SPACETIME PHYSICS

introduction to special relativity

Second Edition

Edwin F. Taylor
Massachusetts Institute of Technology

John Archibald Wheeler
Princeton University and
University of Texas at Austin

W. H. Freeman and Company
New York

Library of Congress Cataloging-in-Publication Data

Taylor, Edwin F.
    Spacetime physics : introduction to special relativity / Edwin F.
Taylor, John Archibald Wheeler. — 2nd ed.
        p.   cm.
    Includes bibliographical references and index.
    ISBN 0-7167-2327-1
    1. Special relativity (Physics)   I. Wheeler, John Archibald,
1911–   . II. Title.
QC173.65T37   1991   92-722
530.1′1 — dc20   CIP

1 2 3 4 5 6 7 8 9 0   KP   9 8 7 6 5 4 3 2

*Both males and females make competent observers. We ordinarily treat the laboratory
observer as male and the rocket observer as female. Beyond this, to avoid alternating
"his" and "her" in a single chapter, we use female pronouns for an otherwise undesig-
nated observer in odd-numbered chapters and male pronouns in even-numbered chapters.*

Epigram, facing page: Einstein remark to his assistant Ernst Straus, quoted in "Mainsprings of Scientific
Discovery" by Gerald Holton in *The Nature of Scientific Discovery,* Owen Gingerich, Editor (Smithsonian
Institution Press, Washington, 1975).

What I'm really interested in is whether God could have made the world in a different way; that is, whether the necessity of logical simplicity leaves any freedom at all.

—Albert Einstein

CONTENTS

Chapter 9 GRAVITY: CURVED SPACETIME IN ACTION 275

Gravity is not a force reaching across space but a distortion — curvature! —
of spacetime experienced right where you are.

CHAPTER 1

SPACETIME: OVERVIEW

> *Our imagination is stretched to the utmost, not, as in fiction, to imagine things which are not really there, but just to comprehend those things which are there.*
>
> Richard P. Feynman

1.1 PARABLE OF THE SURVEYORS

disagree on northward and eastward separations; agree on *distance*

Once upon a time there was a Daytime surveyor who measured off the king's lands. He took his directions of north and east from a magnetic compass needle. Eastward separations from the center of the town square he measured in meters. The northward direction was sacred. He measured northward separations from the town square in a different unit, in miles. His records were complete and accurate and were often consulted by other Daytimers.

A second group, the Nighttimers, used the services of another surveyor. Her north and east directions were based on a different standard of north: the direction of the North Star. She too measured separations eastward from the center of the town square in meters and sacred separations northward in miles. The records of the Nighttime surveyor were complete and accurate. Marked by a steel stake, every corner of a plot appeared in her book, along with its eastward and northward separations from the town square.

Daytimers and Nighttimers did not mix but lived mostly in peace with one another. However, the two groups often disputed the location of property boundaries. Why? Because a given corner of the typical plot of land showed up with different numbers in the two record books for its eastward separation from the town center, measured in meters (Figure 1-1). Northward measurements in miles also did not agree between the two record books. The differences were small, but the most careful surveying did not succeed in eliminating them. No one knew what to do about this single source of friction between Daytimers and Nighttimers.

One fall a student of surveying turned up with novel open-mindedness. Unlike all previous students at the rival schools, he attended both. At Day School he learned

Daytime surveyor uses magnetic north

Nighttime surveyor uses North-Star north

1

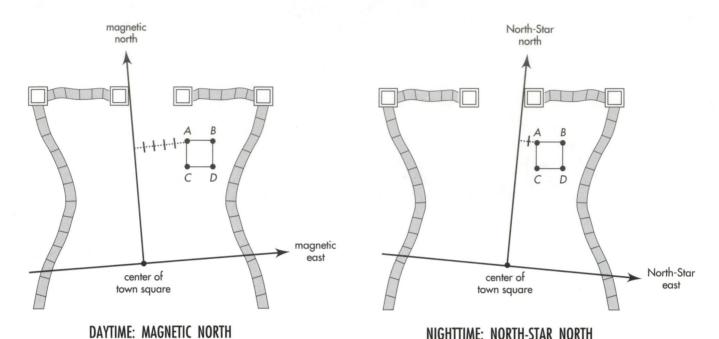

DAYTIME: MAGNETIC NORTH

NIGHTTIME: NORTH-STAR NORTH

FIGURE 1-1. *The town as plotted by Daytime and Nighttime surveyors. Notice that the line of Daytime magnetic north just grazes the* left *side of the north gate, while the line of Nighttime North-Star north just grazes the* right *side of the same gate. Steel stakes A, B, C, D driven into the ground mark the corners of a disputed plot of land. As shown, the eastward separation of stake A from the north—south line measured by the Daytime surveyor is different from that measured by the Nighttime surveyor.*

from one expert his method of recording locations of gates of the town and corners of plots of land based on magnetic north. At Night School he learned the other method, based on North-Star north.

As days and nights passed, the student puzzled more and more in an attempt to find some harmonious relationship between rival ways of recording location. His attention was attracted to a particular plot of land, the subject of dispute between Daytimers and Nighttimers, and to the steel stakes driven into the ground to mark corners of this disputed plot. He carefully compared records of the two surveyors (Figure 1-1, Table 1-1).

Student converts miles to meters

In defiance of tradition, the student took the daring and heretical step of converting northward measurements, previously expressed always in miles, into meters by multiplying with a constant conversion factor k. He found the value of this conversion factor to be $k = 1609.344$ meters/mile. So, for example, a northward separation of 3 miles could be converted to $k \times 3$ miles $= 1609.344$ meters/mile $\times 3$ miles $= 4828.032$ meters. "At last we are treating both directions the same!" he exclaimed.

Next the student compared Daytime and Nighttime measurements by trying various combinations of eastward and northward separation between a given stake and the center of the town square. Somewhere the student heard of the Pythagorean Theorem, that the sum of squares of the lengths of two perpendicular legs of a right triangle equals the square of the length of the hypotenuse. Applying this theorem, he discovered that the expression

$$\left[k \times \begin{pmatrix} \text{northward} \\ \text{separation} \\ \text{(miles)} \end{pmatrix} \right]^2 + \left[\begin{array}{c} \text{eastward} \\ \text{separation} \\ \text{(meters)} \end{array} \right]^2 \qquad \text{(1-1)}$$

Daytime   **Daytime**

───────────────────────⟨ **TABLE 1-1** ⟩───────────────────────

TWO DIFFERENT SETS OF RECORDS; SAME PLOT OF LAND

	Daytime surveyor's axes oriented to magnetic north		*Nighttime surveyor's axes oriented to North-Star north*	
	Eastward (meters)	Northward (miles)	Eastward (meters)	Northward (miles)
Town square	0	0	0	0
Corner stakes:				
Stake *A*	4010.1	1.8330	3950.0	1.8827
Stake *B*	5010.0	1.8268	4950.0	1.8890
Stake *C*	4000.0	1.2117	3960.0	1.2614
Stake *D*	5000.0	1.2054	4960.0	1.2676

based on Daytime measurements of the position of steel stake *C* had exactly the same numerical value as the quantity

$$\left[k \times \begin{pmatrix} \text{Nighttime} \\ \text{northward} \\ \text{separation} \\ \text{(miles)} \end{pmatrix} \right]^2 + \left[\begin{array}{c} \text{Nighttime} \\ \text{eastward} \\ \text{separation} \\ \text{(meters)} \end{array} \right]^2 \tag{1-2}$$

computed from the readings of the Nighttime surveyor for stake *C* (Table 1-2). He

───────────────────────⟨ **TABLE 1-2** ⟩───────────────────────

"INVARIANT DISTANCE" FROM CENTER OF TOWN SQUARE TO STAKE C
(Data from Table 1-1)

Daytime measurements		Nighttime measurements	
Northward separation 1.2117 miles		Northward separation 1.2614 miles	
Multiply by $k = 1609.344$ meters/mile to convert to meters: 1950.0 meters		Multiply by $k = 1609.344$ meters/mile to convert to meters: 2030.0 meters	
Square the value	3,802,500 (meters)^2	Square the value	4,120,900 (meters)^2
Eastward separation 4000.0 meters		Eastward separation 3960.0 meters	
Square the value and add	$+$ 16,000,000 (meters)^2	Square the value and add	$+$ 15,681,600 (meters)^2
Sum of squares	$=$ 19,802,500 (meters)^2	Sum of squares	$=$ 19,802,500 (meters)^2
Expressed as a number squared	$=$ (4450 meters)^2	Expressed as a number squared	$=$ (4450 meters)^2
This is the square of what measurement?	**4450 meters**	This is the square of what measurement?	**4450 meters**

SAME

DISTANCE
from center of Town Square

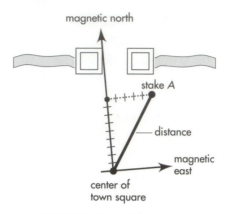

DAYTIME: MAGNETIC NORTH

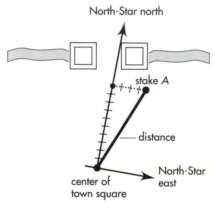

NIGHTTIME: NORTH-STAR NORTH

FIGURE 1-2. *The distance between stake* A *and the center of the town square has the same value for Daytime and Nighttime surveyors, even though the northward and eastward separations, respectively, are not the same for the two surveyors.*

tried the same comparison on recorded positions of stakes *A, B,* and *D* and found agreement here too. The student's excitement grew as he checked his scheme of comparison for all stakes at the corners of disputed plots—and found everywhere agreement.

Flushed with success, the student methodically converted all northward measurements to units of meters. Then the student realized that the quantity he had calculated, the numerical value of the above expressions, was not only the same for Daytime and Nighttime measurements. It was also the square of a length: (meters)². He decided to give this length a name. He called it the **distance** from the center of town.

Discovery: Invariance of distance

$$(\text{distance})^2 = \left[\begin{array}{c} \text{northward} \\ \text{separation} \\ \text{(meters)} \end{array}\right]^2 + \left[\begin{array}{c} \text{eastward} \\ \text{separation} \\ \text{(meters)} \end{array}\right]^2 \tag{1-3}$$

He said he had discovered the **principle of invariance of distance;** he reckoned exactly the same value for distance from Daytime measurements as from Nighttime measurements, despite the fact that the two sets of surveyors' numbers differed significantly (Figure 1-2).

After some initial confusion and resistance, Daytimers and Nighttimers welcomed the student's new idea. The invariance of distance, along with further results, made it possible to harmonize Daytime and Nighttime surveys, so everyone could agree on the location of each plot of land. In this way the last source of friction between Daytimers and Nighttimers was removed. ✐

1.2 SURVEYING SPACETIME

disagree on separations in space and time; agree on spacetime *interval*

The Parable of the Surveyors illustrates the naive state of physics before the discovery of **special relativity** by Einstein of Bern, Lorentz of Leiden, and Poincaré of Paris. Naive in what way? Three central points compare physics at the turn of the twentieth century with surveying before the student arrived to help Daytimers and Nighttimers.

First, surveyors in the mythical kingdom measured northward separations in a sacred unit, the mile, different from the unit used in measuring eastward separations. Similarly, people studying physics measured time in a sacred unit, called the second, different from the unit used to measure space. No one suspected the powerful results of using the same unit for both, or of squaring and combining space and time separations when both were measured in meters. Time in meters is just the time it takes a light flash to go that number of meters. The conversion factor between seconds and meters is the speed of light, $c = 299,792,458$ meters/second. The velocity of light c (in meters/second) multiplied by time t (in seconds) yields ct (in meters).

The second: A sacred unit

The speed of light is the only natural constant that has the necessary units to convert a time to a length. Historically the value of the speed of light was regarded as a sacred number. It was not recognized as a mere conversion factor, like the factor of conversion between miles and meters — a factor that arose out of historical accident in human-kind's choice of units for space and time, with no deeper physical significance.

Speed of light converts seconds to meters

Second, in the parable northward readings as recorded by two surveyors did not differ much because the two directions of north were inclined to one another by only the small angle of 1.15 degrees. At first our mythical student thought that small differences between Daytime and Nighttime northward measurements were due to surveying error alone. Analogously, we used to think of the separation in time between two electric sparks as the same, regardless of the motion of the observer. Only with the publication of Einstein's relativity paper in 1905 did we learn that the separation in time between two sparks really has different values for observers in different states of motion — in different **frames.**

Time between events: Different for different frames

Think of John standing quietly in the front doorway of his laboratory building. Suddenly a rocket carrying Mary flashes through the front door past John, zooms down the middle of the long corridor, and shoots out the back door. An antenna projects from the side of Mary's rocket. As the rocket passes John, a spark jumps across the 1-millimeter gap between the antenna and a pen in John's shirt pocket. The rocket continues down the corridor. A second spark jumps 1 millimeter between the antenna and the fire extinguisher mounted on the wall 2 meters farther down the corridor. Still later other metal objects nearer the rear receive additional sparks from the passing rocket before it finally exits through the rear door.

John and Mary each measure the lapse of time between "pen spark" and "fire-extinguisher spark." They use accurate and fast electronic clocks. John measures this time lapse as 33.6900 thousand-millionths of a second (0.0000000336900 second = 33.6900×10^{-9} second). This equals 33.6900 **nanoseconds** in the terminology of high-speed electronic circuitry. (One nanosecond = 10^{-9} second.) Mary measures a slightly different value for the time lapse between the two sparks, 33.0228 nanoseconds. For John the fire-extinguisher spark is separated in space by 2.0000 meters from the pen spark. For Mary in the rocket the pen spark and fire-extinguisher spark occur at the same place, namely at the end of her antenna. Thus for her their space separation equals zero.

One observer uses laboratory frame

Another observer uses rocket frame

Later, laboratory and rocket observers compare their space and time measurements between the various sparks (Table 1-3). Space locations and time lapses in both frames are measured from the pen spark.

TABLE 1-3

SPACE AND TIME LOCATIONS OF THE SAME SPARKS AS SEEN BY TWO OBSERVERS

| | Distance and time between sparks as measured by observer who is | | | |
| | standing in laboratory (John) | | moving by in rocket (Mary) | |
	Distance (meters)	Time (nanoseconds)	Distance (meters)	Time (nanoseconds)
Reference spark (pen spark)	0	0	0	0
Spark A (fire-extinguisher spark)	2.0000	33.6900	0	33.0228
Spark B	3.0000	50.5350	0	49.5343
Spark C	5.0000	84.2250	0	82.5572
Spark D	8.0000	134.7600	0	132.0915

Discovery: Invariance of spacetime interval

The third point of comparison between the Parable of the Surveyors and the state of physics before special relativity is this: The mythical student's discovery of the concept of distance is matched by the Einstein–Poincaré discovery in 1905 of the **invariant spacetime interval** (formal name **Lorentz interval,** but we often say just **interval**), a central theme of this book. Let each time measurement in seconds be converted to meters by multiplying it by the "conversion factor c," the speed of light:

$$c = 299{,}792{,}458 \text{ meters/second} = 2.99792458 \times 10^8 \text{ meters/second}$$
$$= 0.299792458 \times 10^9 \text{ meters/second} = 0.299792458 \text{ meters/nanosecond}$$

Then the square of the spacetime interval is calculated from the laboratory observer's measurements by *subtracting* the square of the space separation from the square of the time separation. Note the minus sign in equation (1-4).

$$(\text{interval})^2 = \left[c \times \left(\begin{array}{c} \text{time} \\ \text{separation} \\ \text{(seconds)} \end{array} \right) \right]^2 \overset{\text{Laboratory}}{} - \left[\begin{array}{c} \text{space} \\ \text{separation} \\ \text{(meters)} \end{array} \right]^2 \overset{\text{Laboratory}}{} \tag{1-4}$$

The rocket calculation gives exactly the same value of the interval as the laboratory calculation,

$$(\text{interval})^2 = \left[c \times \left(\begin{array}{c} \text{time} \\ \text{separation} \\ \text{(seconds)} \end{array} \right) \right]^2 \overset{\text{Rocket}}{} - \left[\begin{array}{c} \text{space} \\ \text{separation} \\ \text{(meters)} \end{array} \right]^2 \overset{\text{Rocket}}{} \tag{1-5}$$

even though the respective space and time separations are not the same. Two observers find different space and time separations, respectively, between pen spark and fire-extinguisher spark, but when they calculate the spacetime interval between these sparks their results agree (Table 1-4).

The student surveyor found that invariance of distance was most simply written with both northward and eastward separations expressed in the same unit, the meter. Likewise, invariance of the spacetime interval is most simply written with space and

---

TABLE 1-4

"INVARIANT SPACETIME INTERVAL" FROM REFERENCE SPARK TO SPARK A
(Data from Table 1-3)

Laboratory measurements		Rocket measurements	
Time lapse 33.6900×10^{-9} seconds $= 33.6900$ nanoseconds		Time lapse 33.0228×10^{-9} seconds $= 33.0228$ nanoseconds	
Multiply by $c = 0.299792458$ meters per nanosecond to convert to meters: 10.1000 meters		Multiply by $c = 0.299792458$ meters per nanosecond to convert to meters: 9.9000 meters	
Square the value	102.010 (meters)^2	Square the value	98.010 (meters)^2
Spatial separation 2.000 meters		Spatial separation zero	
Square the value and subtract	$-$ 4.000 (meters)^2	Square the value and subtract	$-$ 0
Result of subtaction expressed as a number squared	$= 98.010$ (meters)^2 $= (9.900$ meters)^2	Result of subtaction expressed as a number squared	$= 98.010$ (meters)^2 $= (9.900$ meters)^2
This is the square of what measurement?	**9.900 meters**	This is the square of what measurement?	**9.900 meters**

SAME SPACETIME
INTERVAL
from the reference event

---

time separations expressed in the same unit. Time is converted to meters: t (meters) $= c \times t$ (seconds). Then the interval appears in simplified form:

$$(\text{interval})^2 = \left[\begin{array}{c}\text{time}\\\text{separation}\\\text{(meters)}\end{array}\right]^2 - \left[\begin{array}{c}\text{space}\\\text{separation}\\\text{(meters)}\end{array}\right]^2 \qquad (1\text{-}6)$$

The **invariance of the spacetime interval** — its independence of the state of motion of the observer — forces us to recognize that time cannot be separated from space. Space and time are part of a single entity, **spacetime**. Space has three dimensions: northward, eastward, and upward. Time has one dimension: onward! The interval combines all four dimensions in a single expression. The geometry of spacetime is truly four-dimensional.

Space and time are part of spacetime

To recognize the unity of spacetime we follow the procedure that makes a landscape take on depth — we look at it from several angles. That is why we compare space and time separations between events A and B as recorded by two different observers in relative motion.

Why the minus sign in the equation for the interval? Pythagoras tells us to ADD the squares of northward and eastward separations to get the square of the distance. Who tells us to SUBTRACT the square of the space separation between events from the square of their time separation in order to get the square of the spacetime interval?

Shocked? Then you're well on the way to understanding the new world of very fast motion! This world goes beyond the three-dimensional textbook geometry of Euclid, in which distance is reckoned from a sum of squares. In this book we use another kind of geometry, called **Lorentz geometry,** more real, more powerful than Euclid for the world of the very fast. In Lorentz geometry the squared space separation is combined with the squared time separation in a new way—by *subtraction.* The result is the square of a new unity called the *spacetime interval* between events. The numerical value of this interval is *invariant,* the same for all observers, no matter how fast they are moving past one another. Proof? Every minute of every day an experiment somewhere in the world demonstrates it. In Chapter 3 we derive the invariance of the spacetime interval—with its minus sign—from experiments. They show the finding that no experiment conducted in a closed room will reveal whether that room is "at rest" or "in motion" (Einstein's Principle of Relativity). We won't wait until then to cash in on the idea of interval. We can begin to enjoy the payoff right now.

SAMPLE PROBLEM 1-1

SPARKING AT A FASTER RATE

Another, even faster rocket follows the first, entering the front door, zipping down the long corridor, and exiting through the back doorway. Each time the rocket clock ticks it emits a spark. As before, the first spark jumps the 1 millimeter from the passing rocket antenna to the pen in the pocket of John, the laboratory observer. The second flash jumps when the rocket antenna reaches a doorknob 4.00000000 meters farther along the hall as measured by the laboratory observer, who records the time between these two sparks as 16.6782048 nanoseconds.

- **a.** What is the time between sparks, measured in meters by John, the laboratory observer?

- **b.** What is the value of the spacetime interval between the two events, calculated from John's laboratory measurements?

- **c.** Predict: What is the value of the interval calculated from measurements in the new rocket frame?

- **d.** What is the distance between sparks as measured in this rocket frame?

- **e.** What is the time (in meters) between sparks as measured in this rocket frame? Compare with the time between the same sparks as measured by John in the laboratory frame.

- **f.** What is the speed of this rocket as measured by John in the laboratory?

SOLUTION

- **a.** Time in meters equals time in nanoseconds multiplied by the conversion factor, the speed of light in meters per nanosecond. For John, the laboratory observer,

  16.6782048 nanoseconds $\times$ 0.299792458 meters/nanosecond
  $$= 5.00000000 \text{ meters}$$

- **b.** The square of the interval between two flashes is reckoned by subtracting the square of the space separation from the square of the time separation. Using laboratory figures:

$$(\text{interval})^2 = (\text{laboratory time})^2 - (\text{laboratory distance})^2$$
$$= (5 \text{ meters})^2 - (4 \text{ meters})^2 = 25 \text{ (meters)}^2 - 16 \text{ (meters)}^2$$
$$= 9 \text{ (meters)}^2 = (3 \text{ meters})^2$$

Therefore the interval between the two sparks has the value 3 meters (to nine significant figures).

c. We strongly assert in this chapter that the **spacetime interval is invariant** — has the same value by whomever calculated. Accordingly, the interval between the two sparks calculated from rocket observations has the same value as the interval (3 meters) calculated from laboratory measurements.

d. From the rocket rider's viewpoint, both sparks jump from the same place, namely the end of her antenna, and so distance between the sparks equals zero for the rocket rider.

e. We know the value of the spacetime interval between two sparks as computed in the rocket frame (**c**). And we know that the interval is computed by subtracting the square of the space separation from the square of the time separation in the rocket frame. Finally we know that the space separation in the rocket frame equals zero (**d**). Therefore the rocket time lapse between the two sparks equals the interval between them:

$$(\text{interval})^2 = (\text{rocket time})^2 - (\text{rocket distance})^2$$
$$(3 \text{ meters})^2 = (\text{rocket time})^2 - (\text{zero})^2$$

from which 3 meters equals the rocket time between sparks. Compare this with 5 meters of light-travel time between sparks as measured in the laboratory frame.

f. Measured in the laboratory frame, the rocket moves 4 meters of distance (statement of the problem) in 5 meters of light-travel time (**a**). Therefore its speed in the laboratory is 4/5 light speed. Why? Well, light moves 4 meters of distance in 4 meters of time. The rocket takes longer to cover this distance: 5 meters of time. Suppose that instead of 5 meters of time, the rocket had taken 8 meters of time, twice as long as light, to cover the 4 meters of distance. In that case it would be moving at 4/8 — or half — the speed of light. In the present case the rocket travels the 4 meters of distance in 5 meters of time, so it moves at 4/5 light speed. Therefore its speed equals

$(4/5) \times 2.99792458 \times 10^8$ meters/second
$$= 2.3983397 \times 10^8 \text{ meters/second}$$

1.3 EVENTS AND INTERVALS ALONE!

tools enough to chart matter and motion without any reference frame

In surveying, the fundamental concept is **place.** The surveyor drives a steel stake to mark the corner of a plot of land — to mark a place. A second stake marks another corner of the same plot — another place. Every surveyor — no matter what his or her standard of north — can agree on the value of the distance between the two stakes, between the two places.

Surveying locates a place

Every stake has its own reality. Likewise the *distance* between every pair of stakes also has its own reality, which we can experience directly by pacing off the straight line from one stake to the other stake. The reading on our pedometer — the distance between stakes — is independent of all surveyors' systems, with their arbitrary choice of north.

More: Suppose we have a table of distances between every pair of stakes. That is all we need! From this table and the laws of Euclidean geometry, we can construct the map of every surveyor (see the exercises for this chapter). Distances between stakes: That is all we need to locate every stake, every place on the map.

Physics locates an event

In physics, the fundamental concept is **event.** The collision between one particle and another is an event, with its own location in spacetime. Another event is the emission of a flash of light from an atom. A third is the impact of the pebble that chips the windshield of a speeding car. A fourth event, likewise fixing in and by itself a location in spacetime, is the strike of a lightning bolt on the rudder of an airplane. An event marks a location in spacetime; it is like a steel stake driven into spacetime.

Every laboratory and rocket observer — no matter what his or her relative velocity — can agree on the spacetime interval between any pair of events.

Wristwatch measures interval directly

Every event has its own reality. Likewise the *interval* between every pair of events also has its own reality, which we can experience directly. We carry our wristwatch at constant velocity from one event to the other one. It is not enough just to pass through the two physical locations — we must pass through the actual *events;* we must be at each event precisely when it occurs. Then the space separation between the two events is zero for us — they both occur at our location. As a result, our wristwatch reads directly the spacetime interval between the pair of events:

$$(\text{interval})^2 = \left[\begin{array}{c}\text{time} \\ \text{separation} \\ \text{(meters)}\end{array}\right]^2 - \left[\begin{array}{c}\text{space} \\ \text{separation} \\ \text{(meters)}\end{array}\right]^2$$

$$= \left[\begin{array}{c}\text{time} \\ \text{separation} \\ \text{(meters)}\end{array}\right]^2 - [\text{zero}]^2 = \left[\begin{array}{c}\text{time} \\ \text{separation} \\ \text{(meters)}\end{array}\right]^2 \quad \text{[wristwatch time]}$$

The time read on a wristwatch carried between two events — the interval between those events — is independent of all laboratory and rocket reference frames.

More: To chart all happenings, we need no more than a table of spacetime intervals between every pair of events. That is all we need! From this table and the laws of Lorentz geometry, it turns out, we can construct the space and time locations of events as observed by every laboratory and rocket observer. Intervals between events: That is all we need to specify the location of every event in spacetime.

"Do science" with intervals alone

In brief, we can completely describe and locate events entirely without a reference frame. We can analyze the physical world — we can "do science" — simply by cataloging every event and listing the interval between it and every other event. The unity of spacetime is reflected in the simplicity of entries in our table: intervals only.

Of course, if we want to use a reference frame, we can do so. We then list in our table the individual northward, eastward, upward, and time separations between pairs of events. However, these laboratory-frame listings for a given pair of events will be different from the corresponding listings that our rocket-frame colleague puts in her table. Nevertheless, we can come to agreement if we use the individual separations to reckon the interval between each pair of events:

$$(\text{interval})^2 = (\text{time separation})^2 - (\text{space separation})^2$$

That returns us to a universal, frame-independent description of the physical world.

When two events both occur at the position of a certain clock, that special clock measures directly the interval between these two events. The interval is called the **proper time** (or sometimes the **local time**). The special clock that records the proper time directly has the name **proper clock** for this pair of events. In this book we often call the proper time the **wristwatch time** and the proper clock the **wristwatch** to emphasize that the proper clock is carried so that it is "present" at each of the two events as the events occur.

In Einstein's German, the word for proper time is *Eigenzeit*, or "own-time," implying "one's very own time." The German word provides a more accurate description than the English. In English, the word "proper" has come to mean "following conventional rules." Proper time certainly does not do that!

Hey! I just thought of something: Suppose two events occur at the same time in my frame but very far apart, for example two handclaps, one in New York City and one in San Francisco. Since they are simultaneous in my frame, the time separation between handclaps is zero. But the space separation is not zero — they are separated by the width of a continent. Therefore the square of the interval is a negative number:

$$(interval)^2 = (time\ separation)^2 - (space\ separation)^2$$
$$= (zero)^2 - (space\ separation)^2 = -(space\ separation)^2$$

How can the square of the spacetime interval be negative?

In most of the situations described in the present chapter, there exists a reference frame in which two events occur at the same place. In these cases time separation predominates in all frames, and the interval squared will always be positive. We call these intervals **timelike intervals.**

Euclidean geometry adds squares in reckoning distance. Hence the result of the calculation, distance squared, is always positive, regardless of the relative magnitudes of north and east separations. Lorentz geometry, however, is richer. For your simultaneous handclaps in New York City and San Francisco, space separation between handclaps predominates. In such cases, the interval is called a **spacelike interval** and its form is altered to

$$(interval)^2 = (space\ separation)^2 - (time\ separation)^2 \qquad \text{[when spacelike]}$$

This way, the squared interval is never negative.

The *timelike* interval is measured directly using a wristwatch carried from one event to the other in a special frame in which they occur at the *same place*. In contrast, a *spacelike* interval is measured directly using a rod laid between the events in a special frame in which they occur at the *same time*. This is the frame you describe in your example.

Spacelike interval or timelike interval: In either case the interval is invariant — has the same value when reckoned using rocket measurements as when reckoned using laboratory measurements. You may want to skim through Chapter 6 where timelike and spacelike intervals are described more fully.

1.4 SAME UNIT FOR SPACE AND TIME: METER, SECOND, MINUTE, OR YEAR

meter for particle accelerators; minute for planets; year for the cosmos

The parable of the surveyors cautions us to use the same unit to measure both space and time. So we use meter for both. Time can be measured in meters. Let a flash of light bounce back and forth between parallel mirrors separated by 0.5 meter of

Measure time in meters

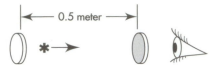

FIGURE 1-3. *This two-mirror "clock" sends to the eye flash after flash, each separated from the next by 1 meter of light-travel time. A light flash (represented by an asterisk) bounces back and forth between parallel mirrors separated from one another by 0.5 meter of distance. The silver coating of the right-hand mirror does not reflect perfectly: It lets 1 percent of the light pass through to the eye each time the light pulse hits it. Hence the eye receives a pulse of light every meter of light-travel time.*

Meter officially defined
using light speed

Measure distance in light-years

distance (Figure 1-3). Such a device is a "clock" that "ticks" each time the light flash arrives back at a given mirror. Between ticks the light flash has traveled a round-trip distance of 1 meter. Therefore we call the stretch of time between ticks **1 meter of light-travel time** or more simply **1 meter of time.**

One meter of light-travel time is quite small compared to typical time lapses in our everyday experience. Light travels nearly 300 million meters per second (300,000,000 meters/second $= 3 \times 10^8$ meters/second, four fifths of the way to Moon in one second). Therefore one second equals 300 million meters of light-travel time. So 1 meter of light-travel time has the small value of one three-hundred-millionth of a second. [How come? Because (1) light goes 300 million meters in one second, and (2) one three-hundred-millionth of that distance (one meter!) is covered in one three-hundred-millionth of that time.] Nevertheless this unit of time is very useful when dealing with light and with high-speed particles. A proton early in its travel through a particle accelerator may be jogging along at "only" one half the speed of light. Then it travels 0.5 meter of distance in 1 meter of light-travel time.

We, our cars, even our jet planes, creep along at the pace of a snail compared with light. We call a deed quick when we've done it in a second. But a second for light means a distance covered of 300 million meters, seven trips around Earth. As we dance around the room to the fastest music, oh, how slow we look to light! Not zooming. Not dancing. Not creeping. Oozing! That long slow ooze racks up an enormous number of meters of light-travel time. That number is so huge that, by the end of one step of our frantic dance, the light that carries the image of the step's beginning is well on its way to Moon.

In 1983 the General Conference on Weights and Measures officially redefined the meter in terms of the speed of light. **The meter is now defined as the distance that light travels in a vacuum in the fraction 1/299,792,458 of a second.** (For the definition of the second, see Box 3-2.) Since 1983 the speed of light is, *by definition,* equal to $c = 299,792,458$ meters/second. This makes official the central position of the speed of light as a conversion factor between time and space.

This official action defines distance (meter) in terms of time (second). Every day we use time to measure distance. "My home is only ten minutes (by car) from work." "The business district is a five-minute walk." Each statement implies a speed — the speed of driving or walking — that converts distance to time. But these speeds can vary — for example, when we get caught in traffic or walk on crutches. In contrast, the speed of light in a vacuum does not vary. It always has the same value when measured over time and the same value as measured by every observer.

We often describe distances to stars and galaxies in a unit of time. These distances we measure in light-years. One light-year equals the distance that light travels in one year. Along with the light-year of space goes the year of time. Here again, space and time are measured in the same units — years. Here again the speed of light is the conversion factor between measures of time and space. From our everyday perspective one light-year of space is quite large, almost 10,000 million million meters: 1 light-year $= 9,460,000,000,000,000$ meters $= 0.946 \times 10^{16}$ meters. Nevertheless it is a convenient unit for measuring distance between stars. For example, the nearest star to our Sun, Proxima Centauri, lies 4.28 light-years away.

Any common unit of space or time may be used as the same unit for both space *and* time. For example, Table 1-5 gives us another convenient measure of time, seconds, compared with time in meters. We can also measure space in the same units, light-seconds. Our Sun is 499 light-seconds — or, more simply, 499 seconds — of distance from Earth. Seconds are convenient for describing distances and times among events that span the solar system. Alternatively we could use minutes of time and light-minutes of distance: Our Sun is 8.32 light-minutes from Earth. We can also use hours of time and light-hours of distance. In all cases, the speed of light is the conversion factor between units of space and time.

◁ TABLE 1-5 ▷

SOME LIGHT-TRAVEL TIMES

	Time in seconds of light-travel time	*Time in meters*
Telephone call one way: New York City to San Francisco via surface microwave link	0.0138	4,139,000
Telephone call one way: New York City to San Francisco via Earth satellite	0.197	59,000,000
Telephone call one way: New York City to San Francisco bounced off Moon	2.51	752,000,000
Flash of light: Emitted by Sun, received on Earth	499.0	149,600,000,000

Expressing time and space in the same unit **meter** is convenient for describing motion of high-speed particles in the confines of the laboratory. Time and space in the same unit **second** (or **minute** or **hour**) is convenient for describing relations among events in our solar system. Time and space in the same unit **year** is convenient for describing relations among stars and among galaxies. In all three arenas spacetime is the stage and special relativity is the spotlight that illuminates the inner workings of Nature.

Use convenient units, the same for space and time

> We are not accustomed to measuring time in meters. So as a reminder to ourselves we add a descriptor: meters *of light-travel time*. But the unit of time is still the meter. Similarly, the added words "seconds *of distance*" and "*light*-years" help to remind us that distance is measured in seconds or years, units we usually associate with time. But this unit of distance is really just second or year. The modifying descriptors are for our convenience only. In Nature, space and time form a unity: spacetime!

The words sound OK. The mathematics appears straightforward. The Sample Problems seem logical. But the ideas are so strange! Why should I believe them? How can invariance of the interval be proved?

No wonder these ideas seem strange. Particles zooming by at nearly the speed of light—how far this is from our everyday experience! Even the soaring jet plane crawls along at less than one-millionth light speed. Is it so surprising that the world appears different at speeds a million times faster than those at which we ordinarily move with respect to Earth?

The notion of *spacetime interval* distills a wealth of real experience. We begin with interval because it endures: It illuminates observations that range from the core of a nucleus to the center of a black hole. Understand the spacetime interval and you vault, in a single bound, to the heart of spacetime.

Chapter 3 presents a logical proof of the invariance of the interval. Chapter 4 reports a knock-down argument about it. Chapters that follow describe many experiments whose outcomes are totally incomprehensible unless the interval is invariant. Real verification comes daily and hourly in the on-going enterprise of experimental physics. ◀▨

SAMPLE PROBLEM 1-2

PROTON, ROCK, AND STARSHIP

a. A proton moving at 3/4 light speed (with respect to the laboratory) passes through two detectors 2 meters apart. Events 1 and 2 are the transits through the two detectors. What are the laboratory space and time separations between the two events, in meters? What are the space and time separations between the events in the proton frame?

b. A speeding rock from space streaks through Earth's outer atmosphere, creating a short fiery trail (Event 1) and continues on its way to crash into Sun (Event 2) 10 minutes later as observed in the Earth frame. Take Sun to be 1.4960×10^{11} meters from Earth. In the Earth frame, what are space and time separations between Event 1 and Event 2 in minutes? What are space and time separations between the events in the frame of the rock?

c. In the twenty-third century a starship leaves Earth (Event 1) and travels at 95 percent light speed, later arriving at Proxima Centauri (Event 2), which lies 4.3 light-years from Earth. What are space and time separations between Event 1 and Event 2 as measured in the Earth frame, in years? What are space and time separations between these events in the frame of the starship?

SOLUTION

a. The space separation measured in the laboratory equals 2 meters, as given in the problem. A flash of light would take 2 meters of light-travel time to travel between the two detectors. Something moving at 1/4 light speed would take four times as long: 2 meters/(1/4) = 8 meters of light-travel time to travel from one detector to the other. The proton, moving at 3/4 light speed, takes 2 meters/(3/4) = 8/3 meters = 2.66667 meters of light-travel time between events as measured in the laboratory.

Event 1 and Event 2 both occur at the position of the proton. Therefore the space separation between the two events equals zero in the proton frame. This means that the spacetime interval — the proper time — equals the time between events in the proton frame.

$$(\text{proton time})^2 - (\text{proton distance})^2 = (\text{interval})^2 = (\text{lab time})^2 - (\text{lab distance})^2$$
$$(\text{proton time})^2 - (\text{zero})^2 = (2.66667 \text{ meters})^2 - (2 \text{ meters})^2$$
$$= (7.1111 - 4) \text{ (meters)}^2$$
$$(\text{proton time})^2 = 3.1111 \text{ (meters)}^2$$

So time between events in the proton frame equals the square root of this, or 1.764 meters of time.

b. Light travels 60 times as far in one minute as it does in one second. Its speed in meters per minute is therefore:

$$2.99792458 \times 10^8 \text{ meters/second} \times 60 \text{ seconds/minute}$$
$$= 1.798754748 \times 10^{10} \text{ meters/minute}$$

So the distance from Earth to Sun is

$$\frac{1.4960 \times 10^{11} \text{ meters}}{1.798754748 \times 10^{10} \text{ meters/minute}} = 8.3169 \text{ light-minutes}$$

This is the distance between the two events in the Earth frame, measured in light-minutes. The Earth-frame time between the two events is 10 minutes, as stated in the problem.

In the frame traveling with the rock, the two events occur at the same place; the time between the two events in this frame equals the spacetime interval — the proper time — between these events:

$$(\text{interval})^2 = (10 \text{ minutes})^2 - (8.3169 \text{ minutes})^2$$
$$= (100 - 69.1708) \text{ (minutes)}^2$$
$$= 30.8292 \text{ (minutes)}^2$$

The time between events in the rest frame of the rock equals the square root of this, or 5.5524 minutes.

c. The distance between departure from Earth and arrival at Proxima Centauri is 4.3 light-years, as given in the problem. The starship moves at 95 percent light speed, or 0.95 light-years/year. Therefore it takes a time 4.3 light-years/(0.95 light-years/year) = 4.53 years to arrive at Proxima Centauri, as measured in the Earth frame.

Starship time between departure from Earth and arrival at Proxima Centauri equals the interval:

$$(\text{interval})^2 = (4.53 \text{ years})^2 - (4.3 \text{ years})^2$$
$$= (20.52 - 18.49) \text{ (years)}^2$$
$$= 2.03 \text{ (years)}^2$$

The time between events in the rest frame of the starship equals the square root of this, or 1.42 years. Compare with the value 4.53 years as measured in the Earth frame. This example illustrates the famous idea that astronaut wristwatch time — proper time — between two events is less than the time between these events measured by any other observer in relative motion. Travel to stay young! This result comes simply and naturally from the invariance of the interval.

1.5 UNITY OF SPACETIME

time and space: equal footing but distinct nature

When time and space are measured in the same unit — whether meter or second or year — the expression for the square of the spacetime interval between two events takes on a particularly simple form:

$$(\text{interval})^2 = (\text{time separation})^2 - (\text{space separation})^2$$
$$= t^2 - x^2 \qquad \text{[same units for time and space]}$$

This formula shows forth the unity of space and time. Impressed by this unity, Einstein's teacher Hermann Minkowski (1864–1909) wrote his famous words, "Henceforth space by itself, and time by itself, are doomed to fade away into mere shadows, and only a union of the two will preserve an independent reality." Today this union of space and time is called spacetime. Spacetime provides the true theater for

Spacetime is a unity

BOX 1-1

PAYOFF OF THE PARABLE

from distance in space to interval in spacetime

DISCUSSION	SURVEYING TOWNSHIP	ANALYZING NATURE
Location marker	Steel stake driven in ground	Collision between two particles Emission of flash from atom Spark jumping from antenna to pen
General name for such a location marker	**Point** or **place**	**Event**
Can its location be staked out for all to see, independent of any scheme of measurement, and independent of all numbers?	Yes	Yes
Simple descriptor of separation between two location markers	**Distance**	**Spacetime interval**
Are there ways directly to measure this separation?	Yes	Yes
With enough markers already staked out, how can we tell someone where we want the next one?	Specify **distances** from other points.	Specify **spacetime intervals** from other events.
Instead of boldly staking out the new marker, or instead of positioning it relative to existing markers, how else can we place the new marker?	By locating point relative to a **reference frame**	By locating event relative to a **reference frame**
Nature of this reference frame?	**Surveyor's grid** yields northward and eastward readings of point (Chapter 1).	**Lattice frame** of rods and clocks yields space and time readings of event (Chapter 2).
Is such a reference frame unique?	No	No
How do two such reference frames differ from one another?	**Tilt** of one surveyor's grid relative to the other	**Uniform velocity** of one frame relative to the other
What are names of two such possible reference frames?	**Daytime grid:** oriented to magnetic north **Nighttime grid:** oriented to North-Star north	**Laboratory frame** **Rocket frame**
What common unit simplifies analysis of the results?	The unit **meter** for both northward and eastward readings	The unit **meter** for both space and time readings
What is the conversion factor from conventional units to meters?	Converting miles to meters: $k = 1609.344$ meters/mile	Converting seconds to meters using the speed of light: $c = 299{,}792{,}458$ meters/second

DISCUSSION	SURVEYING TOWNSHIP	ANALYZING NATURE
For convenience, all measurements are referred to what location?	A common **origin** (center of town)	A common **event** (reference spark)
How do readings for a single marker differ between two reference frames?	Individual northward and eastward readings for one point—for one steel stake—do not have the same values respectively for two surveyors' grids that are tilted relative to one another.	Individual space and time readings for one event—for one spark—do not have the same values respectively for two frames that are in motion relative to one another.
When we change from one marker to two, how do we specify the offset between them in reference-frame language?	**Subtract:** Figure the difference between eastward readings of the two points; also the difference in northward readings.	**Subtract:** Figure the difference between space readings of the two events; also the difference in time readings.
How to figure from offset readings a measure of separation that has the same value whatever the choice of reference frame?	Figure the **distance** between the two points.	Figure the **spacetime interval** between the two events.
Figure how?	$$(\text{distance})^2 = \left(\begin{array}{c}\text{difference in}\\\text{northward readings}\end{array}\right)^2 + \left(\begin{array}{c}\text{difference in}\\\text{eastward readings}\end{array}\right)^2$$	$$(\text{interval})^2 = \left(\begin{array}{c}\text{difference in}\\\text{time readings}\end{array}\right)^2 - \left(\begin{array}{c}\text{difference in}\\\text{space readings}\end{array}\right)^2$$
Result of this reckoning?	Distance between points as figured from readings using one surveyor's grid is the **same** as figured from readings using a second surveyor's grid tilted with respect to first grid.	Interval between events as figured from readings using one latticework frame is the **same** as figured from readings using a second frame in steady straight-line motion relative to first frame.
Phrase to summarize this identity of separation as figured in two reference frames?	**Invariance of the distance** between points	**Invariance of the spacetime interval** between events.
Conclusions from this analysis?	(1) Northward and eastward dimensions are part of a single entity: **space.**	(1) Space and time dimensions are part of a single entity: **spacetime.**
	(2) **Distance** is the simple measure of separation between two **points**, natural because invariant: the same for different surveyor grids.	(2) **Spacetime interval** is the simple measure of separation between two **events**, natural because invariant: the same for different reference frames.

every event in the lives of stars, atoms, and people. Space is different for different observers. Time is different for different observers. Spacetime is the same for everyone.

Minkowski's insight is central to the understanding of the physical world. It focuses attention on those quantities, such as spacetime interval, electrical charge, and particle mass, that are the same for all observers in relative motion. It brings out the merely relative character of quantities such as velocity, momentum, energy, separation in time, and separation in space that depend on relative motion of observers.

Today we have learned not to overstate Minkowski's argument. It is right to say that time and space are inseparable parts of a larger unity. It is wrong to say that time is identical in quality with space.

Difference between time and space

Why is it wrong? Is not time measured in meters, just as space is? In relating the positions of two steel stakes driven into the ground, does not the surveyor measure northward and eastward separations, quantities of identical physical character? By analogy, in locating two events is not the observer measuring quantities of the same nature: space and time separations? How else could it be legitimate to treat these quantities on an equal footing, as in the formula for the interval?

Equal footing, yes; same nature, no. There is a minus sign in the formula for the interval squared = (time separation)^2 — (space separation)^2 that no sleight of hand can ever conjure away. This minus sign distinguishes between space and time. No twisting or turning can ever give the same sign to real space and time separations in the expression for the interval.

The invariance of the spacetime interval evidences the unity of space and time while also preserving—in the formula's minus sign—the distinction between the two.

The principles of special relativity are remarkably simple—simpler than the axioms of Euclidean geometry or the principles of operating an automobile. Yet both Euclid and the automobile have been mastered—perhaps with insufficient surprise —by generations of ordinary people. Some of the best minds of the twentieth century struggled with the concepts of relativity, not because nature is obscure, but because (1) people find it difficult to outgrow established ways of looking at nature, and (2) the world of the very fast described by relativity is so far from common experience that everyday happenings cannot help us develop an intuition for its descriptions.

By now we have won the battle to put relativity in understandable form. The concepts of relativity can now be expressed simply enough to make it easy to think correctly—"to make the bad difficult and the good easy." This leaves only the second difficulty, that of developing intuition—a practiced way of seeing. We understand distance intuitively from everyday experience. Box 1.1 applies our intuition for **distance in space** to help our intuition for **interval in spacetime.**

To put so much into so little, to subsume all of Einstein's teaching on light and motion in the single word *spacetime*, is to cram a wealth of ideas into a small picnic basket that we shall be unpacking throughout the remainder of this book. ✐

REFERENCES

Introductory quote: Richard P. Feynman, *The Character of Physical Law* (MIT Press, Cambridge, Mass., 1967), page 127.

Quote from Minkowski in Section 1.5: H. A. Minkowski, "Space and Time," in H. A. Lorentz *et al., The Principle of Relativity* (Dover Publications, New York, 1952), page 75.

Quote at end of Section 1.5: "to make the bad difficult and the good easy," *"rend le mal difficile et le bien facile."* Einstein, in a similar connection, in a letter to the architect Le Corbusier. Private communication from Le Corbusier.

For an appreciation of Albert Einstein, see John Archibald Wheeler, "Albert Einstein," in *The World Treasury of Physics, Astronomy, and Mathematics,* Timothy Ferris, ed. (Little, Brown, New York, 1991), pages 563–576.

ACKNOWLEDGMENTS

Many students in many classes have read through sequential versions of this text, shared with us their detailed difficulties, and given us advice. We asked students to write down comments, perplexities, and questions as they read and turn in these reading memos for personal response by the teacher. Italicized objections in the text come, in part, from these commentators. Both we who write and you who read are in their debt. Some readers not in classes have also been immensely helpful; among these we especially acknowledge Steven Bartlett. No one could have read the chapters more meticulously than Eric Sheldon, whose wide knowledge has enriched and clarified the presentation. William A. Shurcliff has been immensely inventive in devising new ways of viewing the consequences of relativity; a few of these are specifically acknowledged in later chapters. Electronic-mail courses using this text brought a flood of comments and reading memos from teachers and students around the world. Richard C. Smith originated, organized, and administered these courses, for which we are very grateful. The clarity and simplicity of both the English and the physics were improved by Penny Hull.

Some passages in this text, both brief and extended, have been adapted from the book *A Journey into Gravity and Spacetime* by John Archibald Wheeler (W.H. Freeman, New York, 1990). In turn, certain passages in that book were adapted from earlier drafts of the present text. We have also used passages, logical arguments, and figures from the book *Gravitation* by Charles W. Misner, Kip S. Thorne, and John Archibald Wheeler (W. H. Freeman, New York, 1973).

INTRODUCTION TO THE EXERCISES

Important areas of current research can be analyzed very simply using the theory of relativity. This analysis depends heavily on a physical intuition, which develops with experience. Wide experience is not easy to obtain in the laboratory—simple experiments in relativity are difficult and expensive because the speed of light is so great. As alternatives to experiments, the exercises and problems in this text evoke a wide range of physical consequences of the properties of spacetime. These properties of spacetime recur here over and over again in different contexts:

- paradoxes

- puzzles

- derivations

- technical applications

- experimental results

- estimates

- precise calculations

- philosophical difficulties

The text presents all formal tools necessary to solve these exercises and problems, but intuition — a practiced way of seeing — is best developed without hurry. For this reason we suggest continuing to do more and more of these exercises in relativity after you have moved on to material outside this book. The mathematical manipulations in the exercises and problems are very brief: only a few answers take more than five lines to write down. On the other hand, the exercises require some "rumination time."

In some chapters, exercises are divided into two categories, Practice and Problems. The Practice exercises help you to get used to ideas in the text. The Problems apply these ideas to physical systems, thought experiments, and paradoxes.

WHEELER'S FIRST MORAL PRINCIPLE: *Never make a calculation until you know the answer.* Make an estimate before every calculation, try a simple physical argument (symmetry! invariance! conservation!) before every derivation, guess the answer to every paradox and puzzle. Courage: No one else needs to know what the guess is. Therefore make it quickly, by instinct. A right guess reinforces this instinct. A wrong guess brings the refreshment of surprise. In either case life as a spacetime expert, however long, is more fun!

CHAPTER 1 EXERCISES

PRACTICE

1-1 comparing speeds

Compare the speeds of an automobile, a jet plane, an Earth satellite, Earth in its orbit around Sun, and a pulse of light. Do this by comparing the relative distance each travels in a fixed time. Arbitrarily choose the fixed time to give convenient distances. A car driving at the USA speed limit of 65 miles/hour (105 kilometers/hour) covers 1 meter of distance in about 35 milliseconds = 35×10^{-3} second.

a How far does a commercial jetliner go in 35 milliseconds? (speed: 650 miles/hour = 1046 kilometers/hour)

b How far does an Earth satellite go in 35 milliseconds? (speed: 17,000 miles/hour $\approx$ 27,350 kilometers/hour)

c How far does Earth travel in its orbit around Sun in 35 milliseconds? (speed: 30 kilometers/second)

d How far does a light pulse go in a vacuum in 35 milliseconds? (speed: 3×10^8 meters /second). This distance is roughly how many times the distance from Boston to San Francisco (5000 kilometers)?

1-2 images from Neptune

At 9:00 P.M. Pacific Daylight Time on August 24, 1989, the planetary probe *Voyager II* passed by the planet Neptune. Images of the planet were coded and transmitted to Earth by microwave relay.

It took 4 hours and 6 minutes for this microwave signal to travel from Neptune to Earth. Microwaves (electromagnetic radiation, like light, but of frequency lower than that of visible light), when propagating through interplanetary space, move at the "standard" light speed of one meter of distance in one meter of light-travel time, or 299,792,458 meters/second. In the following, neglect any relative motion among Earth, Neptune, and *Voyager II*.

a Calculate the distance between Earth and Neptune at fly-by in units of minutes, seconds, years, meters, and kilometers.

b Calculate the time the microwave signal takes to reach Earth. Use the same units as in part **a**.

1-3 units of spacetime

Light moves at a speed of 3.0×10^8 meters/second. One mile is approximately equal to 1600 meters. One furlong is approximately equal to 200 meters.

a   How many meters of time in one day?
b   How many seconds of distance in one mile?
c   How many hours of distance in one furlong?
d   How many weeks of distance in one light-year?
e   How many furlongs of time in one hour?

1-4   time stretching and the spacetime interval

A rocket clock emits two flashes of light and the rocket observer records the time lapse (in seconds) between these two flashes. The laboratory observer records the time separation (in seconds) and space separation (in light-seconds) between the same pair of flashes. The results for both laboratory and rocket observers are recorded in the first line of the table.

Now a clock in a different rocket, moving at a different speed with respect to the laboratory, emits a different pair of flashes. The set of laboratory and rocket space and time separations are recorded on the second line of the table. And so on. Complete the table.

1-5   where and when?

Two firecrackers explode at the same place in the laboratory and are separated by a time of 3 years as measured on a laboratory clock.

a   What is the spatial distance between these two events in a rocket in which the events are separated in time by 5 years as measured on rocket clocks?

b   What is the relative speed of the rocket and laboratory frames?

1-6   mapmaking in space

The table shows distances between cities. The units are kilometers. Assume all cities lie on the same flat plane.

a   Use a ruler and a compass (the kind of compass that makes circles) to construct a map of these cities. Choose a convenient scale, such as one centimeter on the map corresponds to ten kilometers on Earth.

Discussion: How to start? With three arbitrary decisions! (1) Choose any city to be at the center of the map. (2) Choose any second city to be "due north" — that is, along any arbitrary direction you select. (3) Even with these choices, there are two places you can locate the third city; choose either of these two places arbitrarily.

b   If you rotate the completed map in its own plane — for example, turning it while keeping it flat on the table — does the resulting map also satisfy the distance entries above?

c   Hold up your map between you and a light, with the marks on the side of the paper facing the

EXERCISE 1-4

SPACE AND TIME SEPARATIONS

	Rocket time lapse (seconds)	Laboratory time lapse (seconds)	Laboratory distance (light-seconds)
Example	20	29	21
a	?	10.72	5.95
b	20	?	99
c	66.8	72.9	?
d	?	8.34	6.58
e	21	22	?

EXERCISE 1-6

DISTANCES BETWEEN CITIES

Distance to city → from city	A	B	C	D	E	F	G	H
A	0	20.0	28.3	28.3	28.3	2.00	28.3	44.7
B		0	20.0	20.0	44.7	40.0	44.7	40.0
C			0	40.0	40.0	44.7	56.6	60.0
D				0	56.6	44.7	40.0	20.0
E					0	20.0	40.0	72.1
F						0	20.0	56.6
G							0	44.7
H								0

light. Does the map you see from the back also satisfy the table entries?

Discussion: In this exercise you use a table consisting only of distances between pairs of cities to construct a map of these cities from the point of view of a surveyor using a given direction for north. In Exercise 5-3 you use a table consisting only of spacetime intervals between pairs of events to draw a "spacetime map" of these events from the point of view of one free-float observer. Exercise 1-7 previews this kind of spacetime map.

1-7   spacetime map

The laboratory space and time measurements of events 1 through 5 are plotted in the figure. Compute the value of the spacetime interval

  a between event 1 and event 2.

  b between event 1 and event 3.

  c between event 1 and event 4.

  d between event 1 and event 5.

  e A rocket moves with constant velocity from event 1 to event 2. That is, events 1 and 2 occur at the same place in this rocket frame. What time lapse is recorded on the rocket clock between these two events?

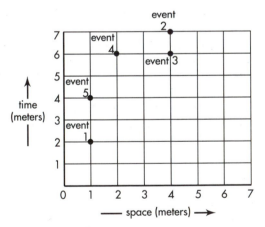

EXERCISE 1-7. *Spacetime map of some events.*

PROBLEMS

1-8   size of a computer

In one second some desktop computers can carry out one million instructions in sequence. One instruction might be, for instance, multiplying two numbers together. In technical jargon, such a computer operates at "one megaflop." Assume that carrying out one instruction requires transmission of data from the memory (where data is stored) to the processor (where the computation is carried out) and transmission of the result back to the memory for storage.

  a What is the maximum average distance between memory and processor in a "one-megaflop" computer? Is this maximum distance increased or decreased if the signal travels through conductors at one half the speed of light in a vacuum?

  b Computers are now becoming available that operate at "one gigaflop," that is, they carry out 10^9 sequential instructions per second. What is the maximum average distance between memory and processor in a "one-gigaflop" machine?

  c Estimate the overall maximum size of a "one-teraflop" machine, that is, a computer that can carry out 10^{12} sequential instructions per second.

  d **Discussion question:** In contrast with most current personal computers, a "parallel processing" computer contains several or many processors that work together on a computing task. One might think that a machine with 10,000 processors would complete a given computation task in 1/10,000 the time. However, many computational problems cannot be divided up in this way, and in any case some fraction of the computing capacity must be devoted to coordinating the team of processors. What limits on physical size does the speed of light impose on a parallel processing computer?

1-9   trips to Andromeda by rocket

The Andromeda galaxy is approximately two million light-years distant from Earth as measured in the Earth-linked frame. Is it possible for you to travel from Earth to Andromeda in your lifetime? Sneak up on the answer to this question by considering a series of trips from Earth to Andromeda, each one faster than the one before. For simplicity, assume the Earth-Andromeda distance to be exactly two million light-years in the Earth frame, treat Earth and Andromeda as points, and neglect any relative motion between Earth and Andromeda.

  a TRIP 1. Your one-way trip takes a time 2.01×10^6 years (measured in the Earth-linked frame) to cover the distance of 2.00×10^6 light-years. How long does the trip last as measured in your rocket frame?

  b What is your rocket speed on Trip 1 as measured in the Earth-linked frame? Express this speed as a decimal fraction of the speed of light. Call this fraction, $v = v_{conv}/c$, where v_{conv} is speed in conventional units, such as meters/second. **Discussion:** If your rocket moves at half the speed of light, it takes

4×10^6 years to cover the distance 2×10^6 light-years. In this case

$$v = \frac{2 \times 10^6 \text{ light-years}}{4 \times 10^6 \text{ years}} = \frac{1}{2}$$

Therefore . . .

c TRIP 2. Your one-way Earth-Andromeda trip takes 2.001×10^6 years as measured in the Earth-linked frame. How long does the trip last as measured in your rocket frame? What is your rocket speed for Trip 2, expressed as a decimal fraction of the speed of light?

d TRIP 3. Now set the rocket time for the one-way trip to 20 years, which is all the time you want to spend getting to Andromeda. In this case, what is your speed as a decimal fraction of the speed of light? **Discussion:** Solutions to many exercises in this text are simplified by using the following approximation, which is the first two terms in the binomial expansion

$$(1 + z)^n \approx 1 + nz \qquad \text{if} \qquad |z| \ll 1$$

Here n can be positive or negative, a fraction or an integer; z can be positive or negative, as long as its magnitude is very much smaller than unity. This approximation can be used twice in the solution to part **d**.

1-10   trip to Andromeda by Transporter

In the *Star Trek* series a so-called Transporter is used to "beam" people and their equipment from a starship to the surface of nearby planets and back. The Transporter mechanism is not explained, but it appears to work only locally. (If it could transport to remote locations, why bother with the starship at all?) Assume that one thousand years from now a Transporter exists that reduces people and things to data (elementary bits of information) and transmits the data by light or radio signal to remote locations. There a Receiver uses the data to reassemble travelers and their equipment out of local raw materials.

One of your descendants, named Samantha, is the first "transporternaut" to be beamed from Earth to the planet Zircon orbiting a star in the Andromeda Nebula, two million light-years from Earth. Neglect any relative motion between Earth and Zircon, and assume: (1) transmission produces a Samantha identical to the original in every respect (except that she is 2 million light-years from home!), and (2) the time required for disassembling Samantha on Earth and reassembling her on Zircon is negligible as measured

in the common rest frame of Transporter and Receiver.

a How much does Samantha age during her outward trip to Zircon?

b Samantha collects samples and makes observations of the Zirconian civilization for one Earth-year, then beams back to Earth. How much has Samantha aged during her entire trip?

c How much older is Earth and its civilization when Samantha returns?

d Earth has been taken over by a tyrant, who wishes to invade Zircon. He sends one warrior and has him duplicated into attack battalions at the Receiver end. How long will the Earth tyrant have to wait to discover whether his ambition has been satisfied?

e A second transporternaut is beamed to a much more remote galaxy that is moving away from Earth at 87 percent of the speed of light. This time, too, the traveler stays in the remote galaxy for one year *as measured by clocks moving with the galaxy* before returning to Earth by Transporter. How much has the transporternaut aged when she arrives back at Earth? (Careful!)

1-11   time stretching with μ-mesons

At heights of 10 to 60 kilometers above Earth, cosmic rays continually strike nuclei of oxygen and nitrogen atoms and produce μ-mesons (mu-mesons: elementary particles of mass equal to 207 electron masses produced in some nuclear reactions). Some of the μ-mesons move vertically downward with a speed nearly that of light. Follow one of the μ mesons on its way down. In a given sample of μ-mesons, half of them decay to other elementary particles in 1.5 microseconds (1.5×10^{-6} seconds), measured with respect to a reference frame in which they are at rest. Half of the remainder decay in the next 1.5 microseconds, and so on. Analyze the results of this decay as observed in two different frames. Idealize the rather complicated actual experiment to the following roughly equivalent situation: All the mesons are produced at the same height (60 kilometers); all have the same speed; all travel straight down; none are lost to collisions with air molecules on the way down.

a Approximately how long a time will it take these mesons to reach the surface of Earth, as measured in the Earth frame?

b If the decay time were the same for Earth observers as for an observer traveling with the μ-mesons, approximately how many half-lives would have passed? Therefore what fraction of those created at a height of 60 kilometers would remain when they reached sea level on Earth? You may express your answer as a power of the fraction $1/2$.

c   An experiment determines that the fraction $1/8$ of the μ-mesons reaches sea level. Call the rest frame of the mesons the rocket frame. In this rocket frame, how many half-lives have passed between creation of a given μ-meson and its arrival as a survivor at sea level?

d   *In the rocket frame,* what is the space separation between birth of a survivor meson and its arrival at the surface of Earth? (Careful!)

e   From the rocket space and time separations, find the value of the spacetime interval between the birth event and the arrival event for a single surviving μ-meson.

Reference: Nalini Easwar and Douglas A. MacIntire, *American Journal of Physics,* Volume 59, pages 589–592 (July 1991).

1-12   time stretching with π^+-mesons

Laboratory experiments on particle decay are much more conveniently done with π^+-mesons (pi-plus mesons) than with μ-mesons, as is seen in the table.

In a given sample of π^+-mesons half will decay to other elementary particles in 18 nanoseconds (18×10^{-9} seconds) measured in a reference frame in which the π^+-mesons are at rest. Half of the remainder will decay in the next 18 nanoseconds, and so on.

a   In a particle accelerator π^+-mesons are produced when a proton beam strikes an aluminum

TIME STRETCHING WITH π^+-MESONS

Particle	Time for half to decay (measured in rest frame)	"Characteristic distance" (speed of light multiplied by foregoing time)
μ-meson (207 times electron mass)	1.5×10^{-6} second	450 meters
π^+-meson (273 times electron mass)	18×10^{-9} second	5.4 meters

target inside the accelerator. Mesons leave this target with nearly the speed of light. If there were no time stretching and if no mesons were removed from the resulting beam by collisions, what would be the greatest distance from the target at which half of the mesons would remain undecayed?

b   The π^+-mesons of interest in a particular experiment have a speed 0.9978 that of light. By what factor is the predicted distance from the target for half-decay increased by time dilation over the previous prediction — that is, by what factor does this dilation effect allow one to increase the separation between the detecting equipment and target?

CHAPTER 2

FLOATING FREE

At that moment there came to me the happiest thought of my life . . . *for an observer falling freely from the roof of a house no gravitational field exists during his fall . . .*

Albert Einstein

2.1 FLOATING TO MOON

will the astronaut stand on the floor — or float?

Less than a month after the surrender at Appomattox ended the American Civil War (1861–1865), the French author Jules Verne began writing *A Trip From the Earth to the Moon* and *A Trip Around the Moon*. Eminent American cannon designers, so the story goes, cast a great cannon in a pit, with cannon muzzle pointing skyward. From this cannon they fire a ten-ton projectile containing three men and several animals (Figure 2-1).

As the projectile coasts outward in unpowered flight toward Moon, Verne says, its passengers walk normally inside the projectile on the end nearer Earth (Figure 2-2). As the trip continues, passengers find themselves pressed less and less against the floor of the spaceship until finally, at the point where Earth and Moon exert equal but opposite gravitational attraction, passengers float free of the floor. Later, as the ship nears Moon, they walk around once again — according to Verne — but now against the end of the spaceship nearer Moon.

Early in the coasting portion of the trip a dog on the ship dies from injuries sustained at takeoff. Passengers dispose of its remains through a door in the spaceship, only to find the body floating outside the window during the entire trip (Figure 2-1).

This story leads to a paradox whose resolution is of crucial importance to relativity. Verne thought it reasonable that Earth's gravitational attraction would keep a passenger pressed against the Earth end of the spaceship during the early part of the trip. He also thought it reasonable that the dog should remain next to the ship, since both ship and dog independently follow the same path through space. But since the dog floats outside the spaceship during the entire trip, why doesn't the passenger float around inside the spaceship? If the ship were sawed in half would the passenger, now "outside," float free of the floor?

Jules Verne:
Passenger stands on floor

Paradox of passenger and dog

IT WAS THE BODY OF SATELLITE.

FIGURE 2-1. *Illustration from an early edition of* A Trip Around the Moon. *Satellite is the name of the unfortunate dog.*

FIGURE 2-2. *Incorrect prediction: Jules Verne believed that a passenger inside a free projectile would stand against the end of the projectile nearest Earth or Moon, whichever had greater gravitational attraction — but that the dog would float along beside the projectile for the entire trip.* **Correct prediction:** *Verne was right about the dog, but a passenger also floats with respect to the free projectile during the entire trip.*

Reality:
Passenger floats in spaceship

Our experience with actual space flights enables us to resolve this paradox (Figure 2-2). Jules Verne was wrong about the passenger's motion inside the unpowered spaceship. Like the dog outside, the passenger inside independently follows the same path through space as the spaceship itself. Therefore he floats freely relative to the ship during the entire trip (after the initial boost inside the cannon barrel). True: Earth's gravity acts on the passenger. But it also acts on the spaceship. In fact, with respect to Earth, gravitational acceleration of the spaceship just equals gravitational acceleration of the passenger. Because of this equality, there is no *relative* acceleration between passenger and spaceship. Thus the spaceship serves as a **reference frame** relative to which the passenger does not experience any acceleration.

To say that acceleration of the passenger relative to the unpowered spaceship equals zero is *not* to say that his velocity relative to it necessarily also equals zero. He may jump from the floor or spring from the side — in which case he hurtles across the spaceship and strikes the opposite wall. However, when he floats with zero initial velocity relative to the ship the situation is particularly interesting, for he will also float with zero velocity relative to it at all later times. He and the ship follow identical paths through space. How remarkable that the passenger, who cannot see outside, nevertheless moves on this deterministic orbit! Without a way to control his motion and even with his eyes closed he will not touch the wall. How could one do better at eliminating detectable gravitational influences? 🪶

2.2 THE INERTIAL (FREE-FLOAT) FRAME

goodbye to the "force of gravity"

It is easy to talk about the simplicity of motion in a spaceship. It is hard to think of conditions being equally simple on the surface of Earth (Figure 2-3). The reason for

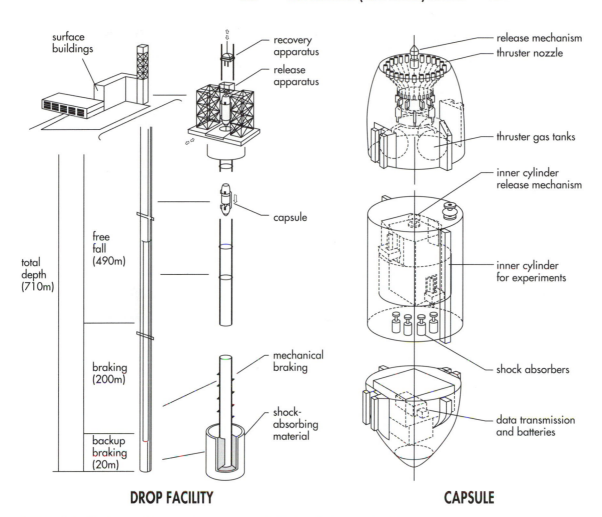

DROP FACILITY

CAPSULE

FIGURE 2-3. *The Japan Microgravity Center (JAMIC) installed in an abandoned coal mine 710 meters deep in the small town of Kamisunagawa on the northern island of Hokkaido, Japan.* The capsule carrying the experimental apparatus provides a free-float frame for 10 seconds as it falls 490 meters through a vertical tube, achieving a maximum velocity of nearly 100 meters/second. It is guided by two contact-free magnetic suspensions along the tube. The vertical tube is not evacuated; downward-thrusting gas jets on the capsule compensate for air drag as the capsule drops. The capsule is slowed down in an additional distance of 200 meters near the bottom of the tube by air resistance after thrusters are turned off, followed by mechanical braking. Twenty meters of cushioning material at the very bottom of the tube provide emergency stopping. The falling capsule is nearly 8 meters long and nearly 2 meters in diameter with a mass of 5000 kilograms, including 1000 kilograms of experimental equipment contained in an inner cylinder 1.3 meters in diameter and 1.8 meters long. The space between capsule and experimental cylinder is evacuated. The inner experimental cylinder is released just before the outer capsule itself. Optical monitoring of the vertical position of the inner cylinder triggers downward-pushing thrusters as needed to overcome air resistance. Thus the experimental cylinder itself acts as an internal "conscience," ensuring that the capsule takes the same course that it would have taken had both resistance and thrust been absent. The result? A nearly free-float frame, with a maximum acceleration of 1.0×10^{-4} g in the experimental capsule, where g is the acceleration of gravity at Earth's surface. Experiments carried out in this facility benefit from conditions of "no air pressure, no heat convection, no floating or sinking buoyancy, no resistance to motion," as well as much lower cost and less environmental damage than those involved in launching and monitoring an Earth satellite. The facility is designed to carry out 400 drops per year, with experiments such as forming large superconducting crystals, creating alloys of materials that do not normally mix, studying transitions between gas and liquid phases, and burning under zero-g. (See also Figure 9-2.)

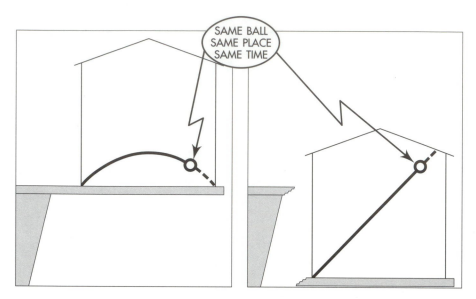

FIGURE 2-4. *Illusion and Reality. The same ball thrown from the same corner of the same room in the same direction with the same speed is seen to undergo very different motions depending on whether it is recorded by an observer with a floor pushing up against his feet or by an observer in "free fall" ("free float") in a house sawed free from the cliff. In both descriptions the ball arrives at the same place—relative to Mother Earth—at the same instant. Let each ball squirt a jet of ink on the wall we are looking at. The resulting record is as crisp for the arc as for the straight line. Is the arc real and the straight line illusion? Or is the straight line real and the arc illusion? Einstein tells us that the two ink trails are equally valid. We have only to be honest and say whether the house, the wall, and the describer of the motion are in free float or whether the describer is continually being driven away from a condition of free float by a push against his feet. Einstein also tells us that physics always looks simplest in a free-float frame. Finally, he tells us that every truly local manifestation of "gravity" can be eliminated by observing motion from a frame of reference that is in free float.*

concern is not far to seek. We experience it every day, every minute, every second. We call it gravity. It shows in the arc of a ball tossed across the room (Figure 2-4, left). How can anyone confront a mathematical curve like that arc and not be trapped again in that tortuous trail of thought that led from ancient Greeks to Galileo to Newton? They thought of gravity as a force acting through space, as something mysterious, as something that had to be "explained."

Einstein put forward a revolutionary new idea. Eliminate gravity!

Where lies the cause of the curved path of the ball? Is it the ball? Is it some mysterious "force of gravity"? Neither, Einstein tells us. It is the fault of the viewers —and the fault of the floor that forces us away from the natural state of motion: the state of **free fall,** or better put, **free float.** Remove the floor and our motion immediately becomes natural, effortless, free from gravitational effects.

Concept of free-float frame

Let the room be cut loose at the moment we throw the ball slantwise upward from the west side at floor level (Figure 2-4, right). The ball has the same motion as it did before. However, the motion looks different. It looks different because we who look at it are in a different frame of reference. We are in a **free-float frame.** In this free-float frame the ball has straight-line motion. What could be simpler?

Even when the room was not cut away from the cliff, the floor did not affect the midair flight of the ball. But the floor did affect us who watched the flight. The floor forced us away from our natural motion, the motion of free fall (free float). We blamed the curved path of the ball on the "force of gravity" acting on the *ball.* Instead we should have blamed the floor for its force acting on *us.* Better yet, get rid of the floor by cutting the house away from the cliff. Then our point of view becomes the natural one: We enter a free-float frame. In our free-float frame the ball flies straight.

What's the fault of the force on my feet?
What pushes my feet down on the floor?
Says Newton, the fault's at Earth's core.
Einstein says, the fault's with the floor;
Remove that and gravity's beat!

— Frances Towne Ruml

How could humankind have lived so many centuries without realizing that the "arc" is an unnecessary distraction, that the idea of local "gravity" is superfluous — the fault of the observer for not arranging to look at matters from a condition of free float?

Even today we recoil instinctively from the experience of free float. We and a companion ride in the falling room, which does not crash on the ground but drops into a long vertical tunnel dug for that purpose along the north–south axis of Earth. Our companion is so filled with consternation that he takes no interest in our experimental findings about free float. He grips the door jamb in terror. "We're falling!" he cries out. His fear turns to astonishment when we tell him not to worry.

"A shaft has been sunk through Earth," we tell him. "It's not the fall that hurts anyone but what stops the fall. All obstacles have been removed from our way, including air. Free fall," we assure him, "is the safest condition there is. That's why we call it free float."

"You may call it float," he says, "but I still call it fall."

"Right now that way of speaking may seem reasonable," we reply, "but after we pass the center of Earth and start approaching the opposite surface, won't the word 'fall' seem rather out of place? Might you not then prefer the word 'float'?" And with "float" our companion at last is happy.

Free-float through Earth

What do we both see? Weightlessness. Free float. Motion in a straight line and at uniform speed for marbles, pennies, keys, and balls in free motion in any direction within our traveling home. No jolts. No shudders. No shakes at any point in all the long journey from one side of Earth to the other.

For our ancestors, travel into space was a dream beyond realization. Equally beyond our reach today is the dream of a house floating along a tunnel through Earth, but this dream nonetheless illuminates the simplicity of motion in a free-float frame. Given the necessary conditions, nothing that we observe inside our traveling room gives us the slightest possibility of discriminating among different free-float frames: one just above Earth's surface, a second passing through Earth's interior, a third in the uttermost reaches of space. Floating inside any of them we find no evidence whatever for the presence of "gravity."

Wait a minute! If the idea of local "gravity" is unnecessary, why does my pencil begin to fall when I hold it in the air and let go? If there is no gravity, my pencil should remain at rest.

And so it does remain at rest — as observed from a free-float frame! The natural motion of your pencil is to remain at rest or to move with constant velocity in a free-float frame. So it is not helpful to ask: "Why does the pencil begin to fall when I let go?" A more helpful question: "Before I let go, why must I apply an upward force to keep the pencil at rest?" Answer: Because you are making observations from an unnatural frame: one held fixed at the surface of Earth. Remove that fixed hold by dropping your room off a cliff. Then for you "gravity" disappears. For you, no force is required to keep the pencil at rest in your free-float frame. ✎

2.3 LOCAL CHARACTER OF FREE-FLOAT FRAME

tidal effects intrude in larger domains

First to strike us about the concept free float has been its paradoxical character. As a first step to explaining gravity Einstein got rid of gravity. There is no evidence of gravity in the freely falling house.

Well, *almost* no evidence. The second feature of free float is its local character. Riding in a very small spaceship (Figure 2-5, left) we find no evidence of gravity. But the enclosure in which we ride—falling near Earth or plunging through Earth— cannot be too large or fall for too long a time without some unavoidable relative changes in motion being detected between particles in the enclosure. Why? Because widely separated particles within a large enclosed space are differently affected by the nonuniform gravitational field of Earth, to use the Newtonian way of speaking. For example, two particles released side by side are both attracted toward the center of Earth, so they move closer together as measured inside a falling long narrow *horizontal* railway coach (Figure 2-5, center).

As another example, think of two particles released far apart vertically but directly above one another in a long narrow *vertical* falling railway coach (Figure 2-5, right). This time their gravitational accelerations toward Earth are in the same direction,

Earth's pull nonuniform: Large spaceship not a free-float frame

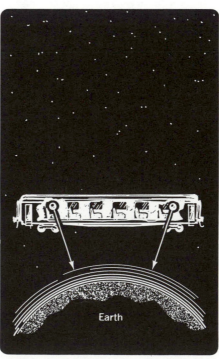
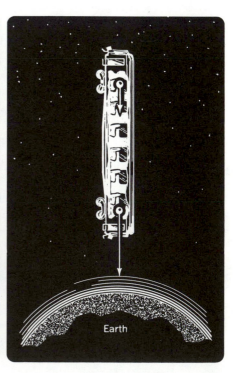

FIGURE 2-5. *Three vehicles in free fall near Earth: small space capsule, Einstein's old-fashioned railway coach in free fall in a* horizontal *orientation, and another railway coach in* vertical *orientation.*

according to the Newtonian analysis. However, the particle nearer Earth is more strongly attracted to Earth and slowly leaves the other behind: the two particles move farther apart as the coach falls. Conclusion: the large enclosure is not a free-float frame.

Even a small room fails to qualify as free-float when we sample it over a long enough time. In the 42 minutes it takes our small room to fall through the tunnel from North Pole to South Pole, we notice relative motion between test particles released initially from rest at opposite sides of the room.

Now, we want the laws of motion to look simple in our floating room. Therefore we want to eliminate all relative accelerations produced by external causes. "Eliminate" means to reduce these accelerations below the limit of detection so that they do not interfere with more important accelerations we wish to study, such as those produced when two particles collide. We eliminate the problem by choosing a room that is sufficiently small. Smaller room? Smaller relative accelerations of objects at different points in the room!

Let someone have instruments for detection of relative accelerations with any given degree of sensitivity. No matter how fine that sensitivity, the room can always be made so small that these perturbing relative accelerations are too small to be detectable. Within these limits of sensitivity our room is a free-float frame. "Official" names for such a frame are the **inertial reference frame** and the **Lorentz reference frame.** Here, however, we often use the name **free-float frame,** which we find more descriptive. These are all names for the same thing.

> A reference frame is said to be an "inertial" or "free-float" or "Lorentz" reference frame in a certain region of space and time when, throughout that region of spacetime — and within some specified accuracy — every free test particle initially at rest with respect to that frame remains at rest, and every free test particle initially in motion with respect to that frame continues its motion without change in speed or in direction.

Wonder of wonders! This test can be carried out entirely within the free-float frame. The observer need not look out of the room or refer to any measurements made external to the room. A free-float frame is "local" in the sense that it is limited in space and time — and also "local" in the sense that its free-float character can be determined from within, locally.

Sir Isaac Newton stated his First Law of Motion this way: "Every body perseveres in its state of rest, or of uniform motion in a right [straight] line, unless it is compelled to change that state by forces impressed upon it." For Newton, **inertia** was a property of objects that described their tendency to maintain their state of motion, whether of rest or constant velocity. For him, objects obeyed the "Law of Inertia." Here we have turned the "Law of Inertia" around: Before we certify a reference frame to be inertial, we *require* observers in that frame to demonstrate that every free particle maintains its initial state of motion or rest. Then Newton's First Law of Motion *defines* a reference frame — an arena or playing field — in which one can study the motion of objects and draft the laws of their motion.

When is the room, the spaceship, or any other vehicle small enough to be called a local free-float frame? Or when is the relative acceleration of two free particles placed at opposite ends of the vehicle too slight to be detected?

"Local" is a tricky word. For example, drop the old-fashioned railway coach in a horizontal orientation from rest at a height of 315 meters onto the surface of Earth (Figure 2-5, center). Time from release to impact equals 8 seconds, or 2400 million meters of light-travel time. At the same instant you drop the coach, release tiny ball bearings from rest — and in midair — at opposite ends of the coach. During the time

Free-float frame is local

Free-float (inertial) frame formally defined

BOX 2-1

THE TIDE-DRIVING POWER OF MOON AND SUN

Note: *Neither astronomers nor newspapers say "the Venus" or "the Mars." All say simply "Venus" or "Mars." Astronomers follow the same snappy practice for Earth, Moon, and Sun. More and more of the rest of the world now follows — as do we in this book — the recommendations of the International Astronomical Union.*

The ocean's rise and fall in a never-ending rhythmic cycle bears witness to the tide-driving power of Moon and Sun. In principle those influences are no different from those that cause relative motion of free particles in the vicinity of Earth. In a free-float frame near Earth, particles separated vertically *increase* their separation with time; particles separated horizontally *decrease* their separation with time (Figure 2-5). More generally, a thin spattering of free-float test masses, spherical in pattern, gradually becomes egg-shaped, with the long axis vertical. Test masses nearer Earth, more strongly attracted than the average, move downward to form the lower bulge. Similarly, test masses farther from Earth, less strongly attracted than the average, move upward to form the upper bulge.

By like action Moon, acting on the waters of Earth — floating free in space — would draw them out into an egg-shaped pattern if there were water everywhere, water of uniform depth. There isn't. The narrow Straits of Gibraltar almost cut off the Mediterranean from the open ocean, and almost kill all tides in it. Therefore it is no wonder that Galileo Galilei, although a great pioneer in the study of gravity, did not take the tides as seriously as the more widely traveled Johannes Kepler, an expert on the motion of Moon and the planets. Of Kepler, Galileo even said, "More than other people he was a person of independent genius . . . [but he] later pricked up his ears and became interested in the action of the moon on the water, and in other occult phenomena, and similar childishness."

Foolishness indeed, it must have seemed, to assign to the tiny tides of the Mediterranean an explanation so cosmic as Moon. But mariners in northern waters face destruction unless they track the tides. For good reason they remember that Moon reaches its summit overhead an average 50.47 minutes later each day. Their own bitter experience tells them that, of the two high tides a day — *two* because there are two projections on an egg — each also comes about 50 minutes later than it did the day before.

Geography makes Mediterranean tides minuscule. Geography also makes tides in the Gulf of Maine and Bay of Fundy the highest in the world. How come? Resonance! The Bay of Fundy and the Gulf of Maine make together a great bathtub in which water sloshes back and forth with a natural period of 13 hours, near to the 12.4-hour timing of Moon's tide-driving power — and to the 12-hour timing of Sun's influence. Build a big power-producing dam in the upper reaches of the Bay of Fundy? Shorten the length of the bathtub? Decrease the slosh time from 13 hours to exact resonance with Moon? Then get one-foot higher tides along the Maine coast!

Want to see the highest tides in the Bay of Fundy? Then choose your visit according to these rules: (1) Come in summer, when this northern body of water tilts most strongly toward Moon. (2) Come when Moon, in its elliptic orbit, is closest to Earth — roughly 10 percent closer than its most distant point, yielding roughly 35 percent greater tide-producing power. (3) Take into account the tide-producing power of Sun, about 45 percent as great as that of Moon. Sun's effect reinforces Moon's influence when Moon is dark, dark because interposed, or almost interposed, between Earth and Sun, so Sun and Moon pull from the same side. But an egg has two projections, so Sun and Moon also assist each other in producing tides when they are on opposite sides of Earth; in this case we see a full Moon.

The result? Burncoat Head in the Minas Basin, Nova Scotia, has the greatest mean range of 14.5 meters (47.5 feet) between low and high tide when Sun and Moon line up. At nearby Leaf Basin, a unique value of 16.6 meters (54.5 feet) was recorded in 1953.

High and low tides witness to the relative accelerations of portions of the ocean separated by the diameter of Earth. High tides show the "stretching" relative acceleration at different radial distances from Moon or Sun. Low tides witness to the "squeezing" relative accelerations at the same radial distance from Moon or Sun but at opposite sides of Earth.

of fall, they move toward each other a distance of 1 millimeter—a thousandth of a meter, the thickness of 16 pages of this book. Why do they move toward one another? Not because of the gravitational attraction between the ball bearings; this is far too minute to bring about any "coming together." Rather, according to Newton's nonlocal view, they are both attracted toward the center of Earth. Their relative motion results from the difference in direction of Earth's gravitational pull on them, says Newton.

As another example, drop the same antique railway coach from rest in a *vertical* orientation, with the lower end of the coach initially 315 meters from the surface of Earth (Figure 2-5, right). Again release tiny ball bearings from rest at opposite ends of the coach. In this case, during the time of fall, the ball bearings move *apart* by a distance of 2 millimeters because of the greater gravitational acceleration of the one nearer Earth, as Newton would put it. This is twice the change that occurs for horizontal separation.

In either of these examples let the measuring equipment in use in the coach be just short of the sensitivity required to detect this relative motion of the ball bearings. Then, with a limited time of observation of 8 seconds, the railway coach—or, to use the earlier example, the freely falling room—serves as a free-float frame.

When the sensitivity of measuring equipment is increased, the railway coach may no longer serve as a local free-float frame unless we make additional changes. Either shorten the 20-meter domain in which observations are made, or decrease the time given to the observations. Or better, cut down some appropriate combination of space and time dimensions of the region under observation. Or as a final alternative, shoot the whole apparatus by rocket up to a region of space where one cannot detect locally the "differential gravitational acceleration" between one side of the coach and another—to use Newton's way of speaking. In another way of speaking, relative accelerations of particles in different parts of the coach must be too small to perceive. Only when these relative accelerations are too small to detect do we have a reference frame with respect to which laws of motion are simple. That's why "local" is a tricky word!

Hold on! You just finished saying that the idea of local gravity is unnecessary. Yet here you use the "differential gravitational acceleration" to account for relative accelerations of test particles and ocean tides near Earth. Is local gravity necessary or not?

Near Earth, two explanations of projectile paths or ocean flow give essentially the same numerical results. Newton says there is a force of gravity, to be treated like any other force in analyzing motion. Einstein says gravity differs from all other forces: Get rid of gravity locally by climbing into a free-float frame. Near the surface of Earth both explanations accurately predict relative accelerations of falling particles toward or away from one another and motions of the tides. In this chapter we use the more familiar Newtonian analysis to predict relative accelerations.

When tests of gravity are very sensitive, or when gravitational effects are large, such as near white dwarfs or neutron stars, then Einstein's predictions are not the same as Newton's. In such cases Einstein's battle-tested 1915 theory of gravity (general relativity) predicts results that are observed; Newton's theory makes incorrect predictions. This justifies Einstein's insistence on getting rid of gravity locally using free-float frames. All that remains of gravity is the relative accelerations of nearby particles—tidal accelerations. ✒

2.4 REGIONS OF SPACETIME

special relativity is limited to free-float frames

"Region of spacetime." What is the precise meaning of this term? The long narrow railway coach in Figure 2-5 probes spacetime for a limited stretch of time and in one or another single direction in space. It can be oriented north–south or east–west or up–down. Whatever its orientation, relative acceleration of the tiny ball bearings released at the two ends can be measured. For all three directions—and for all intermediate directions—let it be found by calculation that the relative drift of two test particles equals half the minimum detectable amount or less. Then throughout a cube of space 20 meters on an edge and for a lapse of time of 8 seconds (2400 million meters of light-travel time), test particles moving every which way depart from straight-line motion by undetectable amounts. In other words, the reference frame is free-float in a local region of spacetime with dimensions

(20 meters $\times$ 20 meters $\times$ 20 meters of space) $\times$ 2400 million meters of time

Notice that this "region of spacetime" is four-dimensional: three dimensions of space and one of time.

"Region of spacetime" is four-dimensional

Why pay so much attention to the small relative accelerations described above? Why not from the beginning consider as reference frames only spaceships very far from Earth, far from our Sun, and far from any other gravitating body? At these distances we need not worry at all about any relative acceleration due to a nonuniform gravitational field, and a free-float frame can be huge without worrying about relative accelerations of particles at the extremities of the frame. Why not study special relativity in these remote regions of space?

Most of our experiments are carried out near Earth and almost all in our part of the solar system. Near Earth or Sun we cannot eliminate relative accelerations of test particles due to nonuniformity of gravitational fields. So we need to know how large a region of spacetime our experiment can occupy and still follow the simple laws that apply in free-float frames.

For some experiments local free-float frames are not adequate. For example, a comet sweeps in from remote distances, swings close to Sun, and returns to deep space. (Consider only the head of the comet, not its 100-million-kilometer-long tail.) Particles traveling near the comet during all those years move closer together or farther apart due to tidal forces from Sun (assuming we can neglect effects of the gravitational field of the comet itself). These relative forces are called **tidal**, because similar differential forces from Sun and Moon act on the ocean on opposite sides of Earth to cause tides (Box 2-1). A frame large enough to include these particles is not free-float. So reduce spatial size until relative motion of encompassed particles is undetectable during that time. The resulting frame is very much smaller than the head of the comet! You cannot analyze the motion of a comet in a frame smaller than the comet. So instead think of a larger free-float frame that surrounds the comet for a limited time during its orbit, so that the comet passes through a series of such frames. Or think of a whole collection of free-float frames plunging radially toward Sun, through which the comet passes in sequence. In either case, motion of the comet over a small portion of its trajectory can be analyzed rigorously with respect to one of these local free-float frames using special relativity. However, questions about the entire trajectory cannot be answered using only one free-float frame; for this we require a series of frames. General relativity—the theory of gravitation—tells how to describe and predict orbits that traverse a string of adjacent free-float frames. Only general relativity can describe motion in unlimited regions of spacetime.

When is general relativity required?

Please stop beating around the bush! In defining a free-float frame, you say that every test particle at rest in such a frame remains at rest "within some specified accuracy." What accuracy? Can't you be more specific? Why do these definitions depend on whether or not we are able to perceive the tiny motion of some test particle? My eyesight gets worse. Or I take my glasses off. Does the world suddenly change, along with the standards for "inertial frame"? Surely science is more exact, more objective than that!

Science can be "exact" only when we agree on acceptable accuracy. A 1000-ton rocket streaks 1 kilometer in 3 seconds; do you want to measure the sequence of its positions during that time with an accuracy of 10 centimeters? An astronaut in an orbiting space station releases a pencil that floats at rest in front of her; do you want to track its position to 1-millimeter accuracy for 2 hours? Each case places different demands on the inertial frame from which the observations are made. Specific figures imply specific requirements for inertial frames, requirements that must be verified by test particles. The astronaut takes off her glasses; then she can determine the position of the pencil with only 3-millimeter accuracy. Suddenly—yes!—requirements on the inertial frame have become less stringent—unless she is willing to observe the pencil over a longer period of time.

2.5 TEST PARTICLE

ideal tool to probe spacetime without affecting it

Test particle defined

"Test particle." How small must a particle be to qualify as a **test particle?** It must have so little mass that, within some specified accuracy, its presence does not affect the motion of other nearby particles. In terms of Newtonian mechanics, gravitational attraction of the test particle for other particles must be negligible within the accuracy specified.

As an example, consider a particle of mass 10 kilograms. A second and less massive particle placed 10 centimeters from it and initially at rest will, in less than 3 minutes, be drawn toward it by 1 millimeter (see the exercises for this chapter). For measurements of this sensitivity or greater sensitivity, the 10-kilogram object is not a test particle. A particle counts as a test particle only when it accelerates as a result of gravitational forces without itself causing measurable gravitational acceleration in other objects—according to the Newtonian way of speaking.

Free-float frame definable because every substance falls with same acceleration

It would be impossible to define a free-float frame were it not for a remarkable feature of nature. Test particles of different size, shape, and material in the same location all fall with the same acceleration toward Earth. If this were not so, an observer inside a falling room would notice that an aluminum object and a gold object accelerate relative to one another, even when placed side by side. At least one of these test particles, initially at rest, would not remain at rest within the falling room. That is, the room would not be a free-float frame according to definition.

How sure are we that particles in the same location but of different substances all fall toward Earth with equal acceleration? John Philoponus of Alexandria argued, in 517 A.D., that when two bodies "differing greatly in weight" are released simultaneously to fall, "the difference in their time [of fall] is a very small one." According to legend Galileo dropped balls made of different materials from the Leaning Tower of Pisa in order to verify this assumption. In 1905 Baron Roland von Eötvös checked that the gravitational acceleration of wood toward Earth is equal to that of platinum within 1 part in 100 million. In the 1960s R. H. Dicke, Peter G. Roll, and Robert V. Krotkov reduced this upper limit on difference in accelerations—for aluminum and gold responding to the gravitational field of Sun—to less than 1 part in 100,000 million (less than 1 in 10^{-11}). This—and a subsequent experiment by Vladimir Braginsky and colleagues—is one of the most sensitive checks of fundamental physical principles in all of science: the equality of acceleration produced by gravity on test particles of every kind.

It follows that a particle made of any material can be used as a test particle to determine whether a given reference frame is free-float. A frame that is free-float for a test particle of one kind is free-float for test particles of all kinds.

2.6 LOCATING EVENTS WITH A LATTICEWORK OF CLOCKS

only the nearest clock records an event

The fundamental concept in physics is **event.** An event is specified not only by a place but also by a time of happening. Some examples of events are emission of a particle or a flash of light (from, say, an explosion), reflection or absorption of a particle or light flash, a collision.

How can we determine the place and time at which an event occurs in a given free-float frame? Think of constructing a frame by assembling meter sticks into a cubical latticework similar to the jungle gym seen on playgrounds (Figure 2-6). At every intersection of this latticework fix a clock. These clocks are identical. They can be constructed in any manner, but their readings are in meters of light-travel time (Section 1.4).

Latticework of rods and clocks

How are the clocks to be set? We want them all to read the "same time" as one another for observers in this frame. When one clock reads midnight (00.00 hours = 0 meters), all clocks in the same frame should read midnight (zero). That is, we want the clocks to be **synchronized** in this frame.

How are the several clocks in the lattice to be synchronized? As follows: Pick one clock in the lattice as the standard and call it the **reference clock.** Start this reference

Synchronizing clocks in lattice

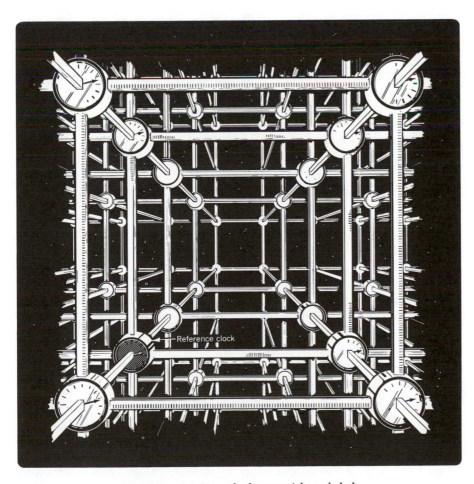

FIGURE 2-6. *Latticework of meter sticks and clocks.*

Reference event defined

clock with its pointer set initially at zero time. At this instant let it send out a flash of light that spreads out as a spherical wave in all directions. Call the flash emission the **reference event** and the spreading spherical wave the **reference flash.**

When the reference flash gets to a slave clock 5 meters away, we want that clock to read 5 meters of light-travel time. Why? Because it takes light 5 meters of light-travel time to travel the 5 meters of distance from reference clock to slave clock. So an assistant sets the slave clock to 5 meters of time long before the experiment begins, holds it at 5 meters, and releases it only when the reference flash arrives. (The assistant has zero reaction time or the slave clock is set ahead an additional time equal to the reaction time.) When assistants at all slave clocks in the lattice follow this prearranged procedure (each setting his slave clock to a time in meters equal to his own distance from the reference clock and starting it when the reference light flash arrives), the lattice clocks are said to be **synchronized.**

This is an awkward way to synchronize lattice clocks with one another. Is there some simpler and more conventional way to carry out this synchronization?

There are other possible ways to synchronize clocks. For example, an extra portable clock could be set to the reference clock at the origin and carried around the lattice in order to set the rest of the clocks. However, this procedure involves a moving clock. We saw in Chapter 1 that the time between two events is not necessarily the same as recorded by clocks in relative motion. The portable clock will not even agree with the reference clock when it is brought back next to it! (This idea is explored more fully in Section 4.6.) However, when we use a moving clock traveling at a speed that is a very small fraction of light speed, its reading is only slightly different from that of clocks fixed in the lattice. In this case the second method of synchronization gives a result nearly equal to the first — and standard — method. Moreover, the error can be made as small as desired by carrying the portable clock around sufficiently slowly.

Locate event with latticework

Use the latticework of synchronized clocks to determine location and time at which any given event occurs. The space position of the event is taken to be the location of the clock nearest the event. The location of this clock is measured along three lattice directions from the reference clock: northward, eastward, and upward. The time of the event is taken to be the time recorded on the same lattice clock nearest the event. The spacetime location of an event then consists of four numbers, three numbers that specify the space position of the clock nearest the event and one number that specifies the time the event occurs as recorded by that clock.

The clocks, when installed by a foresighted experimenter, will be *recording* clocks. Each clock is able to detect the occurrence of an event (collision, passage of light-flash or particle). Each reads into its memory the nature of the event, the time of the event, and the location of the clock. The memory of all clocks can then be read and analyzed, perhaps by automatic equipment.

Why a latticework built of rods that are 1 meter long? What is special about 1 meter? Why not a lattice separation of 100 meters between recording clocks? Or 1 millimeter?

When a clock in the 1-meter lattice records an event, we will not know whether the event so recorded is 0.4 meters to the left of the clock, for instance, or 0.2 meters to the right. The location of the event will be uncertain to some substantial fraction of a meter. The time of the event will also be uncertain with some appreciable fraction of a meter of light-travel time, because it may take that long for a light signal from the event to reach the nearest clock. However, this accuracy of a meter or less is quite adequate for observing the passage of a rocket. It is extravagantly good for measurements on planetary orbits — for a planet it would even be reasonable to increase the lattice spacing from 1 meter to hundreds of meters.

Neither 100 meters nor 1 meter is a lattice spacing suitable for studying the tracks of particles in a high-energy accelerator. There a centimeter or a millimeter would be more appropriate. The location and time of an event can be determined to whatever accuracy is desired by constructing a latticework with sufficiently small spacing.

2.7 OBSERVER

ten thousand local witnesses

In relativity we often speak about the **observer.** Where is this observer? At one place, or all over the place? Answer: **The word "observer" is a shorthand way of speaking about the whole collection of recording clocks associated with one free-float frame.** No one real observer could easily do what we ask of the "ideal observer" in our analysis of relativity. So it is best to think of the observer as a person who goes around reading out the memories of all recording clocks under his control. This is the sophisticated sense in which we hereafter use the phrase "the observer measures such-and-such."

Location and time of each event is recorded by the clock nearest that event. We intentionally limit the observer's report on events to a summary of data collected from clocks. We do not permit the observer to report on widely separated events that he himself views by eye. The reason: travel time of light! It can take a long time for light from a distant event to reach the observer's eye. Even the order in which events are seen by eye may be wrong: Light from an event that occurred a million years ago and a million light-years distant in our frame is just entering our eyes now, after light from an event that occurred on Moon a few seconds ago. We see these two events in the "wrong order" compared with observations recorded by our far-flung latticework of recording clocks. For this reason, we limit the observer to collecting and reporting data from the recording clocks.

The wise observer pays attention only to clock records. Even so, light speed still places limits on how soon he can analyze events after they occur. Suppose that events in a given experiment are widely separated from one another in interstellar space, where a single free-float frame can cover a large region of spacetime. Let remote events be recorded instantly on local clocks and transmitted by radio to the observer's central control room. This information transfer cannot take place faster than the speed of light — the same speed at which radio waves travel. Information on dispersed events is available for analysis at a central location only after light-speed transmission. This information will be full and accurate and in no need of correction — but it will be late. Thus all analysis of events must take place after — sometimes long after! — events are over as recorded in that frame. The same difficulty occurs, in principle, for a free-float frame of any size.

Nature puts an unbreakable speed limit on signals. This limit has profound consequences for decision making and control. A space probe descends onto Triton, a moon of the planet Neptune. The probe adjusts its rocket thrust to provide a slow-speed "soft" landing. This probe must carry equipment to detect its distance from Triton's surface and use this information to regulate rocket thrust on the spot, without help from Earth. Earth is never less than 242 light-minutes away from Neptune, a round-trip radio-signal time of 484 minutes — more than eight hours. Therefore the probe would crash long before probe-to-surface distance data could be sent to Earth and commands for rocket thrust returned. This time delay of information transmission does not prevent a detailed retrospective analysis on Earth of the probe's descent onto Triton — but this analysis cannot take place until at least 242 minutes

Observer defined

Observer limited to clock readings

Speed limit: c
It's the law!

SAMPLE PROBLEM 2-1

METEOR ALERT!

Interstellar Command Center receives word by radio that a meteor has just whizzed past an outpost situated 100 light-seconds distant (a fifth of Earth-Sun distance). The report warns that the meteor is headed directly toward Command Center at one quarter light speed. Assume radio signals travel with light speed. How long do Command Center personnel have to take evasive action?

SOLUTION

The warning radio signal and the meteor leave the outpost at the same time. The radio signal moves with light speed from outpost to Command Center, covering the 100 light-seconds of distance in 100 seconds of time. During this 100 seconds the meteor also travels toward Command Center. The meteor moves at one quarter light speed, so in 100 seconds it covers one quarter of 100 light-seconds, or 25 light-seconds of distance. Therefore, when the warning arrives at Command Center, the meteor is $100 - 25 = 75$ light-seconds away.

The meteor takes an additional 100 seconds of time to move each additional 25 light-seconds of distance. So it covers the remaining 75 light-seconds of distance in an additional time of 300 seconds.

In brief, after receiving the radio warning, Command Center personnel have a relaxed 300 seconds — or five minutes — to stroll to their meteor-proof shelter.

---

after the event. Could we gather last-minute information, make a decision, and send back control instructions? No. Nature rules out micromanagement of the far-away (Sample Problem 2-1).

2.8 MEASURING PARTICLE SPEED

reference frame clocks and rods put to use

The recording clocks reveal particle motion through the lattice: Each clock that the particle passes records the time of passage as well as the space location of this event. How can the path of the particle be described in terms of numbers? By recording locations of these events along the path. Distances between locations of successive events and time lapse between them reveal the particle speed — speed being space separation divided by time taken to traverse this separation.

Speed in meters per meter

The conventional unit of speed is meters per second. However, when time is measured in meters of light-travel time, speed is expressed in meters of distance covered per meter of time. A flash of light moves one meter of distance in one meter of light-travel time: its speed has the value unity in units of meters per meter. In contrast, a particle loping along at half light speed moves one half meter of distance per meter of time; its speed equals one half in units of meter per meter. More generally, particle speed in meters per meter is the ratio of its speed to light speed:

$$(\text{particle speed}) = \frac{(\text{meters of distance covered by particle})}{(\text{meters of time required to cover that distance})}$$

$$= \frac{(\text{particle speed in meters/second})}{(\text{speed of light in meters/second})}$$

In this book we use the letter v to symbolize the speed of a particle in meters of distance per meter of time, or simply meters per meter. Some authors use the lowercase Greek letter beta: β. Let v_{conv} stand for velocity in conventional units (such as meters per second) and c stand for light speed in the same conventional units. Then

$$v = \frac{v_{conv}}{c} \qquad (2\text{-}1)$$

From the motion of test particles through a latticework of clocks—or rather from records of coincidences of these particles with clocks—we determine whether the latticework constitutes a free-float frame. IF records show (a) that—within some specified accuracy—a test particle moves consecutively past clocks that lie in a straight line, (b) that test-particle speed calculated from the same records is constant—again, within some specified accuracy—and, (c) that the same results are true for as many test-particle paths as the most industrious observer cares to trace throughout the given region of space and time, THEN the lattice constitutes a free-float (inertial) frame throughout that region of spacetime.

Test for free-float frame

Particle speed as a fraction of light speed is certainly an unconventional unit of measure. What advantages does it have that justify the work needed to become familiar with it?

The big advantage is that it is a measure of speed independent of units of space and time. Suppose that a particle moves with respect to Earth at half light speed. Then it travels—with respect to Earth—one half meter of distance in one meter of light travel time. It travels one half light-year of distance in a period of one year. It travels one half light-second of distance in a time of one second, one half light-minute in one minute. Units do not matter as long as we use the same units to measure distance and time; the result always equals the same number: 1/2. Another way to say this is that speed is a fraction; same units on top and bottom of the fraction cancel one another. Fundamentally, v is unit-free. Of course, if we wish we can speak of "meters per meter."

2.9 ROCKET FRAME

does it move? or is it the one at rest?

Let two reference frames be two different latticeworks of meter sticks and clocks, one moving uniformly relative to the other, and in such a way that one row of clocks in each frame coincides along the direction of relative motion of the two frames (Figure 2-7). Call one of these frames **laboratory frame** and the other—moving to the right relative to the laboratory frame—**rocket frame.** The rocket is *unpowered* and coasts along with constant velocity relative to the laboratory. Let rocket and laboratory latticeworks be overlapping in the sense that a region of spacetime exists common to both frames. Test particles move through this common region of spacetime. From motion of these test particles as recorded by his own clocks, the laboratory observer verifies that his frame is free-float (inertial). From motion of the same test particles as recorded by her own clocks, the rocket observer verifies that her frame is also free-float (inertial).

Rocket frame defined

Now we can describe the motion of any particle with respect to the laboratory frame. The same particles and—if they collide—the same collisions may be measured and described with respect to the free-float rocket frame as well. These particles, their paths through spacetime, and events of their collisions have an existence inde-

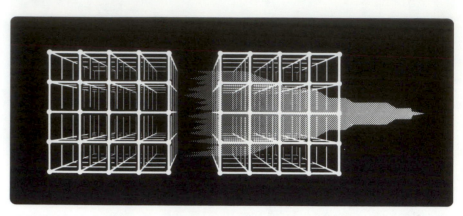

FIGURE 2-7. *Laboratory and rocket frames.* A second ago the two latticeworks were intermeshed.

pendent of any free-float frames in which they are observed, recorded, and described. However, descriptions of these common paths and events are typically different for different free-float frames. For example, laboratory and rocket observers may not agree on the direction of motion of a given test particle (Figure 2-8). Every track that is straight as plotted with respect to one reference frame is straight also with respect to the other frame, because both are free-float frames. This straightness in both frames is possible only because *one free-float frame has uniform velocity relative to any other*

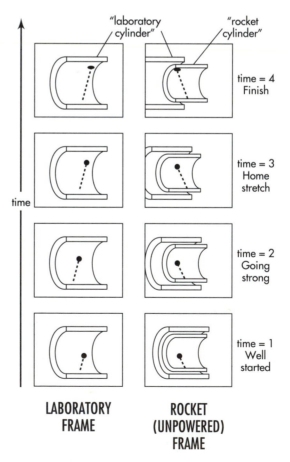

FIGURE 2-8. *A series of "snapshots" of a typical test particle as measured from laboratory and rocket free-float frames, represented by cutaway cylinders. Start at the bottom and read upward (time progresses from bottom to top).*

overlapping free-float frame. However, the direction of this path differs from laboratory to rocket frame, except in the special case in which the particle moves along the line of relative motion of two frames.

How many different free-float rocket frames can there be in a given region of spacetime? An unlimited number! Any unpowered rocket moving through that region in any direction is an acceptable free-float frame from which to make observations. More: There is nothing unique about any of these frames as long as each of them is free-float. All "rocket" frames are unpowered, all are equivalent for carrying out experiments. Even the so-called "laboratory frame" is not unique; you can rename it "Rocket Frame Six" and no one will ever know the difference! All free-float (inertial) frames are equivalent arenas in which to carry out physics experiment. That is the logical basis for special relativity, as described more fully in Chapter 3.

A rocket carries a firecracker. The firecracker explodes. Does this event — the explosion — take place in the rocket frame or in the laboratory frame? Which is the "home" frame for the event? A second firecracker, originally at rest in the laboratory frame, explodes. Does this second event occur in the laboratory frame or in the rocket frame?

Events are primary, the essential stuff of Nature. Reference frames are secondary, devised by humans for locating and comparing events. A given event occurs in both frames — and in all possible frames moving in all possible directions and with all possible constant relative speeds through the region of spacetime in which the event occurs. The apparatus that "causes" the event may be at rest in one free-float frame; another apparatus that "causes" a second event may be at rest in a second free-float frame in motion relative to the first. No matter. Each event has its own unique existence. Neither is "owned" by any frame at all.

A spark jumps 1 millimeter from the antenna of Mary's passing spaceship to a pen in the pocket of John who lounges in the laboratory doorway (Section 1.2). The "apparatus" that makes the spark has parts riding in different reference frames — pen in laboratory frame, antenna in rocket frame. The spark jump — in which frame does this event occur? It is not the property of Mary, not the property of John — not the property of any other observer in the vicinity, no matter what his or her state of motion. The spark-jump event provides data for every observer.

Drive a steel surveying stake into the ground to mark the corner of a plot of land. Is this a "Daytime stake" or a "Nighttime stake"? Neither! It is just a *stake*, marking a location in *space,* the arena of surveying. Similarly an event is neither a "laboratory event" nor a "rocket event." It is just an *event,* marking a location in *spacetime,* the arena of science.

Laboratory frame or rocket frame: Which one is the "primary" free-float frame, the one "really" at rest? There is no way to tell! We apply the names "laboratory" and "rocket" to two free-float enclosures in interstellar space. Someone switches the nameplates while we sleep. When we wake up, there is no way to decide which is which. This realization leads to Einstein's Principle of Relativity and proof of the invariance of the interval, as described in Chapter 3.

Many possible free-float frames

No unique free-float frame

2.10 SUMMARY

what a free-float frame is and what it's good for

The **free-float frame** (also called the **inertial frame** and the **Lorentz frame**) provides a setting in which to carry out experiments without the presence of so-called "gravitational forces." In such a frame, a particle released from rest remains at rest and a particle in motion continues that motion without change in speed or in direction (Section 2.2), as Newton declared in his First Law of Motion.

Where does that frame of reference sit? Where do the east-west, north-south, up-down lines run? We might as well ask where on the flat landscape in the state of Iowa we see the lines that mark the boundaries of the townships. A concrete marker, to be sure, may show itself as a corner marker at a place where a north-south line meets an east-west line. Apart from such on-the-spot evidence, those lines are largely invisible. Nevertheless, they serve their purpose: They define boundaries, settle lawsuits, and fix taxes. Likewise imaginary for the most part are the clock and rod paraphernalia of the idealized inertial reference frame. Work of the imagination though they are, they provide the conceptual framework for everything that goes on in the world of particles and radiation, of masses and motions, of annihilations and creations, of fissions and fusions in every context where tidal effects of gravity are negligible.

Our ability to define a free-float frame depends on the fact that a **test particle** made of any material whatsoever experiences the same acceleration in a given gravitational field (Section 2.5).

Near a massive ("gravitating") body, we can still define a free-float frame. However, in such a frame, free test particles typically accelerate toward or away from one another because of the nonuniform field of the gravitating body (Section 2.3). This limits — in both space and time — the size of a free-float frame, the domain in which the laws of motion are simple. The frame will continue to qualify as free-float and special relativity will continue to apply, provided we reduce the spatial extent, or the time duration of our experiment, or both, until these relative, or **tidal,** motions of test particles cannot be detected in our circumscribed region of spacetime. This is what makes special relativity "special" or limited (French: *relativité restreinte:* "restricted relativity"). General relativity (the theory of gravitation) removes this limitation (Chapter 9).

So there are three central characteristics of a free-float frame. (1) We can "get rid of gravity" by climbing onto (getting into) a free-float frame. (2) The existence of a free-float frame depends on the equal acceleration of all particles at a given location in a gravitational field — in Newton's way of speaking. (3) Every free-float frame is of limited extent in spacetime. All three characteristics appear in a fuller version of the quotation by Albert Einstein that began this chapter:

> At that moment there came to me the happiest thought of my life . . . *for an observer falling freely from the roof of a house no gravitational field exists during his fall* — at least not in his immediate vicinity. That is, if the observer releases any objects, they remain in a state of rest or uniform motion relative to him, respectively, independent of their unique chemical and physical nature. Therefore the observer is entitled to interpret his state as that of "rest." ✍

REFERENCES

Introductory and final quotes: Excerpt from an unpublished manuscript in Einstein's handwriting, dating from about 1919. Einstein is referring to the year 1907. Italics represent material underlined in the original. Quoted by Gerald Holton in *Thematic Origins of Scientific Thought,* Revised Edition (Harvard University Press, Cambridge, Mass.,1988), page 382. Photocopy of the original provided by Professor Holton. Present translation made with the assistance of Peter von Jagow.

Figure 2-1 and Jules Verne story: Jules Verne, *A Trip From the Earth to the Moon* and *A Trip Around the Moon,* paperback edition published by Dover Publications, New York. Hardcover edition published in the Great Illustrated Classics Series by Dodd, Mead and Company, New York, 1962.

Information on Nova Scotia tides in Box 2-1: *Guinness Book of World Records 1988* (Bantam Books, New York, 1987), page 125.

Relative acceleration of different materials, Section 2-5: P. G. Roll, R. Krotkov, and R. H. Dicke, ''The equivalence of inertial and passive gravitational mass,'' *Annals of Physics (USA),* Volume 26, pages 442 – 517 (1964); V. B. Braginsky and V. I. Panov, *Zhurnal Eksperimental'noi i Teoreticheskoi Fiziki,* Volume 61, page 873 (1972) [*Soviet Physics JETP,* Volume 34, page 463 (1972)].

Experimental proofs of Einstein's general relativity theory: Clifford Will, *Was Einstein Right? Putting General Relativity to the Test* (Basic Books, New York, 1986).

CHAPTER 2 EXERCISES

PRACTICE

2-1   human cannonball

A person rides in an elevator that is shot upward out of a cannon. Think of the elevator after it leaves the cannon and is moving freely in the gravitational field of Earth. Neglect air resistance.

   a   While the elevator is still on the way up, the person inside jumps from the ''floor'' of the elevator. Will the person (1) fall back to the ''floor'' of the elevator? (2) hit the ''ceiling'' of the elevator? (3) do something else? If so, what?

   b   The person waits to jump until after the elevator has passed the top if its trajectory and is falling back toward Earth. Will your answers to part **a** be different in this case?

   c   How can the person riding in the elevator tell when the elevator reaches the top if its trajectory?

2-2   free-float bounce

Test your skill as an acrobat and contortionist! Fasten a weight-measuring bathroom scale under your feet and bounce up and down on a trampoline while reading the scale. Describe readings on the scale at different times during the bounces. During what part of each jump will the scale have zero reading? Neglecting air resistance, what is the longest part of the cycle during which you might consider yourself to be in a free-float frame?

2-3   practical synchronization of clocks

You are an observer in the laboratory frame stationed near a clock with spatial coordinates $x = 6$ light-seconds, $y = 8$ light-seconds, and $z = 0$ light-seconds. You wish to synchronize your clock with the one at the origin. Describe in detail and with numbers how to proceed.

2-4   synchronization by a traveling clock

Mr. Engelsberg does not approve of our method of synchronizing clocks by light flashes (Section 2.6).

   a   ''I can synchronize my clocks in any way I choose!'' he exclaims. Is he right?

   Mr. Engelsberg wishes to synchronize two identical clocks, named Big Ben and Little Ben, which are relatively at rest and separated by one million kilometers, which is 10^9 meters or approximately three times the distance between Earth and Moon. He uses a third clock, identical in construction with the first two, that travels with constant velocity between them. As his moving clock passes Big Ben, it is set to read the same time as Big Ben. When the moving clock passes Little Ben, that outpost clock is set to read the same time as the traveling clock.

b "Now Big Ben and Little Ben are synchronized," says Mr. Engelsberg. Is he right?

c How much out of synchronism are Big Ben and Little Ben as measured by a latticework of clocks — at rest relative to them both — that has been synchronized in the conventional manner using light flashes? Evaluate this lack of synchronism in milliseconds when the traveling clock that Mr. Engelsberg uses moves at 360,000 kilometers/hour, or 10^5 meters/second.

d Evaluate the lack of synchronism when the traveling clock moves 100 times as fast.

e Is there any earthly reason — aside from matters of personal preference — why we all should not adopt the method of synchronization used by Mr. Engelsberg?

2-5 Earth's surface as a free-float frame

Many experiments involving fast-moving particles and light itself are observed in earthbound laboratories. Typically these laboratories are not in free fall! Nevertheless, under many circumstances laboratories fixed to the surface of Earth can satisfy the conditions required to be called free-float frames. An example:

a In an earthbound laboratory, an elementary particle with speed $v = 0.96$ passes from side to side through a cubical spark chamber one meter wide. For what length of laboratory time is this particle in transit through the spark chamber? Therefore for how long a time is the experiment "in progress"? How far will a separate test particle, released from rest, fall in this time? [Distance of fall from rest $= (1/2)g t_{sec}^2$, where g = acceleration of gravity ≈ 10 meters/second^2 and t_{sec} is the time of free fall in seconds.] Compare your answer with the diameter of an atomic nucleus (a few times 10^{-15} meter).

b How wide *can* the spark chamber be and still be considered a free-float frame for this experiment? Suppose that by using sensitive optical equipment (an **interferometer**) you can detect a test particle change of position as small as one wavelength of visible light, say 500 nanometers $= 5 \times 10^{-7}$ meter. How long will it take the test particle to fall this distance from rest? How far does the fast elementary particle of part **a** move in that time? Therefore how long can an earthbound spark chamber be and still be considered free-float for this sensitivity of detection?

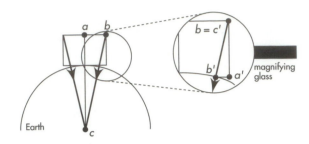

EXERCISE 2-6. *Schematic diagram of two ball bearings falling onto Earth's surface. Not to scale.*

2-6 horizontal extent of free-float frame near Earth

Consider two ball bearings near the surface of Earth and originally separated horizontally by 20 meters (Section 2.3). Demonstrate that when released from rest (relative to Earth) the particles move closer together by 1 millimeter as they fall 315 meters, using the following method of similar triangles or some other method.

Each particle falls from rest toward the center of Earth, as indicated by arrows in the figure. Solve the problem using the ratio of sides of similar triangles abc and $a'b'c'$. These triangles are upside down with respect to each other. However, they are similar because their respective sides are parallel: Sides ac and $a'c'$ are parallel to each other, as are sides bc and $b'c'$ and sides ab and $a'b'$. We know the lengths of some of these sides. Side $a'c' = 315$ meters is the height of fall (greatly exaggerated in the diagram); side ac is effectively equal to the radius of Earth, 6,371,000 meters. Side $ab = (1/2)$ (20 meters) equals half the original separation of the particles. Side $a'b'$ equals HALF their CHANGE in separation as they fall onto Earth's surface. Use the ratio of sides of similar triangles to find this "half-change" and therefore the entire change in separation as two particles initially 20 meters apart horizontally fall from rest 315 meters onto the surface of Earth.

2-7 limit on free-float frame near Earth's Moon

Release two ball bearings from rest a horizontal distance 20 meters apart near the surface of Earth's Moon. By how much does the separation between them decrease as they fall 315 meters? How many seconds elapse during this 315-meter fall? Assume that an initial vertical separation of 20 meters is increased by twice the change in horizontal separation in a fall through the same height. State clearly and completely the dimensions of the region of spacetime in which such a freely falling frame constitutes an inertial frame (to the given accuracy). Moon radius equals

1738 kilometers. Gravitational acceleration at Moon's surface: $g = 1.62$ meters/second^2.

2-8   vertical extent of free-float frame near Earth

Note: This exercise makes use of elementary calculus and the Newtonian theory of gravitation.

A paragraph in Section 2.3 says:

> As another example, drop the same antique [20-meter-long] railway coach from rest in a *vertical* orientation, with the lower end of the coach initially 315 meters from the surface of Earth (Figure 2-5, right). Again release two tiny ball bearings from rest at opposite ends of the coach. In this case, during the time of fall [8 seconds], the ball bearings move *apart* by a distance of two millimeters because of the greater gravitational acceleration of the one nearer Earth, as Newton would put it. This is twice the change that occurs for horizontal separation.

Demonstrate this 2-millimeter increase in separation. The following outline may be useful. Take the gravitational acceleration at the surface of Earth to be $g_0 = 9.8$ meters/second^2 and the radius of Earth to be $r_0 = 6.37 \times 10^6$ meters. More generally, the gravitational acceleration g of a particle of mass m a distance r from the center of Earth (mass M) is given by the expression

$$g = \frac{F}{m} = \frac{GM}{r^2} = \frac{GM}{r_0^2}\frac{r_0^2}{r^2} = \frac{g_0 r_0^2}{r^2}$$

a   Take the differential of this equation for g to obtain an approximate algebraic expression for Δg, the change in g, for a small change Δr in height.

b   Now use $\Delta y = \frac{1}{2}\Delta g t^2$ to find an algebraic expression for increase in distance Δy between ball bearings in a fall that lasts for time t.

c   Substitute numbers given in the quotation above to verify the 2-millimeter change in separation during fall.

2-9   the rising railway coach

You are launched upward inside a railway coach in a horizontal position with respect to the surface of Earth, as shown in the figure. After the launch, but while the coach is still rising, you release two ball bearings at opposite ends of the train and at rest with respect to the train.

a   Riding inside the coach, will you observe the distance between the ball bearings to increase or decrease with time?

b   Now you ride in a second railway coach launched upward in a *vertical* position with respect to

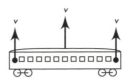

EXERCISE 2-9. *Free-float railway coach rising from Earth's surface, as observed in Earth frame. Two ball bearings were just released from rest with respect to the coach. What will be their subsequent motion as observed from inside the coach? Figure not to scale.*

the surface of Earth (not shown). Again you release two ball bearings at opposite ends of the coach and at rest with respect to the coach. Will you observe these ball bearings to move together or apart?

c   In either of the cases described above, can you, the rider in the railway coach, distinguish whether the coach is rising or falling with respect to the surface of Earth solely by observing the ball bearings from inside the coach? What do you observe at the moment the coach stops rising with respect to Earth and begins to fall?

2-10   test particle?

a   Verify the statement in Section 2.5 that a candidate test particle of mass 10 kilograms placed 0.1 meter from a less massive particle (initially stationary with respect to it), draws the second toward it by 1 millimeter in less than 3 minutes. If this relative motion is detectable by equipment in use at the test site, the result disqualifies the 10-kilogram particle as a "test particle." Assume that both particles are spherically symmetric. Use Newton's Law of Gravitation:

$$F = \frac{GMm}{r^2}$$

where the gravitation constant G has the value $G = 6.670 \times 10^{-11}$ meter^3/(kilogram-second^2). Assume that this force does not change appreciably as the particles decrease separation by one millimeter.

b   Section 2.3 describes two ball bearings released 20 meters apart horizontally in a freely falling railway coach. They move 1 millimeter closer together during 8 seconds of free fall, showing the limitations on this inertial frame. Verify that these ball bearings qualify as test particles by estimating the distance that one will move from rest in 8 seconds under the gravitational attraction of the other, if both were initially at rest in interstellar space far from Earth. Make your own estimate of the mass of each ball bearing.

PROBLEMS

2-11   communications storm!

Sun emits a tremendous burst of particles that travels toward Earth. An astronomer on Earth sees the emission through a solar telescope and issues a warning. The astronomer knows that when the particles arrive, they will wreak havoc with broadcast radio transmission. Communications systems require three minutes to switch from broadcast to underground cable. What is the maximum speed of the particle pulse emitted by Sun such that the switch can occur in time, between warning and arrival of the pulse? Take Sun to be 500 light-seconds from Earth.

2-12   the Dicke experiment

a   The Leaning Tower of Pisa is about 55 meters high. Galileo says, "The variation of speed in air between balls of gold, lead, copper, porphyry, and other heavy materials is so slight that in a fall of 100 cubits [about 46 meters] a ball of gold would surely not outstrip one of copper by as much as four fingers. Having observed this I came to the conclusion that in a medium totally devoid of resistance all bodies would fall with the same speed."

Taking four fingers to be equal to 7 centimeters, find the maximum fractional difference in the acceleration of gravity $\Delta g/g$ between balls of gold and copper that would be consistent with Galileo's experimental result.

b   The result of the more modern Dicke experiment is that the fraction $\Delta g/g$ is not greater than 3×10^{-11}. Assume that the fraction has this more recently determined maximum value. Reckon how far behind the first ball the second one will be when the first reaches the ground if they are dropped simultaneously from the top of a 46-meter vacuum chamber. Under these same circumstances, how far would balls of different materials have to fall in a vacuum in a uniform gravitational field of 10 meters/second/second for one ball to lag behind the other one by a distance of 1 millimeter? Compare this distance with the Earth–Moon separation (3.8×10^8 meters). Clearly the Dicke experiment was not carried out using falling balls!

c   A plumb bob of mass m hangs on the end of a long line from the ceiling of a closed room, as shown in the first figure (left). A very massive sphere at one side of the closed room exerts a horizontal gravitational force mg_s on the plumb bob, where $g_s = GM/R^2$, M is the mass of the large sphere, and R the distance between plumb bob and the center of the sphere. This horizontal force causes a static deflection of the plumb line from the vertical by the small angle ε. (Similar practical example: In northern India the mass of the Himalaya Mountains results in a slight sideways deflection of plumb lines, causing difficulties in precise surveying.) The sphere is now rolled around to a corresponding position on the other side of the room (right), causing a static deflection of the plumb by an angle ε of the same magnitude but in the opposite direction.

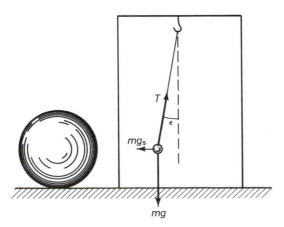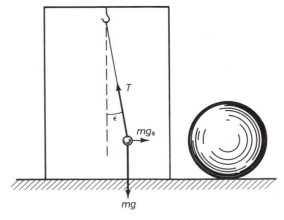

EXERCISE 2-12, first figure. *Left: Nearby massive sphere results in static deflection of plumb line from vertical.* **Right:** *Rolling the sphere to the other side results in static deflection of plumb line in the opposite direction.*

Now the angle ε is very small. (Deflection due to the Himalayas is about 5 seconds of arc, which equals 0.0014 degrees.) However, as the sphere is rolled around and around outside the closed room, an observer inside the room can measure the gravitational field g_s due to the sphere by measuring with greater and greater precision the total deflection angle $2\varepsilon \approx 2 \sin \varepsilon$ of the plumb line, where ε is measured in radians. Derive the equation that we will need in the calculation of g_s.

d We on Earth have a large sphere effectively rolling around us once every day. It is the most massive sphere in the solar system: Sun itself! What is the value of the gravitational acceleration $g_s = GM/R^2$ due to Sun at the position of Earth? (Some constants useful in this calculation appear inside the back cover of this book.)

e One additional acceleration must be considered that, however, will not enter our final comparison of gravitational acceleration g_s for different materials. This additional acceleration is the centrifugal acceleration due to the motion of Earth around Sun. When you round a corner in a car you are pressed against the side of the car on the outward side of the turn. This outward force — called the centrifugal pseudoforce or the centrifugal inertial force — is due to the acceleration of your reference frame (the car) toward the center of the circular turn. This centrifugal inertial force has the value mv_{conv}^2/r, where v_{conv} is the speed of the car in conventional units and r is the radius of the turn. Now Earth moves around Sun in a path that is nearly circular. Sun's gravitational force mg_s acts on a plumb bob in a direction toward Sun; the centrifugal inertial force mv_{conv}^2/R acts in a direction away from Sun. Compare the "centrifugal acceleration" mv_{conv}^2/R at the position of Earth with the oppositely directed gravitational acceleration g_s calculated in part **d**. What is the net acceleration toward or away from Sun of a particle riding on Earth as observed in the (accelerated) frame of Earth?

f Of what use is the discussion thus far? A plumb bob hung near the surface of Earth experiences a gravitational acceleration g_s toward Sun — and an equal but opposite centrifugal acceleration mv_{conv}^2/R away from Sun. Therefore — in the accelerating reference frame of Earth — the bob experiences no net force at all due to the presence of Sun. Indeed this is the method by which we constructed an inertial frame in the first place (Section 2.2): Let the frame be in free fall about the center of gravitational attraction. A particle at rest on Earth's surface is in free fall about Sun and therefore experiences no net force due to Sun. What then does all this have to do with measuring the equality of gravitational acceleration for particles made of different substances — the subject of the

Dicke experiment? Answer: Our purpose is to detect the difference — if any — in the gravitational acceleration g_s toward Sun for different materials. The centrifugal acceleration v^2/R away from Sun is presumably the same for all materials and therefore need not enter any comparison of different materials.

Consider a torsion pendulum suspended from its center by a thin quartz fiber (second figure). A light rod of length L supports at its ends two bobs of equal mass made of different materials — say aluminum and gold. Suppose that the gravitational acceleration g_1 of the gold due to Sun is slightly greater than the acceleration g_2 of the aluminum due to Sun. Then there will be a slight net torque on the torsion pendulum due to Sun. For the position of Sun shown at left in the figure, show that the net torque is counterclockwise when viewed from above. Show also that the magnitude of this net torque is given by the expression

$$\text{torque} = mg_1 \, L/2 - mg_2 \, L/2 = m(g_1 - g_2) \, L/2$$
$$= mg_s(\Delta g/g_s) \, L/2$$

g Suppose that the fraction $(\Delta g/g_s)$ has the maximum value 3×10^{-11} consistent with the results of the final experiment, that L has the value 0.06 meters, and that each bob has a mass of 0.03 kilograms. What is the magnitude of the net torque? Compare this to the torque provided by the added weight of a bacterium of mass 10^{-15} kilogram placed on the end of a meter stick balanced at its center in the gravitational field of Earth.

h Sun moves around the heavens as seen from Earth. Twelve hours later Sun is located as shown at right in the second figure. Show that under these changed circumstances the net torque will have the same magnitude as that calculated in part **g** but now will be clockwise as viewed from above — in a sense opposite to that of part **g**. This change in the sense of the torque every twelve hours allows a small difference $\Delta g = g_1 - g_2$ in the acceleration of gold and aluminum to be detected using the torsion pendulum. As the torsion pendulum jiggles on its fiber because of random motion, passing trucks, Earth tremors and so forth, one needs to consider only those deflections that keep step with the changing position of Sun.

i A torque on the rod causes an angular rotation of the quartz fiber of θ radians given by the formula

$$\text{torque} = k\theta$$

where k is called the **torsion constant** of the fiber. Show that the maximum angular rotation of the torsion pendulum from one side to the other during one

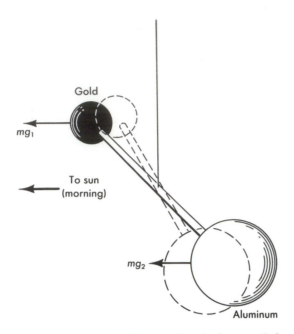

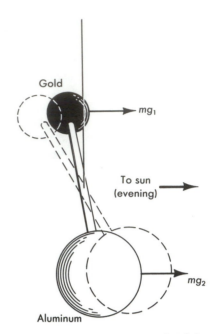

EXERCISE 2-12, second figure. *Schematic diagram of the Dicke experiment.* **Left:** *Hypothetical effect: morning.* **Right:** *Hypothetical effect: evening. Any difference in the gravitational acceleration of Sun for gold and aluminum should result in opposite sense* *of net torque on torsion pendulum in the evening compared with the morning. The large aluminum ball has the same mass as the small high-density gold ball.*

rotation of Earth is given by the expression

$$\theta_{tot} = \frac{mg_sL}{k}\left(\frac{\Delta g}{g_s}\right)$$

j In practice Dicke's torsion balance can be thought of as consisting of 0.030-kilogram gold and aluminum bobs mounted on the ends of a beam 6×10^{-2} meter in length suspended in a vacuum on a quartz fiber of torsion constant 2×10^{-8} newton meter/radian. A statistical analysis of the angular displacements of this torsion pendulum over long periods of time leads to the conclusion that the fraction $\Delta g/g$ for gold and aluminum is less than 3×10^{-11}. To what mean maximum angle of rotation from side to side during one rotation of Earth does this correspond? Random motions of the torsion pendulum — noise! — are of much greater amplitude than this; hence the need for the statistical analysis of the results.

References: R. H. Dicke, "The Eötvös Experiment," *Scientific American,* Volume 205, pages 84–94 (December, 1961). See also P. G. Roll, R. Krotkov, and R. H. Dicke, *Annals of Physics,* Volume 26, pages 442–517 (1964). The first of these articles is a popular exposition written early in the course of the Dicke experiment. The second article reports the final results of the experiment and takes on added interest because of its account of the elaborate precautions required to insure that no influence that might affect the experiment was disregarded. Galileo quote from Galileo Galilei, *Dialogues Concerning Two New Sciences,* translated by Henry Crew and Alfonso de Salvio (Northwestern University Press, Evanston, Illinois, 1950).

2-13   deflection of starlight by Sun

Estimate the deflection of starlight by Sun using an elementary analysis. **Discussion:** Consider first a simpler example of a similar phenomenon. An elevator car of width L is released from rest near the surface of Earth. At the instant of release a flash of light is fired horizontally from one wall of the car toward the other wall. After release the elevator car is an inertial frame. Therefore the light flash crosses the car in a straight line with respect to the car. With respect to Earth, however, the flash of light is falling — because the elevator is falling. Therefore a light flash is deflected in a gravitation field, as Newton would phrase it. (How would Einstein phrase it? See Chapter 9.) As another example, a ray of starlight in its passage tangentially across Earth's surface receives a gravitational deflection (over and above any refraction by Earth's atmosphere). However, the time to cross Earth is so short, and in consequence the deflection so slight, that this effect has not yet been detected on Earth. At the surface of Sun, however, the acceleration of gravity has the much greater value of 275 meters/second/second. Moreover, the time of passage across the surface is much increased because Sun has a greater diameter, 1.4×10^9 meters. In the following, assume that the light just grazes the surface of Sun in passing.

a   Determine an "effective time of fall" from the

diameter of Sun and the speed of light. From this time of fall deduce the net velocity of fall toward Sun produced by the end of the whole period of gravitational interaction. (The maximum acceleration acting for this "effective time" produces the same net effect [calculus proof!] produced by the actual acceleration —changing in magnitude and direction along the path—in the entire passage of the ray through Sun's field of force.)

b Comparing the lateral velocity of the light with its forward velocity, deduce the angle of deflection. The accurate analysis of special relativity gives the same result. However, Einstein's 1915 general relativity predicted a previously neglected effect, associated with the change of lengths in a gravitational field, that produces something like a supplementary refraction of the ray of light and doubles the predicted deflection. [Deflection observed in 1947 eclipse of Sun: $(9.8 \pm 1.3) \times 10^{-6}$ radian; in the 1952 eclipse: $(8.2 \pm 0.5) \times 10^{-6}$ radian.]

CHAPTER 3

SAME LAWS FOR ALL

The name relativity theory was an unfortunate choice: The relativity of space and time is not the essential thing, which is the independence of laws of Nature from the viewpoint of the observer.

Arthur Sommerfeld

3.1 THE PRINCIPLE OF RELATIVITY

fundamental science needs only a closed room

How do you know you are moving? Or at rest? In a car, you pause at a stoplight. You see the car next to you easing forward. With a shock you suddenly realize that, instead, your own car is rolling backward. On an international flight you watch a movie with the cabin shades drawn. Can you tell if the plane is traveling at minimum speed or full speed? In an elaborate joke, could the plane actually be sitting still on the runway, engines running? How would you know?

Everyday observations such as these form the basis for a conjecture that Einstein raised to the status of a postulate and set at the center of the theory of special relativity. He called it the **Principle of Relativity**. Roughly speaking, the Principle of Relativity says that without looking out the window you cannot tell which reference frame you are in or how fast you are moving.

Galileo Galilei made the first known formulation of the Principle of Relativity. Listen to the characters in his book:

Principle of Relativity: With shades drawn you cannot tell your speed

SALVATIUS: Shut yourself up with some friend in the main cabin below decks on some large ship, and have with you there some flies, butterflies, and other small flying animals. Have a large bowl of water with some fish in it; hang up a bottle that empties drop by drop into a wide vessel beneath it. With the ship standing still, observe carefully how the little animals fly with equal speed to all sides of the cabin. The fish swim indifferently in all directions; the drops fall into the vessel beneath; and, in throwing something to your friend, you need throw it no more strongly in one direction than another, the distances being equal; jumping with your feet together, you pass equal spaces in every direction. When you have observed all these things carefully (though there is no doubt that when the ship is standing still everything must happen in this way), have the ship proceed with any speed you like, so long as the motion is uniform and not fluctuating this way and that. You will discover not the least change in all the effects named, nor could you tell from any

Galileo: First known formulation of Principle of Relativity

of them whether the ship was moving or standing still. In jumping, you will pass on the floor the same spaces as before, nor will you make larger jumps toward the stern than toward the prow even though the ship is moving quite rapidly, despite the fact that during the time that you are in the air the floor under you will be going in a direction opposite to your jump. In throwing something to your companion, you will need no more force to get it to him whether he is in the direction of the bow or the stern, with yourself situated opposite. The droplets will fall as before into the vessel beneath without dropping toward the stern, although while the drops are in the air the ship runs many spans. The fish in their water will swim toward the front of their bowl with no more effort than toward the back, and will go with equal ease to bait placed anywhere around the edges of the bowl. Finally the butterflies and flies will continue their flights indifferently toward every side, nor will it ever happen that they are concentrated toward the stern, as if tired out from keeping up with the course of the ship, from which they will have been separated during long intervals by keeping themselves in the air . . .

GALILEO GALILEI
Pisa, February 15, 1564 — Arcetri, near Florence, January 8, 1642

"My portrait is now finished, a very good likeness, by an excellent hand."
— *September 22, 1635*

*   *   *

"If ever any persons might challenge to be signally distinguished for their intellect from other men, Ptolemy and Copernicus were they that had the honor to see farthest into and discourse most profoundly of the World's systems."

*   *   *

"My dear Kepler, what shall we make of all this? Shall we laugh, or shall we cry?"

*   *   *

"When shall I cease from wondering?"

SAGREDUS: Although it did not occur to me to put these observations to the test when I was voyaging, I am sure that they would take place in the way you describe. In confirmation of this I remember having often found myself in my cabin wondering whether the ship was moving or standing still; and sometimes at a whim I have supposed it to be going one way when its motion was the opposite . . .

The Galilean Principle of Relativity is simple in this early formulation, yet not as simple as it might be. In what way is it simple? Physics looks the same in a ship moving uniformly as in a ship at rest. Relative uniform motion of the two ships does not affect the laws of motion in either ship. A ball falling straight down onto one ship appears from the other ship to follow a parabolic course; a ball falling straight down onto that second ship also appears to follow a parabolic course when observed from the first ship. The simplicity of the Galilean Principle of Relativity lies in the equivalence of the two Earthbound frames and the symmetry between them.

In what way is this simplicity not as great as it might be? In Galileo's account the frames of reference are not yet free-float (inertial). To make them so requires only a small conceptual step: from two uniformly moving sea-going ships to two unpowered spaceships. Then up and down, north and south, east and west, all become alike. A ball untouched by force undergoes no acceleration. Its motion with respect to one spaceship is as uniform as it is with respect to the other. This identity of the law of free motion in all inertial reference frames is what one means today by the Galilean Principle of Relativity.

Galileo could not by any stretch of the imagination have asked his hearer to place himself in a spaceship in the year 1632. Yet he could have described the greater simplicity of physics when viewed from such a vantage point. Bottles, drops of water, and all the other test objects float at rest or move at uniform velocity. The zero acceleration of every nearby object relative to the spaceship would have been intelligible to Galileo of all people. Who had established more clearly than he that relative to Earth all nearby objects have a common acceleration?

Einstein's Principle of Relativity is a generalization of such experiments and many other kinds of experiments, involving not only mechanics but also electromagnetism, nuclear physics, and so on.

> **All the laws of physics are the same in every free-float (inertial) reference frame.**

Extension of Galileo's reasoning from ship to spaceship

Principle of Relativity

Einstein's Principle of Relativity says that once the laws of physics have been established in one free-float frame, they can be applied without modification in any other free-float frame. Both the mathematical form of the laws of physics and the numerical values of basic physical constants that these laws contain are the same in every free-float frame. So far as concerns the laws of physics, all free-float frames are equivalent.

We can tell where we are on Earth by looking out of the window. Where we are in the Milky Way we can tell by the configuration of the Big Dipper and other constellations. How fast and in what direction we are going through the larger framework of the universe we measure with a set of microwave horns pointed to pick up the microwave radiation streaming through space from all sides. But now exclude all information from outside. Screen out all radiation from the heavens. Pull down the window shade. Then do whatever experiment we will on the movement and collision of particles and the action of electric and magnetic forces in whatever free-float frame we please. We find not the slightest difference in the fit to the laws of physics between measurements made in one free-float frame and those made in another. We arrive at the Principle of Relativity in its negative form:

> **No test of the laws of physics provides any way whatsoever to distinguish one free-float frame from another.** ✒

Principle of Relativity, negative form

BOX 3-1

THE PRINCIPLE OF RELATIVITY RESTS ON EMPTINESS!

In his paper on special relativity, Einstein says, "We will raise this conjecture (whose intent will from now on be referred to as the 'Principle of Relativity') to a postulate . . . " Is the Principle of Relativity just a postulate? All of special relativity rests on it. How do we know it is true? What lies behind the Principle of Relativity?

This is a philosophical question, not a scientific one. You will have your own opinion; here is ours. We think the Principle of Relativity as used in special relativity rests on one word: emptiness.

Space is empty; there are no kilometer posts or mileposts in space. Do you want to measure distance and time? Then set up a latticework of meter sticks and clocks. Pace off the meter sticks, synchronize the clocks. Use the latticework to carry out your measurements. Discover the laws of physics. This latticework is your construction, not Nature's. Do not ask Nature to choose your latticework in preference to the similar latticework that I have constructed. Why not? Because space is empty. Space accommodates both of us as we go about our constructions and our investigations. But it does not choose either one of us in preference to the other. How can it? Space is empty. Nothing whatever can distinguish your latticework from mine. If we decide in secret to exchange latticeworks, Nature will never be the wiser! It follows that whatever laws of physics you discover employing your latticework must be the same laws of physics I discover using my latticework. The same is true even when our lattices move relative to one another. Which one of us is at rest? There is no way to tell in empty space! This is the Principle of Relativity.

But is space *really* empty? "Definitely not!" says modern quantum physics. "Space is a boiling cauldron of virtual particles. To observe this cauldron,

3.2 WHAT IS *NOT* THE SAME IN DIFFERENT FRAMES

not the same: space separations, time separations, velocities, accelerations, forces, fields

Space and time separations
not the same in different frames

Notice what the Principle of Relativity does *not* say. It does not say that the time between two events is the same when measured from two different free-float frames. Neither does it say that space separation between the two events is the same in the two frames. Ordinarily neither time nor space separations are the same in the two frames.

The catalog of differences between readings in the two frames does not end with laboratory and rocket records of pairs of events. Physics to the Greeks meant the science of change and so it does to us today. Motion gives us a stream of events, for example the blinks of a firefly or the pulses of a sparkplug flashing as it moves. These flashes trace out the sparkplug's trajectory. Record the positions of two sequential

sample regions of space much smaller than the proton. Carry out this sampling during times much shorter than the time it takes light to cross the diameter of the proton.'' These words are familiar or utterly incomprehensible, depending on the amount of our experience with physics. In either case, we can avoid dealing with the ''boiling cauldron of virtual particles'' by observing events that are far apart compared with the dimensions of the proton, events separated from one another by times long compared with the time it takes light to cross the diameter of the proton.

In the realm of classical (nonquantum) physics is space really empty? ''Of course not!'' says modern cosmology. ''Space is full of stars and dust and radiation and neutrinos and white dwarfs and neutron stars and (many believe) black holes. To observe these structures, sample regions of space comparable in size to that of our galaxy. These structures evolve and move with respect to one another in times comparable to millions of years.''

So we choose regions far from massive structures, avoid dust, ignore neutrinos and radiation, and measure events that take place close together in time compared with a million years.

Notice that for the very small and also for the very large, the ''regions'' described span both space and time — they are regions of *spacetime*. ''Emptiness'' refers to spacetime. Therefore we should have said from the beginning, ''*Spacetime* is empty'' — except for us and our apparatus — with limitations described above.

In brief, we can find ''effectively empty'' regions of spacetime of spatial extent quite a few orders of magnitude larger and smaller than dimensions of our bodies and of time spread quite a few orders of magnitude longer and shorter than times that describe our reflexes. In spacetime regions of this general size, empty spacetime can be found. In empty spacetime the Principle of Relativity applies. Where the Principle of Relativity applies, special relativity correctly describes Nature.

spark emissions in the laboratory frame. Record also the laboratory time between these sparks. Divide the change in position by the increase in time, yielding the laboratory-measured velocity of the sparkplug.

Spark events have identities that rise above all differences between reference frames. These events are recorded not only in the laboratory but also by recording devices and clocks in the rocket latticework. From the printouts of the recorders in the rocket frame we read off rocket space and time separations between sequential sparks. We divide. The quotient gives the rocket-measured velocity of the sparkplug. But both the space separation and the time separation between events, respectively, are ordinarily different for the rocket frame than for the laboratory frame. Therefore the rocket-measured velocity of the sparkplug is different from the laboratory-measured velocity of that sparkplug. Same world. Same motion. Different records of that motion. Figures for velocity that differ between rocket and laboratory.

Velocity not the same

Apply force to a moving object: Its velocity changes; it accelerates. Acceleration is the signal that force is being applied. Two events are enough to reveal velocity; three reveal change in velocity, therefore acceleration, therefore force. The laboratory observer reckons velocity between the first and second events, then he reckons velocity

Acceleration not the same

BOX 3-2

THE SPEED OF LIGHT

A "fundamental constant of nature"?
Or a mere factor of conversion between two units of measurement?

METERS AND MILES IN THE PARABLE OF THE SURVEYORS

Meter?
Originally (adopted France, 1799) one ten-millionth of the distance along the surface of Earth from its equator to its pole (in a curved line of latitude passing through the center of Paris).

Mile?
Originally one thousand *paces* — double step: right to left to right — of the Roman soldier.

Modern conversion factor?
1609.344 meters per mile.

Authority for this number?
Measures of equator-to-pole distance eventually (1799 to today) lagged in accuracy compared to laboratory measurement of distance. So the platinum meter rod at Sèvres, Paris, approximating one ten-millionth of that distance, for awhile became — in and by itself — the standard of distance. During that time the British Parliament and the United States Congress redefined the inch to be *exactly* 2.54 centimeters. This decree made the conversion factor (5280 feet/mile) times (12 inches/foot) times (2.54 centimeters/inch) times (1/100 of a meter per centimeter) equal to 1609.344 meters per mile — exactly!

A fundamental constant of nature?
Hardly! Rather, the work of two centuries of committees.

SECONDS AND METERS IN SPACETIME

Second?
Originally 1/24 of 1/60 of 1/60 of the time from high noon one day to high noon the next day. Since 1967, "The second is the duration of 9,192,631,770 periods of the radiation corresponding to the transition between the two hyperfine levels of the fundamental state of the atom cesium 133."

Meter?
Definition evolved from geographic to platinum meter rod to today's "One meter is the distance traveled by light, in vacuum, in the fraction 1/299,792,458 of a second."

Modern conversion factor?
299,792,458 meters per second.

Authority for this number?
Meeting of General Conference on Weights and Measures, 1983. In the accepted definition of the meter important changes took place over the years, and likewise in the definition of the second. With the 1983 definition of the meter these two streams of development have merged. What used to be understood as a measurement of the speed of light is understood today as two ways to measure separation in spacetime.

A fundamental constant of nature?
Hardly! Rather, the work of two centuries of committees.

---

between the second and third events. Subtracting, he obtains the change in velocity. From this change he figures the force applied to the object.

The rocket observer also measures the motion: velocity between the first and second events, velocity between second and third events; from these the change in velocity; from this the force acting on the object. But the rocket-observed velocities are not equal to the corresponding laboratory-observed velocities. The *change* in velocity also

Force not the same   differs in the two frames; therefore the computed *force* on the object is different for

Commentary
Is the distance from Earth's equator to its pole a fundamental constant of nature? No. Earth is plastic and ever changing. Is the distance between the two scratches on the standard meter bar constant? No. Oxidation from decade to decade slowly changes it. Experts in the art and science of measurement move to ever-better techniques. They search out an ever-better object to serve as benchmark. Via experiment after experiment they move from old standards of measurement to new. The goals? Accuracy. Availability. Dependability. Reproducibility.

Make a better measurement of the speed of light. Gain in that way better knowledge about light? No. Win instead an improved value of the ratio between one measure of spacetime interval, the meter, and another such measure, the second—both of accidental and historical origin? Before 1983, yes. Since 1983, no. Today the meter is *defined* as the distance light travels in a vacuum in the fraction 1/299,792,458 of a cesium-defined second. The two great streams of theory, definition, and experiment concerning the meter and the second have finally been unified.

What will be the consequence of a future, still better, measuring technique? Possibly it will shift us from the cesium-atom-based second to a pulsar-based second or to a still more useful standard for the second. But will that improvement in precision change the speed of light? No. Every past International Committee on Weights and Measures has operated on the principle of minimum dislocation of standards; we have to expect that the speed of light will remain at the decreed figure of 299,792,458 meters per second, just as the number of meters in the mile will remain at 1609.344. Through the fixity of this conversion factor c, any substantial improvement in the accuracy of defining the second will bring with it an identical improvement in the accuracy of defining the meter.

Is 299,792,458 a fundamental constant of nature? Might as well ask if 5280 is a fundamental constant of nature!

rocket observer and laboratory observer. The Principle of Relativity does not deny that the force acting on an object is different as reckoned in two frames in relative motion.

An electric field or a magnetic field or some combination of the two, acting on the electron, is the secret of action of many a device doing its quiet duty day after day in home, factory, or car. An electromagnetic force acting on an electron changes its velocity as it moves from event P to event Q and from Q to R. Laboratory and rocket observers do not agree on this change in velocity. Therefore they do not agree on the

Electric and magnetic fields not the same

value of the force that changes that velocity. Nor, finally, do they agree on the magnitudes of the electric and magnetic fields from which the force derives.

In brief, figures for electric and magnetic field strengths, for forces, and for accelerations agree no better between rocket and laboratory observers than do figures for velocity. The Principle of Relativity does not deny these differences. It celebrates them. It explains them. It systematizes them. ➤

3.3 WHAT *IS* THE SAME IN DIFFERENT FRAMES

the same: physical laws, physical constants in those laws

Laws of physics the same in different frames

Different values of some physical quantities between the two frames? Yes, but identical physical *laws!* For example, the relation between the force acting on a particle and the change in velocity per unit time of that particle follows the same law in the laboratory frame as in the rocket frame. The force is not the same in the two frames. Neither is the change in velocity per unit time the same. But the law that relates force and change of velocity per unit time is the same in each of the two frames. All the laws of motion are the same in the one free-float frame as in the other.

Not only the laws of motion but also the laws of electromagnetism and all other laws of physics hold as true in one free-float frame as in any other such frame. This is what it means to say, "No test of the laws of physics provides any way whatsoever to distinguish one free-float frame from another."

Fundamental constants the same

Deep in the laws of physics are numerical values of fundamental physical constants, such as the elementary charge on the electron and the speed of light. The values of these constants must be the same as measured in overlapping free-float frames in relative motion; otherwise these frames could be distinguished from one another and the Principle of Relativity violated.

Speed of light the same

One basic physical constant appears in the laws of electromagnetism: the speed of light in a vacuum, $c = 299,792,458$ meters per second. According to the Principle of Relativity, this value must be the same in all free-float frames in uniform relative motion. Has observation checked this conclusion? Yes, many experiments demonstrate it daily and hourly in every particle-accelerating facility on Earth. Nevertheless, it has taken a long time for people to become accustomed to the apparently absurd idea that there can be one special speed, the speed of light, that has the same value measured in each of two overlapping free-float frames in relative motion.

Values of the speed of light as measured by laboratory and by rocket observer turn out identical. This agreement has cast a new light on light. Its speed rates no longer as a constant of nature. Instead, today the speed of light ranks as mere conversion factor between the meter and the second, like the factor of conversion from the centimeter to the meter. The value of this conversion factor has now been set by decree and the meter defined in terms of it (Box 3.2). This decree *assumes* the invariance of the speed of light. No experimental result contradicts this assumption.

In 1905 the Principle of Relativity was a shocking heresy. It offended most people's intuition and common-sense way of looking at Nature. Consequences of the Principle of Relativity are tried out every day in many experiments where it is continually under severe test. Never has this Principle been verified to lead to a single incorrect experimental prediction. ➤

SAMPLE PROBLEM 3-1

EXAMPLES OF THE PRINCIPLE OF RELATIVITY

Two overlapping free-float frames are in uniform relative motion. According to the Principle of Relativity, which of the quantities on the following list must *necessarily* be the same as measured in the two frames? Which quantities are *not* necessarily the same as measured in the two frames?

a.   numerical value of the speed of light in a vacuum

b.   speed of an electron

c.   value of the charge on the electron

d.   kinetic energy of a proton (the nucleus of a hydrogen atom)

e.   value of the electric field at a given point

f.   time between two events

g.   order of elements in the periodic table

h.   Newton's First Law of Motion ("A particle initially at rest remains at rest, and . . . ")

SOLUTION

a.   The speed of light IS necessarily the same in the two frames. This is one of the central tenets of the Principle of Relativity and a basis of the theory of relativity.

b.   The speed of an electron IS NOT necessarily the same in the two frames. Determining the speed of a particle depends on space and time measurements between events — such as flashes emitted by the particle. Space and time separations between events, respectively, can be measured to be different for observers in relative motion. So the speed — ratio of distance covered to time elapsed — can be different.

c.   The value of the charge on the electron IS necessarily the same in the two frames. Suppose that the charge had one value for the laboratory frame and progressively smaller values for rocket frames moving faster and faster relative to the laboratory frame. Then we could detect the "absolute velocity" of the frame we are in by measuring the charge on the electron. But this violates the Principle of Relativity. Therefore the charge on the electron must have the same value in all free-float frames.

d.   The kinetic energy of a proton IS NOT necessarily the same in the two frames. The value of its kinetic energy depends on the speed of the proton. But speed is not necessarily the same as measured in the two frames (**b**).

e.   The value of the electric field at a given point IS NOT necessarily the same in the two frames. The argument is indirect but inescapable: The electric field is measured by determining the force on a test charge. Force can be measured by change in velocity that the force imparts to a particle of known mass. But the velocity — and the change in velocity — of a particle can be *different* for observers in relative motion (**b**). Therefore the electric field may be different for observers in relative motion.

f.   The time between two events IS NOT necessarily the same in the two frames. This is a direct result of the invariance of the interval (Chapter 1 and Section 3.7).

g. The order of elements in the periodic table by atomic number IS necessarily the same in the two frames. For suppose that the atomic number (the number of protons in the nucleus) were smaller for helium than for uranium in the laboratory frame but greater for helium than for uranium in the rocket frame. Then we could tell which frame we were in by comparing the atomic numbers of helium and uranium.

h. Newton's First Law of Motion IS necessarily the same in the two frames. Newton's First Law is really a definition of the inertial (free-float) frame. We assume that all laboratory and rocket frames are inertial.

3.4 RELATIVITY OF SIMULTANEITY

"same time"? ordinarily true for only one frame!

The Principle of Relativity directly predicts effects that initially seem strange — even weird. Strange or not, weird or not; logical argument demonstrates them and experiment verifies them. One effect has to do with simultaneity: Let two events occur separated in space along the direction of relative motion between laboratory and rocket frames. These two events, even if simultaneous as measured by one observer, cannot be simultaneous as measured by both observers.

Einstein demonstrated the relativity of simultaneity with his famous Train Paradox. (When Einstein developed the theory of special relativity, the train was the fastest common carrier.) Lightning strikes the front and back ends of a rapidly moving train, leaving char marks on the train and on the track and emitting flashes of light that travel forward and backward along the train (Figure 3-1). An observer standing on the ground halfway between the two char marks on the track receives the two light flashes at the same time. He therefore concludes that the two lightning bolts struck the track at the same time — with respect to him they fell simultaneously.

Train Paradox: Two lightning bolts strike simultaneously for ground observer

A second observer rides in the middle of the train. From the viewpoint of the observer on the ground, the train observer moves toward the flash coming from the front of the train and moves away from the flash coming from the rear. Therefore the train observer receives the flash from the front of the train first.

This is just what the train observer finds: The flash from the front of the train arrives at her position first, the flash from the rear of the train arrives later. But she can verify that she stands equidistant from the front and rear of the train, where she sees char marks left by the lightning. Moreover, using the Principle of Relativity, she knows that the speed of light has the same value in her train frame as for the ground observer (Section 3.3 and Box 3-2), and is the same for light traveling in both directions in her frame. Therefore the arrival of the flash first from the front of the train leads her to conclude that the lightning fell first on the front end of the train. For her the lightning bolts did not fall simultaneously. (To allow the train observer to make only measurements with respect to the train, forcing her to ignore Earth, let the train be a cylinder without windows — in other words a spaceship!)

Two lightning bolts do not strike simultaneously for train observer

Did the two lightning bolts strike the front and the back of the train simultaneously? Or did they strike at different times? Decide!

Strange as it seems, there is no unique answer to this question. For the situation described above, the two events are simultaneous as measured in the Earth frame; they

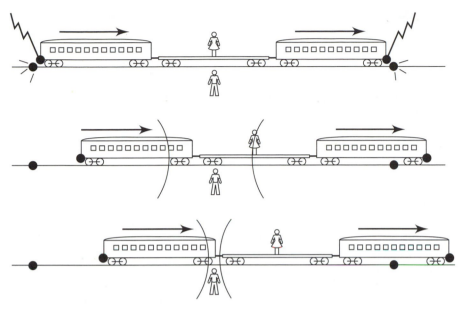

FIGURE 3-1. *Einstein's Train Paradox illustrating the relativity of simultaneity.* ***Top:*** *Light-ning strikes the front and back ends of a moving train, leaving char marks on both track and train. Each emitted flash spreads out in all directions.* ***Center:*** *Observer riding in the middle of the train concludes that the two strokes are not simultaneous. Her argument: "(1) I am equidistant from the front and back char marks on the train. (2) Light has the standard speed in my frame, and equal speed in both directions. (3) The flash arrived from the front of the train first. Therefore, (4) the flash must have left the front of the train first; the front lightning bolt fell before the rear lightning bolt fell. I conclude that the lightning strokes were not simultaneous."* ***Bottom:*** *Observer standing by the tracks halfway between the char marks on the tracks concludes that the strokes were simultaneous,* since the flashes from the strokes reach him at the same time.

are not simultaneous as measured in the train frame. We say that the simultaneity of events is, in general, *relative,* different for different frames. Only in the special case of two or more events that occur at the same point (or in a plane perpendicular to the line of relative motion at that point — see Section 3.6) does simultaneity in the laboratory frame mean simultaneity in the rocket frame. When the events occur at different locations along the direction of relative motion, they cannot be simultaneous in both frames. This conclusion is called the **relativity of simultaneity.**

The relativity of simultaneity is a difficult concept to understand. Almost without exception, every puzzle and apparent paradox used to "disprove" the theory of relativity hinges on some misconception about the relativity of simultaneity. ✒

Simultaneity is relative

3.5 LORENTZ CONTRACTION OF LENGTH

space separation between two length-measuring events? disagreement!

How do we measure the length of a moving rod — the distance between one end and the other end? One way is to use our latticework of clocks to mark the location of the two ends at the same time. But when the rod lies along the direction of relative motion, someone riding with the rod does not agree that our marking of the positions of the two ends occurs at the same time (Section 3.4). The relativity of simultaneity tells us

Length of a rod = separation between simultaneous sparks at its two ends that rocket and laboratory observers disagree about the simultaneity of two events (firecrackers exploding at the two ends of the rod) that occur at different locations along the direction of relative motion. Therefore the two observers disagree about whether or not a valid measurement of length has taken place.

Go back to the Train Paradox. For the observer standing on the ground, the two lightning bolts strike the front and back of the train at the same time. Therefore for him the distance between the char marks on the track constitutes a valid measure of the length of the train. In contrast, the observer riding on the train measures the front lightning bolt to strike first, the rear bolt later. The rider on the train exclaims to her Earth-based colleague, "See here! Your front mark was made before the back mark — since the flash from the front reached me (at the middle of the train) before the flash from the back reached me. Of course the train moved during the time lapse between these two lightning strikes. By the time the stroke fell at the back of the train, the front of the train had moved well past the front char mark on the track. Therefore your measurement of the length of the train is too small. The train is really longer than you measured."

Disagree about simultaneity? Then disagree about length.

There are other ways to measure the length of a moving rod. Many of these methods lead to the same result: the space separation between the ends of the rod is less as measured in a frame in which the rod is moving than as measured in a frame in which the rod is at rest. This effect is called **Lorentz contraction.** Section 5.8 examines the Lorentz contraction quantitatively.

Suppose we agree to measure the length of a rod by determining the position of its two ends at the same time. Then an observer for whom the rod is at rest measures the rod to be longer than does any other observer. This "rest length" of the rod is often called its **proper length.**

You keep using the word "measure." Occasionally you say "observe." You never talk about that most delicate, sensitive, and refined of our five senses: sight. Why not just look and *see these remarkable relativistic effects?*

We have been careful to say that the relativity of simultaneity and the Lorentz contraction are *measured,* not *seen* with the eye. *Measurement* employs the latticework of rods and clocks that constitutes a free-float frame. As mentioned in Chapter 2, seeing with the eye leads to confused images due to the finite speed of light. Stand in an open field in the southern hemisphere as Sun sets in the west and full Moon rises in the east: You see Moon as it was 1.3 seconds ago, Sun as it was eight minutes ago, the star Alpha Centauri (nearest star visible to the naked eye) as it was 4.34 years ago, the Andromeda nebula as it was *2 million years ago* — you see them all *now.* Similarly, light from the two separated ends of a speeding rod typically takes different times to reach your eye. This relative time delay results in visual distortion that is avoided when the location of each end is recorded locally, with zero or minimal delay, by the nearest lattice clock. Visual appearance of rapidly moving objects is itself an interesting study, but for most scientific work it is an unnecessary distraction. To avoid this kind of confusion we set up the free-float latticework of synchronized recording clocks and insist on its use — at least in principle!

Aha! Then I have caught you in a contradiction. Figure 3-1 shows lightning flashes and trains. Is this not a picture of what we would see with our eyes?

No. Strictly speaking, each of the three "pictures" in Figure 3-1 summarizes where parts of the train are as recorded by the Earth latticework of clocks at a given instant of Earth time. The position of each light flash at this instant is also recorded by the clocks in the lattice. The summary of data is then given to a draftsman, who draws the picture for that Earth time. To distinguish such a drafted picture from the visual view, we will often refer to it as a **plot.** For example, Figure 3-1 (top) is the Earth plot at the time when lightning bolts strike the two ends of the train.

Actually, all three plots in Figure 3-1 show approximately what you see through a telescope when you are very far from the scene in a direction perpendicular to the direction of motion of the train and at a position centered on the action. At such a remote location, light from all parts of the scene takes approximately equal times to reach your eye, so you would see events and objects at approximately the same time according to Earth clocks. Of course, you receive this information later than it actually occurs because of the time it takes light to reach you. ✐

3.6 INVARIANCE OF TRANSVERSE DIMENSION

"faster" does not mean "thinner" or "fatter"

A rocket ship makes many trips past the laboratory observer, each at successively higher speed. For each new and greater speed of the rocket, the laboratory observer measures its length to be shorter than it was on the trip before. This observed contraction is **longitudinal**—along its direction of motion. Does the laboratory observer also measure contraction in the **transverse** dimension, perpendicular to the direction of relative motion? In brief, is the rocket measured to get thinner as well as shorter as it moves faster and faster?

The answer is No. This is confirmed experimentally by observing the width of electron and proton beams traveling in high-energy accelerators. It is also easily demonstrated by simple thought experiments.

Speeding-Train Thought Experiment: Return to Einstein's high-speed railroad train seen end-on (Figure 3-2). Suppose the Earthbound observer measures the train to get thinner as it moves faster. Then for the Earth observer the right and left wheels of the train would come closer and closer together as the train speeds up, finally slipping off *between the tracks* to cause a terrible wreck. In contrast, the train observer regards herself as at rest and the tracks as speeding by in the opposite direction. If she

Transverse dimension same for laboratory and rocket observers

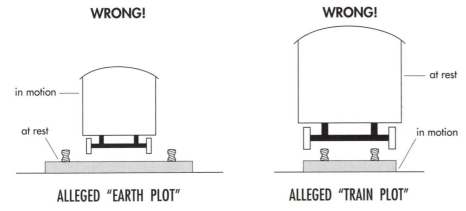

ALLEGED "EARTH PLOT"   ALLEGED "TRAIN PLOT"

FIGURE 3-2. *Two possible alternatives (both wrong!) if the moving train is measured to shrink transverse to its direction of motion. The "Earth plot" assumes the speeding train to be measured as getting thinner with increasing speed. The train's wheels would slip off between the tracks. The "train plot" of the same circumstance assumes the speeding rails to be measured as getting closer together. In this case the wheels would slip off outside the tracks. But this is a contradiction. Therefore the wheel separation —and the transverse dimensions of train and track—must be invariant, the same for all free-float observers moving along the track. (If you think that the actual transverse contraction might be too small to cause a wreck for the train shown, assume that both the wheels and the track are knife edges; the same argument still applies.)*

measures the speeding tracks to get closer together as they move faster and faster, the train wheels will slip off *outside the tracks,* also resulting in a wreck. But this is absurd: the wheels cannot end up between the tracks and outside the tracks under the same circumstances. Conclusion: High speed leads to no measured change in transverse dimensions — no observed thinning or fattening of fast objects. We are left with the conclusion that high relative speed affects the measured values of longitudinal dimensions but not transverse dimension: a welcome simplification!

Speeding-Pipes Thought Experiment: Start with a long straight pipe. Paint one end with a checkerboard pattern and the other end with stripes. Cut out and discard the middle of the pipe, leaving only the painted ends. Now hurl the ends toward each other, with their cylindrical axes lying along a common line parallel to the direction of relative motion (Figure 3-3). Suppose that a moving object is measured to be thinner. Then someone riding on the checkerboard pipe will observe the striped pipe to pass inside her cylinder. *All* observers — everyone looking from the side — will see a checkerboard pattern. In contrast, someone riding on the striped pipe will observe the checkerboard pipe to pass inside his cylinder. In this case, all observers will see a striped pattern. Again, this is absurd: All observers must see stripes, or all must see checkerboard. The only tenable conclusion is that speed has no measurable effect on transverse dimensions and the pipe segments will collide squarely edge on.

Thought experiments demonstrate invariance of transverse dimension

A simple question leads to an even more fundamental argument against the difference of transverse dimensions of a speeding object as observed by different free-float observers in relative motion: *About what axis* does the contraction take place?

We try to define an "axis of shrinkage" parallel to the direction of relative motion. Can we claim that a speeding pipe gets thinner by shrinking uniformly toward an "axis of shrinkage" lying along its center? Then what happens when two pipe segments move along their lengths, side by side as a pair? Does each pipe shrink separately, causing the clear space between them to *increase?* Or does the combination of both pipes contract toward the line midway between them, causing the clear space between them to *decrease?* Is the answer different if one pipe is made of lead and the other one of paper? Or if one pipe is entirely in our imagination?

There is no logically consistent way to define an "axis of shrinkage." Given the direction of relative motion of two objects, we cannot select uniquely an "axis of shrinkage" from the infinite number of lines that lie parallel in this direction. For each different choice of axis a different pattern of distortions results. But this is logically intolerable. The only way out is to conclude that there is no transverse shrinkage at all (and, by a similar argument, no transverse expansion).

The above analysis leads to conclusions about events as well as about objects. A set of explosions occurs around the perimeter of the checkerboard pipe. More: These explosions occur simultaneously in this checkerboard frame. Then these events are simultaneous also in the striped frame. How do we know? By symmetry! For suppose the explosions were *not* simultaneous in the striped frame. Then which one of these

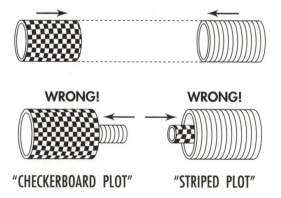

"CHECKERBOARD PLOT"   "STRIPED PLOT"

FIGURE 3-3. *Two identical-size pipe segments hurtle toward each other along a common centerline. What will happen when they meet? Here are two possible alternatives (both wrong!) if a moving object is observed to shrink transverse to direction of motion. Which pipe passes inside the other? The impossibility of a consistent answer to this question leads to the conclusion that neither pipe can be measured to change transverse dimension.*

events would occur first in the striped frame? The one on the right side of the pipe or the one on the left side of the pipe? But "left" and "right" cannot be distinguished by means of any physical effect: Each pipe is cylindrically symmetric. Moreover, space is the same in all directions—space is **isotropic,** the same to right as to left. So neither the event on the right side nor the event on the left side can be first. They must be simultaneous. The same argument can be made for events at the "top" and "bottom" of the pipe, and for every other pair of events on opposite sides of the pipe. Conclusion: If the explosions are simultaneous in the checkerboard frame, they must also be simultaneous in the striped frame.

We make the following summary conclusions about dimensions transverse to the direction of relative motion:

> Dimensions of moving objects transverse to the direction of relative motion are measured to be the same in laboratory and rocket frames (invariance of transverse distance).
>
> Two events with separation only transverse to the direction of relative motion and simultaneous in either laboratory or rocket frame are simultaneous in both. ✐

"Same time" agreed on for events separated only transverse to relative motion

3.7 INVARIANCE OF THE INTERVAL PROVED

laboratory and rocket observers agree on something important

The Principle of Relativity has a major consequence. It demands that the spacetime interval have the same value as measured by observers in every overlapping free-float frame; in brief, it demands "invariance of the interval." Proof? Plan of attack: Determine the separation in space and the separation in time between two events, E and R, in the rocket frame. Then determine the quite different space and time separations between the same two events as measured in a free-float laboratory frame. Then look for—and find—what is invariant. It is the "interval." Now for the details (Figures 3-4 and 3-5).

Event E we take to be the reference event, the emission of a flash of light from the central laboratory and rocket reference clocks as they coincide at the zero of time (Section 2.6). The path of this flash is tracked by the recording clocks in the rocket lattice. Riding with the rocket, we examine that portion of the flash that flies straight "up" 3 meters to a mirror. There it reflects straight back down to the photodetector located at our rocket reference clock, where it is received and recorded. The act of reception constitutes the second event we consider. This event, R, is located at the rocket space origin, at the same location as the emission event E. Therefore, for the rocket observer, the space separation between event E and event R equals zero.

What is the time separation between events E and R in the rocket frame? The light travels 3 meters up to the mirror and 3 meters back down again, a total of 6 meters of distance. At the "standard" light speed of 1 meter of distance per meter of light-travel time, the flash takes a total of 6 meters of time to complete the round trip. In summary, for the rocket observer the event of reception, R, is separated from the event of emission, E, by zero meters in space and 6 meters in time.

What are the space and time separations of events E and R measured in the free-float laboratory frame? As measured in the laboratory, the rocket moves at high speed to the right (Figures 3-4 and 3-5). The rocket goes so fast that the simple

Principle of Relativity leads to invariance of spacetime interval

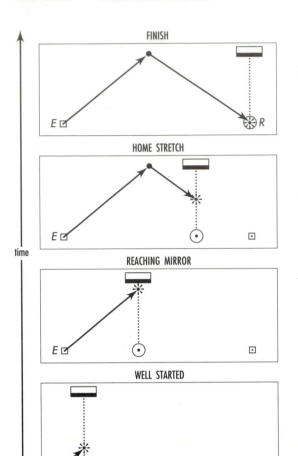

FIGURE 3-4. *Plot of the flash path as recorded in the laboratory frame. Time progresses from bottom to top:* **Well started:** *The flash (represented as an asterisk) has been emitted (event E) from a moving rocket clock (shown as a circle) that coincided with a laboratory clock (shown as a square).* **Reaching mirror** *and* **Home stretch:** *The flash reaches a mirror and reflects from it. The mirror moves along in step with the rocket clock.* **Finish:** *The flash is received (event R) back at the same rocket clock, which has moved in the laboratory frame to coincide with a second laboratory clock. Figure 3-5 shows the trajectory of the same flash in three different free-float frames.*

Greater distance of travel for light flash: longer time!

up-down track of the light in the rocket frame appears in the laboratory to have the profile of a tent, with its right-hand corner — the place of reception of the light — 8 meters to the right of the starting point.

When does the event of reception, *R*, take place as registered in the laboratory frame? Note that it occurs at the time 6 meters in the rocket frame. All we know about everyday events urges us to say, "Why, obviously it occurs at 6 meters of time in the laboratory frame too." But no. More binding than preconceived expectations are the demands of the Principle of Relativity. Among those demands none ranks higher than this: The speed of light has the standard value 1 meter of distance in 1 meter of light-travel time in every free-float frame.

Figure 3-6 punches us in the eye with this point: The light flash travels *farther* as recorded in the laboratory frame than as recorded in the rocket frame. The perpendicular "altitude" of the mirror from the line along which the rocket reference clock moves has the same value in laboratory frame as in rocket frame no matter how fast the rocket — as shown in Section 3.6. Therefore on its slanted path toward and away from the mirror the flash must cover more distance in the laboratory frame than it does in the rocket frame. More distance covered means more time required at the "standard" light speed. We conclude that the time between events *E* and *R* is greater in the laboratory frame than in the rocket frame — a staggering result that stood physics on its ear when first proposed. There is no way out.

In the laboratory frame the flash has to go "up" 3 meters, as before, and "down" again 3 meters. But in addition it has to go 8 meters to the right: 4 meters to the right while rising to hit the mirror, and 4 meters more to the right while falling again to the receptor. The Pythagorean Theorem, applied to the right triangles of Figure 3-6, tells

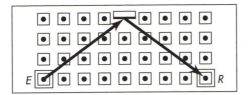

LABORATORY PLOT

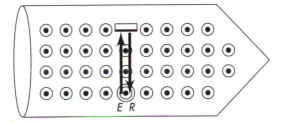

ROCKET PLOT

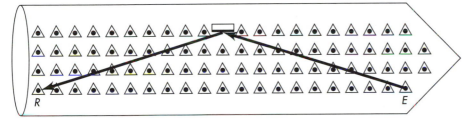

SUPER-ROCKET PLOT

FIGURE 3-5. *Plots of the path in space of a reflected flash of light as measured in three different frames, showing event* **E,** *emission of the flash, and event* **R,** *its reception after reflection. Squares, circles, and triangles represent latticeworks of recording clocks in laboratory, rocket, and super-rocket frames, respectively. The super-rocket frame moves to the right with respect to the rocket, and with such relative speed that the event of reception,* R, *occurs to the left of the event of emission,* E, *as measured in the super-rocket frame. The reflecting mirror is fixed in the rocket, hence appears to move from left to right in the laboratory and from right to left in the super-rocket.*

FIGURE 3-6. *Laboratory plot of the path of the light flash. The flash rises 3 meters while it moves to the right 4 meters. Then it falls 3 meters as it moves an additional 4 meters to the right. From the Pythagorean Theorem, the total length of the flash path equals 5 meters plus 5 meters or 10 meters. Therefore 10 meters of light-travel time is the separation in time between emission event* E *and reception event* R *as measured in the laboratory frame.*

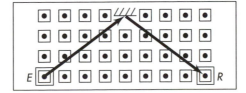

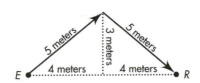

us that each slanted leg of the trip has length 5 meters:

$$(3 \text{ meters})^2 + (4 \text{ meters})^2 = (5 \text{ meters})^2$$

Thus the total length of the trip equals 10 meters, definitely longer than the length of the round trip, 6 meters, as observed in the rocket frame. Moreover, the light can cover that slanted and greater distance only at the standard rate of 1 meter of distance in 1 meter of light-travel time. Therefore there is no escape from saying that the time of reception as recorded in the laboratory frame equals 10 meters. Thus there is a great variance between what is recorded in the two frames (Figure 3-5, Laboratory plot and Rocket plot): separation in time and in space between the emission E of a pulse of light and its reception R after reflection.

In spite of the difference in space separation between events E and R and the difference in time lapse between these events as measured in laboratory and rocket frames, there exists a measure of their separation that has the same value for both observers. This is the interval calculated from the difference of squares of time and space separations (Table 3-1). For both observers the interval has the value 6 meters. The interval is an **invariant** between free-float frames.

Two central results are to be seen here, one of variance, the other of invariance. We discover first that typically there is not and cannot be an absolute time difference between two events. The difference in time depends on our choice of the free-float frame, which inertial frame we use to record events. There is no such thing as a simple concept of universal and absolute separation in time.

Between events: No absolute time, but invariant interval

Second, despite variance between the laboratory frame and the rocket frame in the values recorded for time and space separations individually, the difference between the squares of those separations is identical, that is, invariant with respect to choice of reference frame. The difference of squares obtained in this way defines the square of the interval. The invariant interval itself has the value 6 meters in this example.

◁ **TABLE 3-1** ▷

RECKONING THE SPACETIME INTERVAL FROM ROCKET AND LABORATORY MEASUREMENTS

	Rocket measurements		Laboratory measurements
Time from emission of the flash to its reception	6 meters	← DIFFERENT! →	10 meters
Distance from the point of emission of the flash to its point of reception	0 meters	← DIFFERENT! →	8 meters
Square of time	36 (meters)^2		100 (meters)^2
Square distance and subtract	-0 (meters)^2		-64 (meters)^2
Result of subtraction	36 (meters)^2		36 (meters)^2
This is the square of what measurement?	6 meters		6 meters

SAME SPACETIME
INTERVAL

3.8 INVARIANCE OF THE INTERVAL FOR *ALL* FREE-FLOAT FRAMES

super-rocket observer joins the agreement

The interval between two events has the same value for *all possible* relative speeds of overlapping free-float frames. As an example of this claim, consider a third free-float frame moving at a different speed with respect to the laboratory frame—a speed different from that of the rocket frame.

We now measure the same events of emission and reception from a "super-rocket frame" moving faster than the rocket (but not faster than light!) along the line between events E and R (Figure 3-5, Super-rocket plot). For convenience we arrange that the reference clock of this frame also coincides with reference clocks of the other two frames at event E.

Events E and R occur at the same place in the rocket frame. Between these two events the super-rocket moves to the *right* with respect to the rocket. As a result, the super-rocket observer records event R as occurring to the *left* of the emission event. How far to the left? That depends on the relative speed of the super-rocket frame.

The super-rocket is not super-size; rather it has super-speed. We adjust this super-speed so that the reception occurs 20 meters to the left of the emission for the super-rocket observer. Then the flash of light that rises vertically in the rocket must travel the same 3 meters upward in the super-rocket but also 10 meters to the left as it slants toward the mirror. Hence the distance it travels to the mirror in the super-rocket frame is the length of a hypotenuse, 10.44 meters:

$$(3 \text{ meters})^2 + (10 \text{ meters})^2 = 9 \text{ meters}^2 + 100 \text{ meters}^2 = 109 \text{ meters}^2$$
$$= (10.44 \text{ meters})^2$$

It must travel another 10.44 meters as it slants downward and leftward to the event of reception. The total distance traveled equals 20.88 meters. It follows that the total time lapse between E and R equals 20.88 meters of light-travel time for the super-rocket observer.

The speed of the super-rocket is very high. As a result the space separation between emission and reception is very great. But then the time separation is also very great. Moreover, the magnitude of the time separation is perfectly tailored to the size of the space separation. In consequence, the particular quantity equal to the difference of their squares has the value $(6 \text{ meters})^2$, no matter how great the space separation and time separation individually may be. For the super-rocket frame:

$$(20.88 \text{ meters})^2 - (20 \text{ meters})^2 = 436 \text{ meters}^2 - 400 \text{ meters}^2 = 36 \text{ meters}^2$$
$$= (6 \text{ meters})^2$$

In spite of the difference in space separation observed in the three frames (0 meters for the rocket, 8 meters for the laboratory, 20 meters for the super-rocket) and the difference in time separation (6 meters for the rocket, 10 meters for the laboratory, 20.88 meters for the super-rocket), the interval between the two events has the same value for all three observers:

In general: (time separation)^2 — (space separation)^2 = (interval)^2

Rocket frame: $(6 \text{ meters})^2 - (0 \text{ meters})^2 = (6 \text{ meters})^2$

Laboratory frame: $(10 \text{ meters})^2 - (8 \text{ meters})^2 = (6 \text{ meters})^2$

Super-rocket frame: $(20.88 \text{ meters})^2 - (20 \text{ meters})^2 = (6 \text{ meters})^2$

Super-rocket: Same interval between events

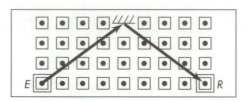

FIGURE 3-6 (repeated). *Laboratory plot of the path of the light flash.*

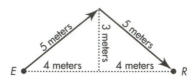

The laboratory observer clocks the time between the flash and its reception as 10 meters, in total disagreement with the 6 meters of timelike interval he figures between those two events. The observer in the super-rocket frame marks an even greater discrepancy, 20.88 meters of her time versus the 6 meters of timelike interval. Only for the rocket observer does clock time agree with interval. Why? Because only she sees reception at the same place as emission.

Invariance of interval from invariance of transverse dimension

The invariance of the interval can be seen at a glance in Figure 3-6. The hypotenuse of the first right triangle has a length equal to half the time separation between E and R. Its base has a length equal to half the space separation. To say that (time separation)^2 — (space separation)^2 has a standard value, and consequently to state that (half the time separation)^2 — (half the space separation)^2 has a standard value, is simply to say that the altitude of this right triangle has a fixed magnitude (3 meters in the diagram) for rocket and all super-rocket frames, no matter how fast they move. And this altitude has a length equal to half the interval between these two events.

SAMPLE PROBLEM 3-2

THE K^+ MESON

SOLUTION

A beam of (unstable) K^+ mesons, traveling at a speed of $v = 0.868$, passes through two counters 9 meters apart. The particles suffer negligible loss of speed and energy in passing through the counters but give electrical pulses that can be counted. The first counter records 1000 pulses (1000 passing particles); the second records 250 counts (250 passing particles). This decrease arises almost entirely from decay of particles in flight. Determine the half-life of the K^+ meson in its own rest frame.

Unstable particles of different kinds decay at different rates. By definition, the half-life of unstable particles of a particular species measures the particle wristwatch time during which — on the average — half of the particles decay. Half of the remaining particles decay in an additional time lapse equal to the same half-life, and so forth. In this case, one quarter of the K^+ particles remain after passage from counter to counter. Therefore the particles that survive experience the passage of two half-lives between counter and counter. We make the interval between those two passages, those two events, the center of our attention, because it has the same value in the laboratory frame where we do our measuring as it does in the free-float frame of the representative particle.

The keystone of the argument establishing the invariance of the interval between two events for all free-float frames? The Principle of Relativity, according to which there is no difference in the laws of physics between one free-float frame and another. This principle showed here in two very different ways. First, it said that distances at right angles to the direction of relative motion are recorded as of equal magnitude in the laboratory frame and the rocket frame (Section 3.6). Otherwise one frame could be distinguished from the other as the one with the shorter perpendicular distances.

Second, the Principle of Relativity demanded that the speed of light be the same in the laboratory frame as in the rocket frame. The speed being the same, the fact that the light-travel path in the laboratory frame (the hypotenuse of two triangles) is longer than the simple round-trip path in the rocket frame (the altitudes of these two triangles: up 3 meters and down again) directly implies a longer time in the laboratory frame than in the rocket frame.

In brief, one elementary triangle in Figure 3-6 displays four great ideas that underlie all of special relativity: invariance of perpendicular distance, invariance of the speed of light, dependence of space and time separations upon the frame of reference, and invariance of the interval.

Basis of invariance of interval: Principle of Relativity

3.9 SUMMARY

same laws for all; invariant interval for all

The **Principle of Relativity** says that the laws of physics are the same in every inertial (free-float) reference frame (Section 3.1). This simple principle has important consequences. Specifically:

$$(\text{interval})^2 = \left(\begin{array}{c}\text{separation}\\\text{in lab}\\\text{time}\end{array}\right)^2 - \left(\begin{array}{c}\text{separation}\\\text{in lab}\\\text{position}\end{array}\right)^2 = \left(\begin{array}{c}\text{separation}\\\text{in moving-}\\\text{particle time}\end{array}\right)^2 - \left(\begin{array}{c}\text{separation}\\\text{in moving-}\\\text{particle position}\end{array}\right)^2$$

$$= \left(\frac{9 \text{ meters of distance}}{0.868 \text{ meters of distance per meter of time}}\right)^2 - \left(\begin{array}{c}9 \text{ meters}\\\text{of distance}\end{array}\right)^2 = (2 \text{ half-lives})^2 - \left(\begin{array}{c}\text{zero separation}\\\text{in space (in}\\\text{particle frame)}\\\text{between those}\\\text{two events}\end{array}\right)^2$$

$$= \left(\begin{array}{c}10.368 \text{ meters}\\\text{of light-travel time}\end{array}\right)^2 - \left(\begin{array}{c}9 \text{ meters}\\\text{of distance}\end{array}\right)^2 = (2 \text{ half-lives})^2$$

A little arithmetic tells us that two half-lives total 5.15 meters of light-travel time. Consequently the K^+ half-life itself is 2.57 meters of time or $(2.57 \text{ meters})/(3.00 \times 10^8$ meters/second$) = 8.5 \times 10^{-9}$ second or 8.5 nanoseconds.

1. Two events that lie along the direction of relative motion between two frames cannot be simultaneous as measured in both frames **(relativity of simultaneity).** (Section 3.4)
2. An object in high-speed motion is measured to be shorter along its direction of motion than its **proper length,** measured in its rest frame **(Lorentz contraction).** (Section 3.5)
3. The dimensions of moving objects transverse to their direction of relative motion are measured to be the same, whatever the relative speed **(invariance of transverse distances).** (Section 3.6)
4. Two events with separation only transverse to the direction of relative motion and simultaneous in either frame are simultaneous in both. (Section 3.6)

BOX 3-3

FASTER THAN LIGHT?

We always want to go faster. Faster than what? Faster than anything has gone before. What is our greatest possible speed, according to the theory of relativity? The speed of light in a vacuum! How do we know that this is the greatest possible speed that we can travel? Many lines of evidence reach this conclusion. Rocket speed greater than the speed of light would lead to the destruction of the essential relation between cause and effect, a result explored in Special Topic: Lorentz Transformation (especially Box L-1) and in Chapter 6. In particular, we could find a frame in which a faster-than-light object arrives before it starts! Moreover, in particle accelerators built over several decades we have spent hundreds of millions of dollars effectively trying to accelerate electrons and protons to the greatest possible speed — which by experiment never exceeds light speed.

The conclusion that no thing can move faster than light arises also from the invariance of the interval. To see this, let a rocket emit two flashes of light a time t' apart as measured in the rocket frame. (Use a prime to distinguish rocket measurements from laboratory measurements.) In the rocket frame the two emissions occur at the same place: the separation x' between them equals zero. Let t and x be the corresponding separations in time and space as measured in the laboratory frame. Then the invariance of the interval tells us that the three quantities t', t, and x are related by the equation

$$(t')^2 - (x')^2 = (t')^2 - (0)^2 = t^2 - x^2$$

whence

$$(t')^2 = t^2 - x^2 \tag{3-1}$$

In the laboratory frame the rocket is moving with some speed; give this speed the symbol v. The distance x between emissions is just the distance that the rocket moves in time t in the laboratory frame. The relation between

5. The spacetime interval between two events is invariant—it has the same value in laboratory and rocket frames (Sections 3.7 and 3.8):

$$(\text{interval})^2 = \left(\begin{array}{c}\text{Laboratory}\\\text{time}\\\text{separation}\end{array}\right)^2 - \left(\begin{array}{c}\text{Laboratory}\\\text{space}\\\text{separation}\end{array}\right)^2$$

$$= \left(\begin{array}{c}\text{Rocket}\\\text{time}\\\text{separation}\end{array}\right)^2 - \left(\begin{array}{c}\text{Rocket}\\\text{space}\\\text{separation}\end{array}\right)^2$$

6. In any free-float frame, no object moves with a speed greater than the speed of light (Box 3-3). 🍃

distance, time, and speed is

$$x = vt \qquad (3\text{-}2)$$

Substitute this into equation (3-1) to obtain $(t')^2 = t^2 - (vt)^2 = t^2[1 - v^2]$, or

$$t' = t(1 - v^2)^{1/2} \qquad (3\text{-}3)$$

Now, v is the speed of the rocket. How large can that speed be? Equation (3-3) makes sense for any rocket speed less than the speed of light, or when v has a value less than one.

Suppose we try to force the rocket to move faster than the speed of light. If we should succeed, v would have a value greater than one. Then v^2 also would have a value greater than one. But in this case the expression $1 - v^2$ would have a negative value and its square root would have no physical meaning. In a formal mathematical sense, the rocket time t' would be an imaginary number for the case of rocket speed greater than the speed of light. But clocks do not read imaginary time; they read real time—three hours, for example. Therefore a rocket speed greater than the speed of light leads to an impossible consequence.

Equation (3-3) does not forbid a rocket to go as close to the speed of light as we wish, as long as this speed remains less than the speed of light. For v very close to the speed of light, equation (3-3) tells us that the rocket time can be very much smaller than the laboratory time. Now suppose that emission of the first flash occurs when the rocket passes Earth on its outward trip to a distant star. Let emission of the second flash occur as the rocket *arrives* at that distant star. No matter how long the laboratory time t between these two events, we can find a rocket speed, v, such that the rocket time t' is as small as we wish. This means that in principle we can go to any remote star in as short a rocket time as we want. In brief, although our speed is limited to less than the speed of light, the distance we can travel in a lifetime has no limitation. We can go anywhere! This result is explored further in Chapter 4.

BOX 3-4

DOES A MOVING CLOCK *REALLY* "RUN SLOW"?

You keep saying, "The time between clock-ticks is shorter as MEASURED in the rest frame of the clock than as MEASURED in a frame in which the clock is moving." I am interested in reality, not someone's measurements. Tell me what really happens!

What is reality? You will have your own opinion and speculations. Here we pose two related scientific questions whose answers may help you in forming your opinion.

Are differences in clock rates *really* verified by experiment?
Different values of the time between two events as observed in different frames? Absolutely! Energetic particles slam into solid targets in accelerators all over the world, spraying forward newly created particles, some of which decay in very short times as measured in their rest frames. But these "short-lived" particles survive much longer in the laboratory frame as they streak from target to detector. In consequence, the detector receives a much larger fraction of the undecayed fast-moving particles than would be predicted from their decay times measured at rest. This result has been tested thousands of times with many different kinds of particles. Such experiments carried out over decades lead to dependable, consistent, repeatable results. As far as we can tell, they are correct, true, and reliable and cannot effectively be denied. If that is what you personally mean by "real," then these results are "what really happens."

Does something about a clock *really* change when it moves, resulting in the observed change in tick rate?
Absolutely not! Here is why: Whether a free-float clock is at rest or in motion in the frame of the observer is controlled by the observer. You want the clock

REFERENCES

Introductory quote: A. Sommerfeld, *Naturwissenschaft Rundschau,* Volume 1, pages 97–100, reprinted in *Gesammelte Schriften* (Vieweg, Braunschweig, 1968), Volume IV, pages 640–643.

Galileo quote, Section 3.1: Galileo Galilei, *Dialogue Concerning the Two Chief World Systems — Ptolemaic and Copernican,* first published February 1632; the translation quoted here is by Stillman Drake (University of California Press, Berkeley, 1962), pages 186ff. Galileo's writings, along with those of Dante, by reason of their strength and aptness, are treasures of human thought, studied today in Italy by secondary school students as part of a great literary heritage.

Einstein quote, Box 3-1: Albert Einstein, "On the Electrodynamics of Moving Bodies," *Annalen der Physik,* Volume 17, pages 891–921 (1905), translated by Arthur I. Miller in *Albert Einstein's Special Theory of Relativity* (Addison-Wesley, Reading, Mass., 1981), page 392.

to be at rest? Move along with it! Now do you want the clock to move? Simply change your own velocity! This is true even when you and the clock are separated by the diameter of the solar system. The magnitude of the clock's steady velocity is entirely under your control. Therefore the time between its ticks as measured in your frame is determined by your actions. How can your change of motion affect the inner mechanism of a distant clock? It cannot and does not.

Every time you change your motion on Earth — and even when you sit down, letting the direction of your velocity change as Earth rotates — you change the rate at which the planets revolve around Sun, as measured in your frame. (You also change the shape of planetary orbits, contracting them along the direction of your motion relative to Sun.) Do you think this change on your velocity really affects the workings of the "clock" we call the solar system? If so, what about a person who sits down on the other side of Earth? That person moves in the opposite direction around the center of Earth, so the results are different from yours. Are each of you having a different effect on the solar system? And are there still different effects — different solar-system clocks — for observers who could in principle be scattered on other planets?

We conclude that free-float motion does not affect the structure or operation of clocks (or rods). If this is what you mean by reality, then there are *really* no such changes due to uniform motion.

Is there some unity behind these conflicting measurements of time and space? Yes! The interval: the proper time (wristwatch time) between ticks of a clock as measured in a frame in which ticks occur at the same place, in which the clock is at rest. Proper time can also be calculated by all free-float observers, whatever their state of motion, and all agree on its value. Behind the confusing clutter of conflicting measurements stands the simple, consistent, powerful view provided by spacetime.

ACKNOWLEDGMENTS

The idea for Box 3-1 was suggested by Kenneth L. Laws. Box 3-4 and the argument for Section 3.6, Invariance of Transverse Dimension, is adapted from material by William A. Shurcliff, private communications. Sample Problem 3-2 is adapted from A. P. French, *Special Relativity* (W. W. Norton, New York, 1968), page 121.

CHAPTER 3 EXERCISES

PRACTICE

3-1 relativity and swimming

The idea here is to illustrate how remarkable is the invariance of the speed of light (light speed same in all free-float frames) by contrasting it with the case of a swimmer making her way through water.

Light goes through space at 3×10^8 meters/second, and the swimmer goes through the water at 1 meter/second. "But how can there otherwise be any difference?" one at first asks oneself.

For a light flash to go down the length of a 30-meter spaceship and back again takes

time = (distance)/(speed)
    = $2 \times (30 \text{ meters})/(3 \times 10^8 \text{ meters/second})$
    = 2×10^{-7} second

as measured in the spaceship, regardless of whether the ship is stationary at the spaceport or is zooming past it at high speed.

Check how very different the story is for the swimmer plowing along at 1 meter/second with respect to the water.

a How long does it take her to swim down the length of a 30-meter pool and back again?

b How long does it take her to swim from float A to float B and back again when the two floats, A and B, are still 30 meters apart, but now are being towed through a lake at 1/3 meter/second? **Discussion:** When the swimmer is swimming in the same direction in which the floats are being towed, what is her speed relative to the floats? And how great is the distance she has to travel expressed in the "frame of reference" of the floats? So how long does it take to travel that leg of her trip? Then consider the same three questions for the return trip.

c Is it true that the total time from A to B and back again is independent of the reference system ("stationary" pool ends vs. moving floats)?

d Express in the cleanest, clearest, sharpest one-sentence formulation you can the difference between what happens for the swimmer and what happens for a light flash.

3-2 Einstein puzzler

When Albert Einstein was a boy of 16, he mulled over the following puzzler: A runner looks at herself in a mirror that she holds at arm's length in front of her. If she runs with nearly the speed of light, will she be able to see herself in the mirror? Analyze this question using the Principle of Relativity.

3-3 construction of clocks

For the measurement of time, we have made no distinction among spring clocks, quartz crystal clocks, biological clocks (aging), atomic clocks, radioactive clocks, and a clock in which the ticking element is a pulse of light bouncing back and forth between two mirrors (Figure 1-3). Let all these clocks be adjusted by the laboratory observer to run at the same rate when at rest in the laboratory. Now let the clocks all be accelerated gently to a high speed in a rocket, which then turns off its engines. Make a simple but powerful argument that the free-float rocket observer will also measure these different clocks all to run at the same rate as one another. Does it follow that the (common) clock rate of these clocks measured by the rocket observer is the same as their (common) rate measured by the laboratory observer as they pass by in the rocket?

3-4 the Principle of Relativity

Two overlapping free-float frames are in uniform relative motion. On the following list, mark with a "yes" the quantities that must *necessarily* be the same as measured in the two frames. Mark with a "no" the quantities that are *not* necessarily the same as measured in the two frames.

a time it takes for light to go one meter of distance in a vacuum
b spacetime interval between two events
c kinetic energy of an electron
d value of the mass of the electron
e value of the magnetic field at a given point
f distance between two events
g structure of the DNA molecule
h time rate of change of momentum of a neutron

3-5 many unpowered rockets

In the laboratory frame, event 1 occurs at $x = 0$ light-years, $t = 0$ years. Event 2 occurs at $x = 6$ light-years, $t = 10$ years. In all rocket frames, event 1 also occurs at the position 0 light-years and the time 0 years. The y- and z-coordinates of both events are zero in both frames.

a In rocket frame A, event 2 occurs at time $t' = 14$ years. At what position x' will event 2 occur in this frame?

b In rocket frame B, event 2 occurs at position x'' = 5 light-years. At what time t'' will event 2 occur in this frame?

c How fast must rocket frame C move if events 1 and 2 occur at the same place in this rocket frame?

d What is the time between events 1 and 2 in rocket frame C of part c?

3-6 down with relativity!

Mr. Van Dam is an intelligent and reasonable man with a knowledge of high school physics. He has the following objections to the theory of relativity. Answer each of Mr. Van Dam's objections decisively—without criticizing him. If you wish, you may present a single connected account of how and why one is driven to relativity, in which these objections are all answered.

a "Observer A says that B's clock goes slow, and observer B says that A's clock goes slow. This is a logical contradiction. Therefore relativity should be abandoned."

b "Observer A says that B's meter sticks are contracted along their direction of relative motion, and observer B says that A's meter sticks are contracted. This is a logical contradiction. Therefore relativity should be abandoned."

c "Relativity does not even have a unique way to *define* space and time coordinates for the instantaneous position of an object. Laboratory and rocket observers typically record different coordinates for this position and time. Therefore anything relativity says about the velocity of the object (and hence about its motion) is without meaning."

d "Relativity postulates that light travels with a standard speed regardless of the free-float frame from which its progress is measured. This postulate is certainly wrong. Anybody with common sense knows that that travel at high speed in the direction of a receding light pulse will decrease the speed with which the pulse recedes. Hence a flash of light *cannot* have the same speed for observers in relative motion. With this disproof of the basic postulate, all of relativity collapses."

e "There isn't a single experimental test of the *results* of special relativity."

f "Relativity offers no way to describe an event without coordinates—and no way to speak about coordinates without referring to one or another particular reference frame. However, physical events have an existence independent of all choice of coordinates and all choice of reference frame. Hence relativity—with its coordinates and reference frames—cannot provide a valid description of these events."

g "Relativity is preoccupied with how we *observe* things, not what is *really* happening. Hence it is not a scientific theory, since science deals with reality."

PROBLEMS

3-7 space war

Two rockets of equal rest length are passing "head on" at relativistic speeds, as shown in the figure (left). Observer o has a gun in the tail of her rocket pointing perpendicular to the direction of relative motion

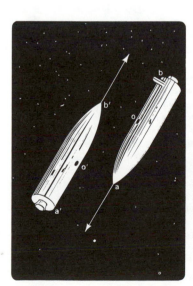 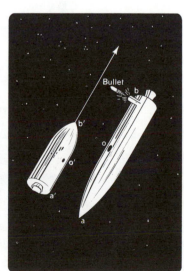 

EXERCISE 3-7. *Left:* Two rocket ships passing at high speed. *Center:* In the frame of o one expects a bullet fired when a coincides with a' to miss the other ship. *Right:* In the frame of o' one expects a bullet fired when a coincides with a' to hit the other ship.

(center). She fires the gun when points *a* and *a'* coincide. In her frame the other rocket ship is Lorentz contracted. Therefore *o* expects her bullet to miss the other rocket. But in the frame of the other observer *o'* it is the rocket ship of *o* that is measured to be Lorentz contracted (right). Therefore when points *a* and *a'* coincide, observer *o'* should observe a hit.

Does the bullet actually hit or miss? Pinpoint the looseness of the language used to state the problem and the error in one figure. Show that your argument is consistent with the results of the Train Paradox (Section 3.4).

3-8 Čerenkov radiation

No particle has been observed to travel faster than the speed of light in a vacuum. However particles have been observed that travel in a material medium faster than the speed of light in that medium. When a charged particle moves through a medium faster than light moves in that medium, it radiates coherent light in a cone whose axis lies along the path of the particle. (Note the rough similarity to waves created by a motorboat speeding across calm water and the more exact similarity to the "cone of sonic boom" created by a supersonic aircraft.) This is called Čerenkov radiation (Russian *Č* is pronounced as "ch"). Let *v* be the speed of the particle in the medium and v_{light} be the speed of light in the medium.

a From this information use the first figure to show that the half-angle ϕ, of the light cone is given by the expression

$$\cos \phi = v_{light}/v$$

b Consider the plastic with the trade name Lucite, for which $v_{light} = 2/3$. What is the minimum velocity that a charged particle can have if it is to produce Čerenkov radiation in Lucite? What is the *maximum* angle ϕ at which Čerenkov radiation can be produced in Lucite? Measurement of the angle provides a good way to measure the velocity of the particle.

c In water the speed of light is approximately $v_{light} = 0.75$. Answer the questions of part **b** for the case of water. See the second figure for an application of Čerenkov radiation in water.

3-9 aberration of starlight

A star lies in a direction generally perpendicular to Earth's direction of motion around Sun. Because of Earth's motion, the star appears to an Earth observer to lie in a slightly different direction than it would

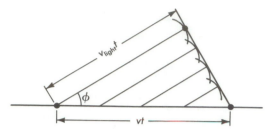

EXERCISE 3-8, first figure. *Calculation of Čerenkov angle ϕ.*

EXERCISE 3-8, second figure. *Use of Čerenkov radiation for indirect detection of neutrinos in the Deep Underwater Muon and Neutrino Detector (DUMAND) 30 kilometers off Keahole Point on the island of Hawaii. Neutrinos have no electric charge and their mass, if any, has so far escaped detection (Box 8-1). Neutrinos interact extremely weakly with matter, passing through Earth with almost no collisions. Indeed, the DUMAND detector array selects for analysis only neutrinos that come upward through Earth. In this way Earth itself acts as a shield to eliminate all other cosmic-ray particles.*

What are possible sources for these neutrinos? Theory predicts the emission of very high-energy (greater than 10^{12} electron-volt) neutrinos from matter plunging toward a black hole. Black holes may be the energy sources for extra-bright galactic nuclei and for quasars —small, distant, enigmatic objects shining with the light of hundreds of galaxies (Section 9.8). Information about conditions deep within these astronomical structures may be carried by neutrinos as they pierce Earth and travel upward through the DUMAND detector array.

In a rare event, a neutrino moving through the ocean slams into one of the quarks that make up a proton or a neutron in, say, an oxygen nucleus in the water, creating a burst of particles. All of these particles are quickly absorbed by the surrounding water except a stable negatively charged muon, 207 times the mass of the electron (thus sometimes called a "fat electron"). This muon streaks through the water in the same direction as the neutrino that created it and at a speed greater than that of light in water, thus emitting Čerenkov radiation. The Čerenkov radiation is detected by photomultiplier tubes in an array anchored to the ocean floor.

Photomultipliers are strung along 9 vertical cables, 8 cables spaced around a circle 100 meters in diameter on the ocean floor, the ninth cable rising from the center of the circle. Each cable is 335 meters long and holds 24 glass spheres positioned 10 meters apart on the top 230 meters of its length. There are no detectors on the bottom 110 meters, in order to avoid any cloud of sediments from the bottom. Above the bottom, the water is so clear and modern photodetectors so sensitive that Čerenkov radiation can be detected from a muon that passes within 40 meters of a detector.

Photomultipliers in the glass spheres detect Čerenkov radiation from the passing muons, transmitting this signal through underwater optical fibers to computers on the nearby island of Hawaii. The computers select for examination only those events in which (1) several optical sensors detect bursts that are (2) within 40 meters or so of a straight line, (3) spaced in time to show that the particle is moving at essentially the speed of light in a vacuum, and (4) from a particle moving upward through the water. A system of sonar beacons and hydrophones tracks the locations of the photomultipliers as the strings sway with the slow ocean currents. As a result, the direction of motion of the original neutrino can be recorded to an accuracy of one degree.

The DUMAND facility is designed to create a new sky map of neutrino sources to supplement our knowledge of the heavens, so far obtained primarily from the electromagnetic spectrum (radio, infrared, optical, ultraviolet, X-ray, gamma ray).

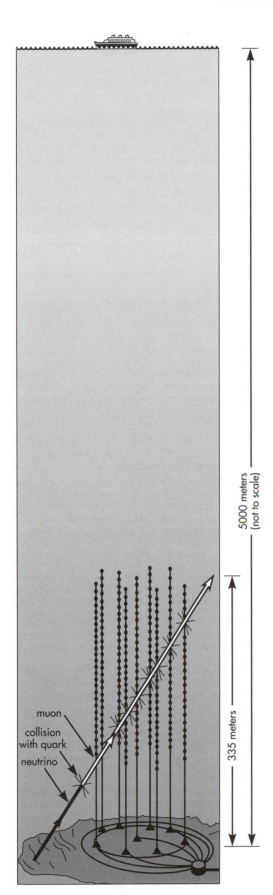

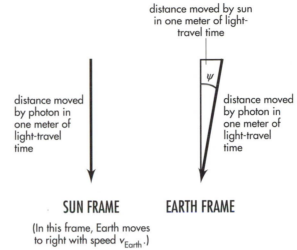

SUN FRAME

(In this frame, Earth moves
to right with speed v_{Earth}.)

EARTH FRAME

EXERCISE 3-9. *Aberration of starlight. Not to scale.*

appear to an observer at rest relative to Sun. This
effect is called **aberration.** Using the diagram, find
this apparent difference of direction.

a   Find a trigonometric expression for the aberration angle ψ shown in the figure.

b   Evaluate your expression using the speed of
Earth around Sun, $v_{\text{Earth conv}} = 30$ kilometers/second.
Find the answer in radians and in seconds of arc. (One
degree equals 60 minutes of arc; one minute equals
60 seconds of arc.) This change in apparent position
can be detected with sensitive equipment.

c   The nonrelativistic answer to this problem—
the answer using nonrelativistic physics—is $\tan \psi =
v_{\text{Earth}}$ (in meters/meter). Do you think that the exper-
imental difference between relativistic and nonrelativistic answers for stellar aberration observed from
Earth can be the basis of a crucial experiment to decide
between the correctness of the two theories?

Discussion: Of course we cannot climb off Earth
and view the star from the Sun frame. But Earth
reverses direction every six months (with respect to
what?), so light from a "transverse star" viewed in,
say, July will appear to be shifted through twice the
aberration angle calculated in part **b** compared with
the light from the same star in January. New question: Since the background of stars behind the one
under observation also shifts due to aberration, how
can the effect be measured at all?

d   A rocket in orbit around Earth suddenly
changes its velocity from a very small fraction of the
speed of light to $v = 0.5$ with respect to Sun, moving
in the same direction as Earth is moving around Sun.
In what direction will the rocket astronaut now see the
star of parts **a** and **b**?

3-10 the expanding universe

a A giant bomb explodes in otherwise empty space. What is the nature of the motion of one fragment relative to another? And how can this relative motion be detected? **Discussion:** Imagine each fragment equipped with a beacon that gives off flashes of light at regular, known intervals $\Delta\tau$ of time as measured in its own frame of reference (proper time!). Knowing this interval between flashes, what method of detection can an observer on one fragment employ to determine the velocity v — relative to her — of any other fragment? Assume that she uses, in making this determination, (1) the known proper time $\Delta\tau$ between flashes and (2) the time $\Delta t_{reception}$ between the arrival of consecutive flashes at her position. (This is *not* equal to the time Δt in her frame between the emission of the two flashes from the receding emitter; see the figure.) Derive a formula for v in terms of proper time lapse $\Delta\tau$ and $\Delta t_{reception}$. How will the measured recession velocity depend on the distance from one's own fragment to the fragment at which one is looking? Hint: In any given time in any given frame, fragments evidently travel distances in that frame from the point of explosion that are in direct proportion to their velocities in that frame.

b How can observation of the light from stars be used to verify that the universe is expanding? **Discussion:** Atoms in hot stars give off light of different frequencies characteristic of these atoms ("spectral lines"). The observed period of the light in each spectral line from starlight can be measured on Earth. From the pattern of spectral lines the kind of atom emitting the light can be identified. The same kind of atom can then be excited in the laboratory to emit light while at rest and the proper period of the light in any spectral line can be measured. Use the results of part **a** to describe how the observed period of light in one spectral line from starlight can be compared to the proper period of light in the same spectral line from atoms at rest in the laboratory to give the velocity of recession of the star that emits the light. This observed change in period due to the velocity of the source is called the Doppler shift. (For a more detailed treatment of Doppler shift, see the exercises for Chapters 5 and 8.) If the universe began in a gigantic explosion, how must the observed velocities of recession of different stars at different distances compare with one another? Slowing down during expansion — by gravitational attraction or otherwise — is to be neglected here but is considered in more complete treatments.

c The brightest steadily shining objects in the heavens are called quasars, which stands for "quasistellar objects." A single quasar emits more than 100 times the light of our entire galaxy. One possible source of quasar energy is the gravitational energy released as material falls into a black hole (Section 9.8). Because they are so bright, quasars can be observed at great distances. As of 1991, the greatest observed quasar red shift $\Delta t_{reception}/\Delta\tau$ has the value 5.9. According to the theory of this exercise, what is the velocity of recession of this quasar, as a fraction of the speed of light?

3-11 law of addition of velocities

In a spacebus a bullet shoots forward with speed 3/4 that of light as measured by travelers in the bus. The spacebus moves forward with speed 3/4 light speed as measured by Earth observers. How fast does the bullet move as measured by Earth observers: $3/4 + 3/4 = 6/4 = 1.5$ times the speed of light? No! Why not? Because (1) special relativity predicts that noth-

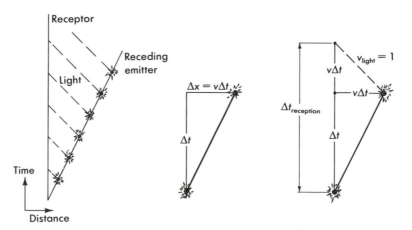

EXERCISE 3-10. *Calculation of the time $t_{reception}$ between arrival at observer of consecutive flashes from receding emitter. Light moves one meter of distance in one meter of time, so lines showing motion of light are tilted at $\pm 45°$ from the vertical.*

ing can travel faster than light, and (2) hundreds of millions of dollars have been spent accelerating particles ("bullets") to the fastest possible speed without anyone detecting a single particle that moves faster than light in a vacuum. Then where is the flaw in our addition of velocities? And what is the correct law of addition of velocities? These questions are answered in this exercise.

a First use Earth observers to record the the motions of the spacebus (length L measured in the Earth frame, speed v_{rel}) and the streaking bullet (speed v_{bullet}). The bullet starts at the back of the bus. To give it some competition, let a light flash (speed = 1) race the bullet from the back of the bus toward the front. The light flash wins, of course, reaching the front of the bus in time $t_{forward}$. And $t_{forward}$ is also equal to the distance that the light travels in this time. Show that this distance (measured in the Earth frame) equals the length of the bus plus the distance the bus travels in the same time:

$$t_{forward} = L + v_{rel}\, t_{forward} \quad \text{or} \quad t_{forward} = \frac{L}{1 - v_{rel}} \quad \text{(1)}$$

b In order to to rub in its advantage over the bullet, the light flash reflects from the front of the bus and moves backward until, after an additional time $t_{backward}$, it rejoins the forward-plodding bullet. This meeting takes place next to the seat occupied by Fred, who sits a distance fL behind the front of the bus, where f is a fraction of the bus length L. Show that for this leg of the trip the Earth-measured distance $t_{backward}$ traveled by the light flash can also be expressed as

$$t_{backward} = fL - v_{rel}\, t_{backward} \quad \text{or}$$
$$t_{backward} = \frac{fL}{1 + v_{rel}} \quad \text{(2)}$$

c The light flash has moved forward and then backward with respect to Earth. What is the *net* forward distance covered by the light flash at the instant it rejoins the bullet? Equate this with the forward distance moved by the bullet (at speed v_{bullet}) to obtain the equation

$$v_{bullet}(t_{forward} + t_{backward}) = t_{forward} - t_{backward}$$

or

$$(1 + v_{bullet})\, t_{backward} = (1 - v_{bullet})\, t_{forward} \quad \text{(3)}$$

d What are we after? We want a relation between the bullet speed v_{bullet} as measured in the Earth

frame and the bullet speed, call it v'_{bullet} (with a prime), as measured in the spacebus frame. The times given in parts **a**, **b**, and **c** are of no use to this end. Worse, we already know that times between events are typically different as measured in the spacebus frame than times between the same events measured in the Earth frame. So get rid of these times! Moreover, the Lorentz-contracted length L of the spacebus itself as measured in the Earth frame will be different from its rest length measured in the bus frame (Section 3.5). So get rid of L as well. Equations (1), (2), and (3) can be treated as three equations in the three unknowns $t_{forward}$, $t_{backward}$, and L. Substitute equations for the times (1) and (2) into equation (3). Lucky us: The symbol L cancels out of the result. Show that this result can be written

$$f = \frac{(1 - v_{bullet})}{(1 + v_{bullet})}\frac{(1 + v_{rel})}{(1 - v_{rel})} \quad \text{(4)}$$

e Now repeat the development of parts **a** through **d** for the spacebus frame, with respect to which the spacebus has its rest length L' and the bullet has speed v'_{bullet} (both with primes). Show that the result is:

$$f = \frac{(1 - v'_{bullet})}{(1 + v'_{bullet})} \quad \text{(5)}$$

Discussion: Instead of working hard, work smart! Why not use the old equations (1) through (4) for the spacebus frame? Because there is no relative velocity v_{rel} in the spacebus frame; the spacebus is at rest in its own frame! No problem: Set $v_{rel} = 0$ in equation (4), replace v_{bullet} by v'_{bullet} and obtain equation (5) directly from equation (4). If this is too big a step, carry out the derivation from the beginning in the spacebus frame.

f Do the two fractions f in equations (4) and (5) have the same value? In equation (4) the number f locates Fred's seat in the bus as a fraction of the total length of the bus in the Earth frame. In equation (5) the number f locates Fred's seat in the bus as a fraction of the total length of the bus in the bus frame. But this fraction must be the same: Fred cannot be halfway back in the Earth frame and, say, three quarters of the way back in the spacebus frame. Equate the two expressions for f given in equations (4) and (5) and solve for v_{bullet} to obtain the Law of Addition of Velocities:

$$v_{bullet} = \frac{v'_{bullet} + v_{rel}}{1 + v'_{bullet}\, v_{rel}} \quad \text{(6)}$$

g Explore some consequences of the Law of Addition of Velocities.

(1) An express bus on Earth moves at 108 kilometers/hour (approximately 67 miles/hour or 30 meters per second). A bullet moves forward with speed 600 meters/second with respect to the bus. What are the values of v_{rel} and v'_{bullet} in meters/meter? What is the value of their product in the denominator of equation (6)? Does this product of speeds increase the value of the denominator significantly over the value unity? Therefore what approximate form does equation (6) take for everyday speeds? Is this the form you would expect from your experience?

(2) Analyze the example that began this exercise: Speed of bullet with respect to spacebus $v'_{bullet} = 3/4$; speed of spacebus with respect to Earth $v_{rel} = 3/4$. What is the speed of the bullet measured by Earth observers?

(3) Why stop with bullets that saunter along at less than the speed of light? Let the bullet itself be a flash of light. Then the bullet speed as measured in the bus is $v'_{bullet} = 1$. For $v_{rel} = 3/4$, with what speed does this light flash move as measured in the Earth frame? Is this what you expect from the Principle of Relativity?

(4) Suppose a light flash is launched from the front of the bus directed toward the back ($v'_{bullet} = -1$). What is the velocity of this light flash measured in the Earth frame? Is this what you expect from the Principle of Relativity?

Reference: N. David Mermin, *American Journal of Physics,* Volume 51, pages 1130–1131 (1983).

3-12 Michelson–Morley experiment

a An airplane moves with air speed c (not the speed of light) from point A to point B on Earth. A stiff wind of speed v is blowing from B toward A. (In this exercise only, the symbol v stands for velocity in conventional units, for example meters/second.) Show that the time for a round trip from A to B and back to A under these circumstances is greater by a factor $1/(1 - v^2/c^2)$ than the corresponding round trip time in still air. Paradox: The wind helps on one leg of the flight as well as hinders on the other. Why, therefore, is the round-trip time not the same in the presence of wind as in still air? Give a simple physical reason for this difference. What happens when the wind speed is nearly equal to the speed of the airplane?

b The same airplane now makes a round trip between A and C. The distance between A and C is the same as the distance from A to B, but the line from A to C is perpendicular to the line from A to B, so that in moving between A and C the plane flies across the wind. Show that the round-trip time between A and C under these circumstances is greater by a factor $1/(1 - v^2/c^2)^{1/2}$ than the corresponding round-trip time in still air.

c Two airplanes with the same air speed c start from A at the same time. One travels from A to B and back to A, flying first against and then with the wind (wind speed v). The other travels from A to C and back to A, flying across the wind. Which one will arrive home first, and what will be the difference in their arrival times? Using the first two terms of the binomial theorem,

$$(1 + z)^n \approx 1 + nz \qquad \text{for } |z| \ll 1$$

show that if $v \ll c$, then an approximate expression for this time difference is $\Delta t \approx (L/2c)(v/c)^2$, where L is the round-trip distance between A and B (and between A and C).

d The South Pole Air Station is the supply depot for research huts on a circle of 300-kilometer radius centered on the air station. Every Monday many supply planes start simultaneously from the station and fly radially in all directions at the same altitude. Each plane drops supplies and mail to one of the research huts and flies directly home. A Fussbudget with a stopwatch stands on the hill overlooking the air station. She notices that the planes do not all return at the same time. This discrepancy perplexes her because she knows from careful measurement that (1) the distance from the air station to every research hut is the same, (2) every plane flies with the same air speed as every other plane — 300 kilometers/hour — and (3) every plane travels in a straight line over the ground from station to hut and back. The Fussbudget finally decides that the discrepancy is due to the wind at the high altitude at which the planes fly. With her stopwatch she measures the time from the return of the first plane to the return of the last plane to be 4 seconds. What is the wind speed at the altitude where the planes fly? What can the Fussbudget say about the direction of this wind?

e In their famous experiment Michelson and Morley attempted to detect the so-called **ether drift** — the motion of Earth through the "ether," with respect to which light was supposed to have the velocity c. They compared the round-trip times for light to travel equal distances parallel and perpendicular to the direction of motion of Earth around Sun. They reflected the light back and forth between nearly parallel mirrors. (This would correspond to part **c** if each airplane made repeated round trips.) By this means they were able to use a total round-trip length of 22 meters for each path. If the "ether" is at rest with respect to Sun, and if Earth moves at 30×10^3 meters/second in its path around Sun, what is the approximate difference in time of return between light flashes that are emitted simultaneously and travel along the two perpendicular paths? Even with the instruments of today, the difference predicted by the ether-drift hypothesis would be too small to measure directly, and the following method was used instead.

f   The original Michelson–Morley interferometer is diagrammed in the figure. Nearly monochromatic light (light of a single frequency) enters through the lens at a. Some of the light is reflected by the half-silvered mirror at b and the rest of the light continues toward d. Both beams are reflected back and forth until they reach mirrors e and e_1 respectively, where each beam is reflected back on itself and re-

traces its path to mirror b. At mirror b parts of each beam combine to enter telescope f together. The transparent piece of glass at c, of the same dimensions as the half-silvered mirror b, is inserted so that both beams pass the same number of times (three times) through this thickness of glass on their way to telescope f. Suppose that the perpendicular path lengths are exactly equal and the instrument is at rest with respect to the ether. Then monochromatic light from the two paths that leave mirror b in some relative phase will return to mirror b in the same phase. Under these circumstances the waves entering telescope f will add crest to crest and the image in this telescope will be bright. On the other hand, if one of the beams has been delayed a time corresponding to one half period of the light, then it will arrive at mirror b one half period later and the waves entering the telescope will cancel (crest to trough), so the image in the telescope will be dark. If one beam is retarded a time corresponding to one whole period, the telescope image will be bright, and so forth. What time corresponds to

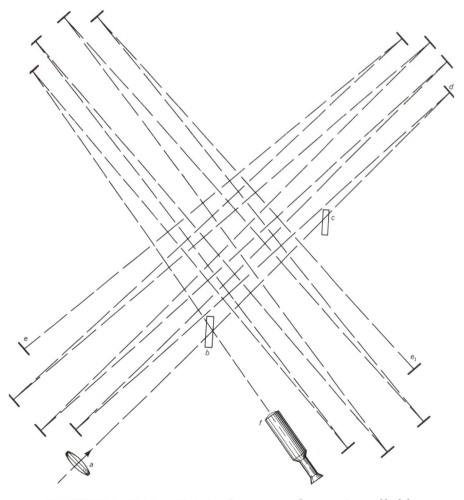

EXERCISE 3-12. *Michelson–Morley interferometer mounted on a rotating marble slab.*

one period of the light? Michelson and Morley used sodium light of wavelength 589 nanometers (one nanometer is equal to 10^{-9} meter). Use the equations $f\lambda = c$ and $f = 1/T$ that relate frequency f, period T, wavelength λ, and speed c of an electromagnetic wave. Show that one period of sodium light corresponds to about 2×10^{-15} seconds.

Now there is no way to "turn off" the alleged ether drift, adjust the apparatus, and then turn the alleged ether drift on again. Instead of this, Michelson and Morley floated their interferometer in a pool of mercury and rotated it slowly about its center like a phonograph record while observing the image in the telescope (see the figure). In this way if light is delayed on either path when the instrument is oriented in a certain direction, light on the other path will be delayed by the same amount of time when the instrument has rotated 90 degrees. Hence the total change in delay time between the two paths observed as the interferometer rotates should be twice the difference calculated using the expression derived in part **c**. By refinements of this method Michelson and Morley were able to show that the time change between the two paths as the instrument rotated corresponded to less than one one-hundredth of the shift from one dark image in the telescope to the next dark image. Show that this result implies that the motion of the ether at the surface of Earth — if it exists at all — is less than one sixth of the speed of Earth in its orbit. In order to eliminate the possibility that the ether was flowing past Sun at the same rate as Earth was moving its orbit, they repeated the experiment at intervals of three months, always with negative results.

g Discussion question: Does the Michelson–Morley experiment, by itself, disprove the theory that light is propagated through an ether? Can the ether theory be modified to agree with the results of this experiment? How? What further experiment can be used to test the modified theory?

Reference: A. A. Michelson and E. W. Morley, *American Journal of Science,* Volume 134, pages 333–345 (1887).

3-13 the Kennedy–Thorndike experiment

Note: Part **d** of this exercise uses elementary calculus.

The Michelson–Morley experiment was designed to detect any motion of Earth relative to a hypothetical fluid — the ether — a medium in which light was supposed to move with characteristic speed c. No such relative motion of earth and ether was detected. Partly as a result of this experiment the concept of ether has since been discarded. In the modern view, light requires no medium for its transmission. What significance does the negative result of the Michelson–Morley experiment have for us who do not believe in the ether theory of light propagation? Simply this: (1) The round-trip speed of light measured on earth is the same in every direction — the speed of light is isotropic. (2) The speed of light is isotropic not only when Earth moves in one direction around Sun in, say, January (call Earth with this motion the "laboratory frame"), but also when Earth moves in the opposite direction around Sun six months later, in July (call Earth with this motion the "rocket frame"). (3) The generalization of this result to any pair of inertial frames in relative motion is contained in the statement, The round-trip speed of light is isotropic both in the laboratory frame and in the rocket frame. This result leaves an important question unanswered: Does the round-trip speed of light — which is isotropic in both laboratory and rocket frames — also have the same numerical value in laboratory and rocket frames? The assumption that this speed has the same numerical value in both frames played a central role in demonstrating the invariance of the interval (Section 3.7). But is this assumption valid?

a An experiment to test the assumption of the equality of the round-trip speed of light in two inertial frames in relative motion was conducted in 1932 by Roy J. Kennedy and Edward M. Thorndike. The experiment uses an interferometer with arms of unequal length (see the figure). Assume that one arm of the interferometer is Δl longer than the other arm. Show that a flash of light entering the apparatus will take a time $2\Delta l/c$ longer to complete the round trip along the longer arm than along the shorter arm. The difference in length Δl used by Kennedy and Thorndike was approximately 16 centimeters. What is the approximate difference in time for the round trip of a light flash along the alternative paths?

b Instead of a pulse of light, Kennedy and Thorndike used continuous monochromatic light of period $T = 1.820 \times 10^{-15}$ seconds ($\lambda = 546.1$ nanometers $= 546.1 \times 10^{-9}$ meters) from a mercury source. Light that traverses the longer arm of the interferometer will return approximately how many periods n later than light that traverses the shorter arm? If in the actual experiment the number of periods is an integer, the reunited light from the two arms will add (crest-to-crest) and the field of view seen through the telescope will be bright. In contrast, if in the actual experiment the number of periods is a half-integer, the reunited light from the two arms will cancel (crest-to-trough) and the field of view of the telescope will be dark.

c Earth continues on its path around Sun. Six months later Earth has reversed the direction of its velocity relative to the fixed stars. In this new frame of

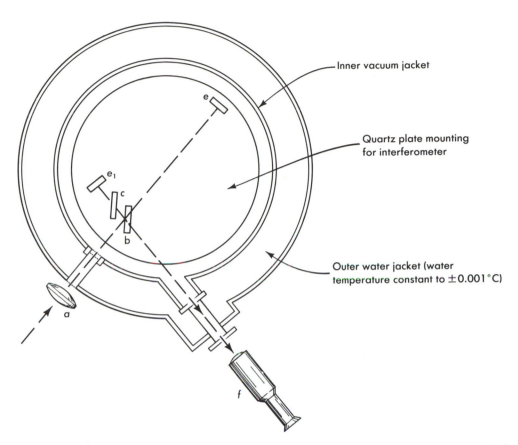

Inner vacuum jacket

Quartz plate mounting
for interferometer

Outer water jacket (water
temperature constant to ±0.001°C)

EXERCISE 3-13. *Schematic diagram of apparatus used for the Kennedy–Thorndike experiment. Parts of the interferometer have been labeled with letters corresponding to those used in describing the Michelson–Morley interferometer (Exercise 3-12). The experimenters went to great lengths to insure the optical and mechanical stability of their apparatus. The interferometer is mounted on a plate of quartz, which changes dimension very little when temperature changes. The interferometer is enclosed in a vacuum jacket so that changes in atmospheric pressure will not alter the effective optical path length of the interferometer arms (slightly different speed of light at different atmospheric pressure). The inner vacuum jacket is surrounded by an outer water jacket in which the water is kept at a temperature that varies less than ±0.001 degrees Celsius. The entire apparatus shown in the figure is enclosed in a small darkroom (not shown) maintained at a temperature constant within a few hundredths of a degree. The small darkroom is in turn enclosed in a larger darkroom whose temperature is constant within a few tenths of a degree. The overall size of the apparatus can be judged from the fact that the difference in length of the two arms of the interferometer (length eb compared with length e_1b) is 16 centimeters.*

reference will the round-trip speed of light have the same numerical value c as in the original frame of reference? One can rewrite the answer to part **b** for the original frame of reference in the form

$$c = (2/n)(\Delta l/T)$$

where Δl is the difference in length between the two interferometer arms, T is the time for one period of the atomic light source, and n is the number of periods that elapse between the return of the light on the shorter path and the return of the light on the longer path. Suppose that as Earth orbits Sun no shift is observed in the telescope field of view from, say, light toward dark. This means that n is observed to be constant. What would this hypothetical result tell about the numerical value c of the speed of light?

Point out the standards of distance and time used in determining this result, as they appear in the equation. Quartz has the greatest stability of dimension of any known material. Atomic time standards have proved to be the most dependable earth-bound timekeeping mechanisms.

d In order to carry out the experiment outlined in the preceding paragraphs, Kennedy and Thorndike would have had to keep their interferometer operating perfectly for half a year while continuously observing the field of view through the telescope. Uninterrupted operation for so long a time was not feasible. The actual durations of their observations varied from eight days to a month. There were several such periods of observation at three-month time separations. From the data obtained in these periods, Kennedy and Thorndike were able to estimate that over a single six-month observation the number of periods n of relative delay would vary by less than the fraction 3/1000 of one period. Take the differential of the equation in part **c** to find the largest fractional change dc/c of the round-trip speed of light between the two frames consistent with this estimated change in n (frame 1 — the "laboratory" frame — and frame 2 — the "rocket" frame — being in the present analysis Earth itself at two different times of year, with a relative velocity twice the speed of Earth in its orbit: 2×30 kilometers/second).

Historical note: At the time of the Michelson–Morley experiment in 1887, no one was ready for the idea that physics — including the speed of light — is the same in every inertial frame of reference. According to today's standard Einstein interpretation it seems obvious that both the Michelson–Morley and the Kennedy–Thorndike experiments should give null results. However, when Kennedy and Thorndike made their measurements in 1932, two alternatives to the Einstein theory were open to consideration (designated here as theory A and theory B). Both A and B assumed the old idea of an absolute space, or "ether," in which light has the speed c. Both A and B explained the zero fringe shift in the Michelson–Morley experiment by saying that all matter that moves at a velocity v (expressed as a fraction of lightspeed) relative to "absolute space" undergoes a shrinkage of its space dimensions in the direction of motion to a new length equal to $(1 - v^2)^{1/2}$ times the old length ("Lorentz-FitzGerald contraction hypothesis"). The two theories differed as to the effect of "motion through absolute space" on the running rate of a clock. Theory A said, No effect. Theory B said that a standard seconds clock moving through absolute space at velocity v has a time between ticks of $(1 - v^2)^{1/2}$ seconds. In theory B the ratio $\Delta l / T$ in the equation in part **b** will not be affected by the velocity of the clock, and the Kennedy–Thorndike experiment will give a null result, as observed ("complicated explanation for simple effect"). In theory A the ratio $\Delta l / T$ in the equation will be multiplied by the factor $(1 - v_1^2)^{1/2}$ at a time of year when the "velocity of Earth relative to absolute space" is v_1 and multiplied by $(1 - v_2^2)^{1/2}$ at a time of year when this velocity is v_2. Thus the fringes should shift from one time of year ($v_1 = v_{orbital} + v_{Sun}$) to another time of year ($v_2 = v_{orbital} - v_{Sun}$) unless by accident Sun happened to have "zero velocity relative to absolute space" — an accident judged so unlikely as not to provide an acceptable explanation of the observed null effect. Thus the Kennedy–Thorndike experiment ruled out theory A (length contraction alone) but allowed theory B (length contraction plus time contraction) — and also allowed the much simpler

Einstein theory of equivalence of all inertial reference frames.

The "sensitivity" of the Kennedy–Thorndike experiment depends on the theory under consideration. In the context of theory A the observations set an upper limit of about 15 kilometers/second to the "speed of Sun through absolute space" (sensitivity reported in the Kennedy–Thorndike paper). In the context of Einstein's theory the observations say that the round-trip speed of light has the same numerical magnitude — within an error of about 3 meters/second — in inertial frames of reference having a relative velocity of 60 kilometers/second.

Reference: R. J. Kennedy and E. M. Thorndike, *Physical Review*, Volume 42, pages 400–418 (1932).

3-14   things that move faster than light

Can "things" or "messages" move faster than light? Does relativity really say "No" to this possibility? Explore these questions further using the following examples.

a   **The Scissors Paradox.** A very long straight rod, inclined at an angle θ to the x-axis, moves downward with uniform speed v_{rod} as shown in the figure. Find the speed v_A of the point of intersection A of the lower edge of the stick with the x-axis. Can this speed be greater than the speed of light? If so, for what values of the angle θ and v_{rod} does this occur? Can the motion of intersection point A be used to transmit a message faster than light from someone at the origin to someone far out on the x-axis?

b   **Transmission of a Hammer Pulse.** Suppose the same rod is initially at rest in the laboratory with the point of intersection initially at the origin. The region of the rod centered at the origin is struck sharply with the downward blow of a hammer. The point of intersection moves to the right. Can this motion of the point of intersection be used to transmit a message faster than the speed of light?

c   **Searchlight Messenger?** A very powerful searchlight is rotated rapidly in such a way that its beam sweeps out a flat plane. Observers A and B are at rest on the plane and each the same distance from the searchlight but not near each other. How far from the searchlight must A and B be in order that the searchlight beam will sweep from A to B faster than a light signal could travel from A to B? Before they took their positions, the two observers were given the following instruction:

To A: "When you see the searchlight beam, fire a bullet at B."
To B: "When you see the searchlight beam, duck because A has fired a bullet at you."

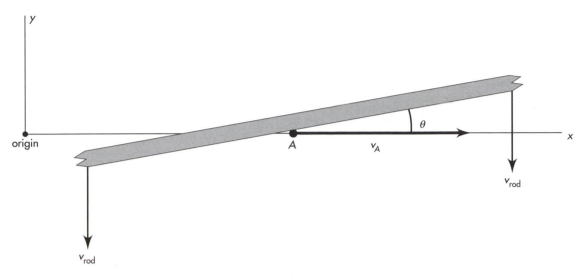

EXERCISE 3-14. *Can the point of intersection* A *move with a speed* v_A *greater than the speed of light?*

Under these circumstances, has a warning message traveled from *A* to *B* with a speed faster than that of light?

d Oscilloscope Writing Speed. The manufacturer of an oscilloscope claims a writing speed (the speed with which the bright spot moves across the screen) in excess of the speed of light. Is this possible?

3-15 four times the speed of light?

We look westward across the United States and see the rocket approaching us at four times the speed of light.

How can this be, since nothing moves faster than light?

We did not say the rocket *moves* faster than light; we said only that we *see* it moving faster than light.

Here is what happens: The rocket streaks under the Golden Gate Bridge in San Francisco, emitting a flash of light that illuminates the rocket, the bridge, and the surroundings. At time Δt later the rocket threads the Gateway Arch in St. Louis that commemorates the starting point for covered wagons. The arch and the Mississippi riverfront are flooded by a second flash of light. The top figure is a visual summary of measurements from our continent-spanning latticework of clocks taken at this moment.

Now the rocket continues toward us as we stand in New York City. The center figure summarizes data taken as the first flash is about to enter our eye. Flash 1 shows us the rocket passing under the Golden Gate Bridge. An instant later flash 2 shows us the rocket passing through the Gateway Arch.

a Answer the following questions using symbols from the first two figures. The images carried by the two flashes show the rocket how far apart in space? What is the time lapse between our reception of these two images? Therefore, what is the apparent speed of the approaching rocket we see? For what speed v of the rocket does the apparent speed of approach equal four times the speed of light? For what rocket speed do we see the approaching rocket to be moving at 99 times the speed of light?

b Our friend in San Francisco is deeply disappointed. Looking eastward, she sees the retreating rocket traveling at less than half the speed of light (bottom figure). She wails, "Which one of us is wrong?" "Neither one." we reply. "No matter how high the speed v of the rocket, you will never see it moving directly away from you at a speed greater than half the speed of light."

Use the bottom figure to derive an expression for the apparent speed of recession of the rocket. When we in New York see the rocket approaching at four times the speed of light, with what speed does our San Francisco friend see it moving away from her? When we see a faster rocket approaching at 99 times the speed of light, what speed of recession does she behold?

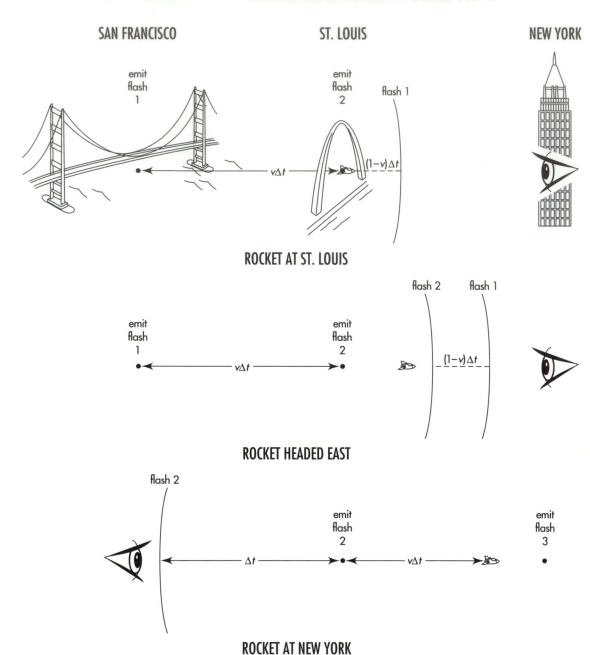

EXERCISE 3-15. *Top: Rocket headed east, shown at the instant it passes under the Gateway Arch in St. Louis and emits flash 2. The rocket is chasing flash 1, emitted earlier as it passed under the Golden Gate Bridge in San Francisco.* **Center:** *The two image-carrying flashes are close together, so they enter the eye in rapid succession. This gives the viewer the visual impression that the rocket moved from San Francisco to St. Louis in a very short time.*

Bottom: Rocket headed east, shown at the instant it approaches the Empire State Building in New York City and emits flash 3. When the rocket moves away from the viewer, the distance of rocket travel is added to the separation between flashes. This increases the apparent time between flashes, giving the viewer the impression that the rocket moved from St. Louis to New York at less than one half light-speed.

3-16 superluminal expansion of quasar 3C273?

The most powerful sources of energy we know or conceive or see in all the universe are so-called quasi-stellar objects, or **quasars,** starlike sources of light located billions of light-years away. Despite being far smaller than any galaxy, the typical quasar manages to put out more than 100 times as much energy as our own Milky Way, with its hundred billion stars. Quasars, unsurpassed in brilliance and remoteness, we count today as lighthouses of the heavens.

One of the major problems associated with quasars is that some are composed of two or more components

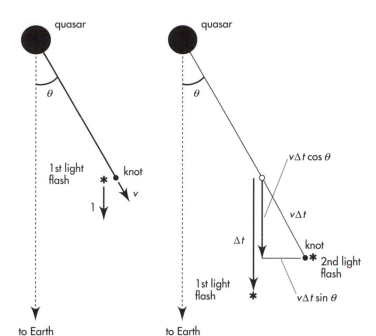

EXERCISE 3-16, first figure. *Left: Bright "knot" of plasma ejected from a quasar at high speed* v *emits a first flash of light toward Earth.* **Right:** *The knot emits a second light flash toward Earth a time* Δt *later. This time* Δt *is measured locally near the knot using the Earth-linked latticework of rods and clocks (har! har!).*

that appear to be separating from each other with relative velocity greater than the speed of light ("superluminal" velocity). One theory that helps explain this effect pictures the quasar as a core that ejects a jet of plasma at relativistic speed. Disturbances or instabilities in such a jet appear as discrete "knots" of plasma. The motion and light emission from a knot may account for its apparent greater-than-light speed, as shown using the first figure.

a   The first figure shows two Earth-directed light flashes emitted from the streaking knot. The time between emissions is Δt as measured locally near the knot using the Earth-linked latticework of rods and clocks. Of course the clock readings on this portion of the Earth-linked latticework are not available to us on Earth; therefore we cannot measure Δt directly. Rather, we see the time separation between the arrivals of the two flashes at Earth. From the figure, show that this Earth-seen time separation Δt_{seen} is given by the expression

$$\Delta t_{\text{seen}} = \Delta t(1 - v \cos \theta)$$

b   We have another disability in viewing the knot from Earth. We do not see the motion of the knot toward us, only the apparent motion of the knot across our field of view. Find an expression for this transverse motion (call it Δx_{seen}) between emissions of the two light flashes in terms of Δt.

c   Now calculate the speed v^x_{seen} of the rightward motion of the knot as seen on Earth. Show that the result is

$$v^x_{\text{seen}} = \frac{\Delta x_{\text{seen}}}{\Delta t_{\text{seen}}} = \frac{v \sin \theta}{1 - v \cos \theta}$$

d   What is the value of v^x_{seen} when the knot is emitted in the direction exactly toward Earth? when it is emitted perpendicular to this direction? Find an expression that gives the range of angles θ for which v^x_{seen} is greater than the speed of light. For $\theta = 45$ degrees, what is the range of knot speeds v such that v^x_{seen} is greater than the speed of light?

e   If you know calculus, find an expression for the angle θ_{max} at which v^x_{seen} has its maximum value for a given knot speed v. Show that this angle satisfies the equation $\cos \theta_{\text{max}} = v$. Whether or not you derive this result, use it to show that the maximum apparent transverse speed is seen as

$$v^x_{\text{seen, max}} = \frac{v}{(1 - v^2)^{1/2}}$$

f   What is this maximum transverse speed seen on Earth when $v = 0.99$?

g   The second figure shows the pattern of radio emission from the quasar 3C273. The decreased pe-

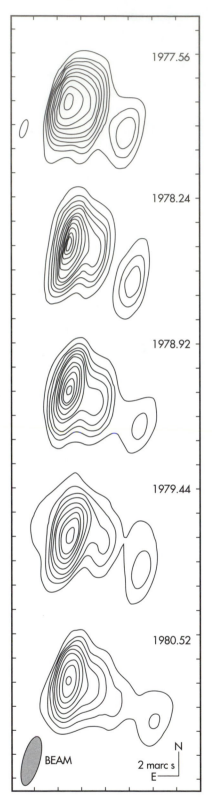

1977.56

1978.24

1978.92

1979.44

1980.52

BEAM

N

2 marc s

E

EXERCISE 3-16, second figure. *Contour lines of radio emission from the quasar 3C273 showing a bright "knot" of plasma apparently moving away from it at a speed greater than the speed of light. The time of each image is given as calendar year and decimal fraction. Horizontal scale divisions are in units of 2 milli arc-seconds. (1 milli arc-second = $10^{-3}/3600$ degree = 4.85×10^{-9} radian)*

riod of radiation from this source (Exercise 3-10) shows that it is approximately 2.6×10^9 light-years from Earth. A secondary source is apparently moving away from the central quasar. Take your own measurements on the figure. Combine this with data from the figure caption to show that the apparent speed of separation is greater than 9 times the speed of light.

Note: As of 1990, apparent greater-than-light-speed ("superluminal") motion has been observed in approximately 25 different sources.

References: Analysis and first figure adapted from Denise C. Gabuzda, *American Journal of Physics,* Volume 55, pages 214–215 (1987). Second figure and data taken from T. J. Pearson, S. C. Unwin, M. H. Cohen, R. P. Linfield, A. C. S. Readhead, G. A. Seielstad, R. S. Simon, and R. C. Walker, *Nature,* Volume 290, pages 365–368 (2 April 1981).

3-17   contraction or rotation?

A cube at rest in the rocket frame has an edge of length 1 meter in that frame. In the laboratory frame the cube is Lorentz contracted in the direction of motion, as shown in the figure. Determine this Lorentz contraction, for example, from locations of four clocks at rest and synchronized in the laboratory lattice with which the four corners of the cube, *E, F, G, H,* coincide when all four clocks read the same time. This latticework measurement eliminates time lags in the travel of light from different corners of the cube.

Now for a different observing procedure! Stand in the laboratory frame and look at the cube with one eye as the cube passes overhead. What one sees at any time is light that enters the eye at that time, even if it left the different corners of the cube at different times. Hence, what one sees visually may not be the same as what one observes using a latticework of clocks. If the cube is viewed from the bottom then the distance *GO* is equal to the distance *HO,* so light that leaves *G* and *H* simultaneously will arrive at *O* simultaneously. Hence, when one sees the cube to be overhead one will see the Lorentz contraction of the bottom edge.

a   Light from *E* that arrives at *O* simultaneously with light from *G* will have to leave *E* earlier than light from *G* left *G.* How much earlier? How far has the cube moved in this time? What is the value of the distance *x* in the right top figure?

b   Suppose the eye interprets the projection in in the figures as a rotation of a cube that is not Lorentz contracted. Find an expression for the angle of apparent rotation ϕ of this uncontracted cube. Interpret this expression for the two limiting cases of cube speed in the laboratory frame: $v \rightarrow 0$ and $v \rightarrow 1$.

c   **Discussion question:** Is the word "really" an appropriate word in the following quotations?

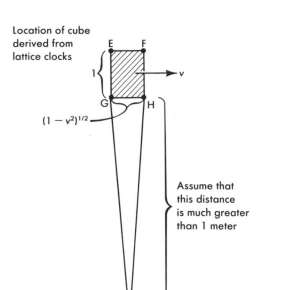

Location of cube derived from lattice clocks

Assume that this distance is much greater than 1 meter

Position of observer's eye

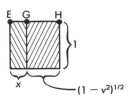

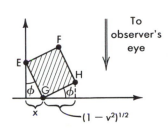

To observer's eye

EXERCISE 3-17. *Left: Position of eye of visual observer watching cube pass overhead. **Right top:** What the visual observer sees as she looks up from below. **Right bottom:** How the visual observer can interpret the projection of the second figure.*

(1) An observer using the rocket latticework of clocks says, "The stationary cube is really neither rotated nor contracted."

(2) Someone riding in the rocket who looks at the stationary cube agrees, "The cube is really neither rotated nor contracted."

(3) An observer using the laboratory latticework of clocks says, "The passing cube is really Lorentz contracted but not rotated."

(4) Someone standing in the laboratory frame looking at the passing cube says, "The cube is really rotated but not Lorentz contracted."

What can one rightfully say — in a sentence or two — to make each observer think it reasonable that the other observers should come to different conclusions?

d The analysis of parts **b** and **c** assumes that the visual observer looks with one eye and has no depth perception. How will the cube passing overhead be perceived by the viewer with accurate depth perception?

Reference: For a more complete treatment of this topic, see Edwin F. Taylor, *Introductory Mechanics* (John Wiley and Sons, New York, 1963), pages 346–360.

LORENTZ TRANSFORMATION

L.1 LORENTZ TRANSFORMATION: USEFUL OR NOT?

related events or lonely events?

Events, and the intervals between events, define the layout of the physical world. No latticework of clocks there! Only events and the relation between event and event as expressed in the interval. That's spacetime physics, lean and spare, as it offers itself to us to meet the needs of industry, science, and understanding.

There's another way to express the same information and use it for the same purposes: Set up a free-float latticework of recording clocks, or the essential rudiments of such a latticework. The space and time coordinates of that Lorentz frame map each event as a lonesome individual, with no mention of any connection, any spacetime interval, to any other event.

This lattice-based method for doing spacetime physics has the advantage that it can be mechanized and applied to event after event, wholesale. These regimented space and time coordinates then acquire full usefulness only when we can translate them from the clock-lattice frame used by one analyst to the clock-lattice frame used by another.

This scheme of translation has acquired the name "Lorentz transformation." Its usefulness depends on the user. Some never need it because they deal always with intervals. Others use it frequently because it regiments records and standardizes analysis. For their needs we insert this Special Topic on the Lorentz transformation. The reader may wish to read it now, or skip it altogether, or defer it until after Chapter 4, 5, or 6. The later the better, in our opinion. 🪶

Events and intervals only: Spacetime lean and spare

Or isolated events described using latticework

Lorentz transformation: Translate event description from lattice to lattice

L.2 FASTER THAN LIGHT?

a reason to examine the Lorentz transformation

No object travels faster than light.

So YOU say, but watch ME: I travel in a rocket that you observe to move at 4/5 light speed. Out the front of my rocket I fire a bullet that I observe to fly forward at 4/5 light speed. Then you measure this bullet to streak forward at 4/5 + 4/5 = 8/5 = 1.6 light speed, which is greater than the speed of light. There!

   No!

Why not? Is it not true that 4/5 + 4/5 = 1.6?

Velocities do not add

As a mathematical abstraction: always true. As a description of the world: only sometimes true! Example 1: Add 4/5 liter of alcohol to 4/5 liter of water. The result? Less than 8/5 = 1.6 liter of liquid! Why? Molecules of water interpenetrate molecules of alcohol to yield a combined volume less that the sum of the separate volumes. Example 2: Add the speed you measure for the bullet (4/5) to the speed I measure for your rocket (4/5). The result? The speed I measure for the bullet is 40/41 = 0.9756. This remains less than the speed of light.

Why? And where did you get that number 40/41 for the bullet speed you measure?

I got the number from the Lorentz transformation, the subject of this Special Topic. The Lorentz transformation embodies a central feature of relativity: Space and time separations typically do not have the same values as observed in different frames.

Space and time separations between what?

Between events.

What events are we talking about here?

Event 1: You fire the bullet out the front of your rocket. Event 2: The bullet strikes a target ahead of you.

What do these events have to do with speed? We are arguing about speed!

Events define velocities

Let the bullet hit the target four meters in front of you, as measured in your rocket. Then the space separation between event 1 and event 2 is 4 meters. Suppose the time of flight is 5 meters as measured by your clocks, the time separation between the two events. Then your bullet speed measurement is (4 meters of distance)/(5 meters of time) = 4/5, as you said.

And what do YOU measure for the space and time separations in your laboratory frame?

For that we need the **Lorentz coordinate transformation equations.**

Phooey! I know how to reckon spacetime separations in different frames. We have been doing it for several chapters! From measurements in one frame we figure the spacetime interval, which has the same value in all frames. End of story.

No, not the end of the story, but at least its beginning. True, the invariant interval has the same value as derived from measurements in every frame. That allows you to predict the time between firing and impact as measured by the passenger riding on the bullet—and measured directly by the bullet passenger alone.

Interval: Only a start in reckoning spacetime separations in different frames

Predict how?

You know your space separation $x' = 4$ meters (primes for rocket measurements), and your time separation, $t' = 5$ meters. You know the space separation for the bullet rider, $x'' = 0$ (double primes for bullet measurements), since she is present at both the firing and the impact. From this you can use invariance of the interval to determine the wristwatch time between these events for the bullet rider:

$$(t'')^2 - (x'')^2 = (t')^2 - (x')^2$$

or

$$(t'')^2 - (0)^2 = (5 \text{ meters})^2 - (4 \text{ meters})^2 = (3 \text{ meters})^2$$

so that $t'' = 3$ meters. This is the proper time, agreed on by all observers but measured directly only on the wristwatch of the bullet rider.

Fine. Can't we use the same procedure to determine the space and time separations between these events in your laboratory frame, and thus the bullet speed for you?

Unfortunately not. We do reckon the same value for the interval. Use unprimed symbols for laboratory measurements. Then $t^2 - x^2 = (3 \text{ meters})^2$. That, however, is not sufficient to determine x or t separately. Therefore we cannot yet find their ratio x/t, which determines the bullet's speed in our frame.

Need more to compare velocities in different frames

So how can we reckon these x and t separations in your laboratory frame, thereby allowing us to predict the bullet speed you measure?

Use the Lorentz transformation. This transformation reports that our laboratory space separation between firing and impact is $x = 40/3$ meters and the time separation is slightly greater: $t = 41/3$ meters. Then bullet speed in my laboratory frame is predicted to be $v = x/t = 40/41 = 0.9756$. The results of our analysis in three reference frames are laid out in Table L-1.

Compare velocities using Lorentz transformation

Is the Lorentz transformation generally useful, beyond the specific task of reckoning speeds as measured in different frames?

Oh yes! Generally, we insert into the Lorentz transformation the coordinates x', t' of an event determined in the rocket frame. The Lorentz transformation then grinds and whirs, finally spitting out the coordinates x, t of the same event measured in the laboratory frame. Following are the Lorentz transformation equations. Here v_{rel} is the relative velocity between rocket and laboratory frames. For our convenience we lay the positive x-axis along the direction of motion of the rocket as observed in the laboratory frame and choose a common reference event for the zero of time and space for both frames.

$$\boxed{\text{TABLE L-1}}$$

HOW FAST THE BULLET?

	Bullet fired (coordinates of this event)	Bullet hits (coordinates of this event)	Speed of bullet (computed from frame coordinates)
Rocket frame (moves at $v_{rel} = 4/5$ as measured in laboratory)	$x' = 0$ $t' = 0$	$x' = 4$ meters $t' = 5$ meters	as measured in rocket frame: $v' = 4/5 = 0.8$
Bullet frame (moves at $v' = 4/5$ as measured in rocket)	$x'' = 0$ $t'' = 0$	$x'' = 0$ $t'' = 3$ meters (from invariance of the interval)	as measured in bullet frame: $v'' = 0$
Laboratory frame	$x = 0$ $t = 0$	$x = 40/3$ meters $t = 41/3$ meters (from Lorentz transformation)	as measured in laboratory frame: $v = 40/41 = 0.9756$

Lorentz transformation previewed

$$x = \frac{x' + v_{rel}\, t'}{(1 - v_{rel}^2)^{1/2}}$$

$$t = \frac{v_{rel}\, x' + t'}{(1 - v_{rel}^2)^{1/2}}$$

$$y = y' \qquad \text{and} \qquad z = z'$$

Check for yourself that for the impact event of bullet with target (rocket coordinates: $x' = 4$ meters, $t' = 5$ meters; rocket speed in laboratory frame: $v_{rel} = 4/5$) one obtains laboratory coordinates $x = 40/3$ meters and $t = 41/3$ meters. Hence $v = x/t = 40/41 = 0.9756$.

You say the Lorentz transformation is general. If it is so important, then why is this a special topic rather than a regular chapter?

Lorentz transformation: Useful but not fundamental

The Lorentz transformation is powerful; it brings the technical ability to transform coordinates from frame to frame. It helps us predict how to add velocities, as outlined here. It describes the Doppler shift for light (see the exercises for this chapter). On the other hand, the Lorentz transformation is not fundamental; it does not expose deep new features of spacetime. But no matter! Physics has to get on with the world's work. One uses the method of describing separation best suited to the job at hand. On some occasions the useful fact to give about a luxury yacht is the 50-meter distance between bow and stern, a distance independent of the direction in which the yacht is headed. On another occasion it may be much more important to know that the bow is 30 meters east of the stern and 40 meters north of it as observed by its captain, who uses North-Star north.

What does the Lorentz transformation rest on? On what foundations is it based?

Two foundations of Lorentz transformation

On two foundations: (1) The equations must be linear. That is, space and time coordinates enter the equations to the first power, not squared or cubed. This results from the requirement that you may choose any event as the zero of space and time.

(2) The spacetime interval between two events must have the same value when computed from laboratory coordinate separations as when reckoned from rocket coordinate separations.

All right, I'll reserve judgment on the validity of what you claim, but show me the derivation itself.

Read on! ✐

L.3 FIRST STEPS

invariance of the interval gets us started

Recall that the coordinates y and z transverse to the direction of relative motion between rocket and laboratory have the same values in both frames (Section 3.6):

$$y = y'$$
$$z = z' \qquad \text{(L-1)}$$

where primes denote rocket coordinates. A second step makes use of the difference in observed clock rates when the clock is at rest or in motion (Section 1.3 and Box 3-3). Think of a sparkplug at rest at the origin of a rocket frame that moves with speed v_{rel} relative to the laboratory. The sparkplug emits a spark at time t' as measured in the rocket frame. The sparkplug is at the rocket origin, so the spark occurs at $x' = 0$.

Where and when (x and t) does this spark occur in the laboratory? That depends on how fast, v_{rel}, the rocket moves with respect to the laboratory. The spark must occur at the location of the sparkplug, whose position in the laboratory frame is given by

Derive difference in clock rates

$$x = v_{rel}t$$

Now the invariance of the interval gives us a relation between t and t',

$$(t')^2 - (x')^2 = (t')^2 - (0)^2 = t^2 - x^2 = t^2 - (v_{rel}t)^2 = t^2(1 - v_{rel}^2)$$

from which

$$t' = t\,(1 - v_{rel}^2)^{1/2}$$

or

$$t = \frac{t'}{(1 - v_{rel}^2)^{1/2}} \qquad \text{[when } x' = 0\text{]} \quad \text{(L-2)}$$

The awkward expression $1/(1 - v_{rel}^2)^{1/2}$ occurs often in what follows. For simplicity, this expression is given the symbol Greek lower-case gamma: γ.

$$\gamma \equiv \frac{1}{(1 - v_{rel}^2)^{1/2}}$$

Because it gives the ratio of observed clock rates, γ is sometimes called the **time stretch factor** (Section 5.8). Strictly speaking, we should use the symbol γ_{rel}, since the value of γ is determined by v_{rel}. For simplicity, however, we omit the subscript in the hope that this will cause no confusion. With this substitution, equation (L-2) becomes

Time stretch factor defined

$$t = \gamma t' \qquad \text{[when } x' = 0\text{]} \quad \text{(L-3)}$$

Substitute this into the equation x = v_{rel} t above to find laboratory position in terms of rocket measurements:

$$x = v_{rel}\gamma t'$$

[when x' = 0]   (L-4)

Equations (L-1), (L-3), and (L-4) give the first answer to the question, "If we know the space and time coordinates of an event in one free-float frame, what are its space and time coordinates in some other overlapping free-float frame?" These equations are limited, however, since they apply only to a particular situation: one in which both events occur at the same place ($x' = 0$) in the rocket. ✐

L.4 FORM OF THE LORENTZ TRANSFORMATION

any event can be reference event? then transformation is linear

What general form does the Lorentz transformation have? It has the form that mathematicians call a **linear transformation.** This means that laboratory coordinates x and t are related to linear (first) power of rocket coordinates x' and t' by equations of the form

Lorentz transformation:
Linear equations

$$t = Bx' + Dt'$$
$$x = Gx' + Ht'$$

(L-5)

where our task is to find expressions for the coefficients B, D, G, and H that do not depend on either the laboratory or the rocket coordinates of a particular event, though they do depend on the relative speed v_{rel}.

Why must these transformations be linear? Because we are free to choose any event as our reference event, the common origin $x = y = z = t = 0$ in all reference frames. Let our rocket sparkplug emit the flashes at $t' = 1$ and 2 and 3 meters. These are equally spaced in rocket time. According to equation (L-3) these three events occur at laboratory times $t = 1\gamma$ and 2γ and 3γ meters of time. These are equally spaced in laboratory time. Moving the reference event to the first of these events still leaves them equally spaced in time for both observers: $t' = 0$ and 1 and 2 meters in the rocket and $t = 0$ and 1γ and 2γ in the laboratory.

In contrast, suppose that equation (L-3) were not linear, reading instead $t = Kt'^2$, where K is some constant. Rocket times $t' = 1$ and 2 and 3 meters result in laboratory times $t = 1K$ and $4K$ and $9K$ meters. These are not equally spaced in time for the laboratory observer. Moving the reference event to the first event would result in rocket times $t' = 0$ and 1 and 2 meters as before, but in this case laboratory times $t = 0$ and $1K$ and $4K$ meters, with a completely different spacing. But the choice of reference event is arbitrary: Any event is as qualified to be reference event as any other. A clock that runs steadily as observed in one frame must run steadily in the other, independent of the choice of reference event. We conclude that the relation between t and t' must be a linear one. A similar argument requires that events equally separated in space in the rocket must also be equally separated in space as measured in the laboratory. Hence the Lorentz transformation must be linear in both space and time coordinates. ✐

Arbitrary event as reference event?
Then Lorentz transformation
must be linear.

L.5 COMPLETING THE DERIVATION

invariance of the interval completes the story

Equations (L-3) and (L-4) provide coefficients D and H called for in equation (L-5):

$$t = Bx' + \gamma t'$$
$$x = Gx' + v_{rel}\gamma t' \tag{L-6}$$

About the two constants B and G we know nothing, for an elementary reason. All events so far considered occured at point $x' = 0$ in the rocket. Therefore the two coefficients B and G could have any finite values whatever without affecting the numerical results of the calculation. To determine B and G we turn our attention from an $x' = 0$ event to a more general event, one that occurs at a point with arbitrary rocket coordinates x' and t'. Then we demand that the spacetime interval have the same numerical value in laboratory and rocket frames for any event whatever:

Demanding invariance of interval . . .

$$t^2 - x^2 = t'^2 - x'^2$$

Substitute expressions for t and x from equation (L-6):

$$(Bx' + \gamma t')^2 - (Gx' + v_{rel}\gamma t')^2 = t'^2 - x'^2$$

On the left side, multiply out the squares. This leads to the rather cumbersome result

$$B^2 x'^2 + 2B\gamma x't' + \gamma^2 t'^2 - G^2 x'^2 - 2Gv_{rel}\gamma x't' - v_{rel}^2\gamma^2 t'^2 = t'^2 - x'^2$$

Group together coefficients of t'^2, coefficients of x'^2, and coefficients of the cross-term $x't'$ to obtain

$$\gamma^2(1 - v_{rel}^2)t'^2 + 2\gamma(B - v_{rel}G)\,x't' - (G^2 - B^2)x'^2 = t'^2 - x'^2 \tag{L-7}$$

Now, t' and x' can each take on any value whatsoever, since they represent the coordinates of an arbitrary event. Under these circumstances, it is impossible to satisfy equation (L-7) with a single choice of values of B and G unless they are chosen in a very special way. The quantities B and G must first be such as to make the coefficient of $x't'$ on the left side of equation (L-7) vanish as it does on the right:

. . . between any pair of events whatsoever . . .

$$2\gamma(B - v_{rel}G) = 0$$

But γ can never equal zero. The value of $\gamma = 1/(1 - v_{rel}^2)^{1/2}$ equals unity when $v_{rel} = 0$ and is greater than this for any other values of v_{rel}. Hence the left side of this equation can be zero only if

$$(B - v_{rel}G) = 0 \qquad \text{or} \qquad B = v_{rel}G \tag{L-8}$$

Second, B and G must be such as to make the coefficient of x'^2 equal on the left and right of equation (L-7); hence

$$G^2 - B^2 = 1 \tag{L-9}$$

Substitute B from equation (L-8) into equation (L-9):

. . . leads to completed form of Lorentz transformation.

$$G^2 - (v_{rel}G)^2 = 1 \qquad \text{or} \qquad G^2(1 - v_{rel}^2) = 1$$

Divide through by $(1 - v_{rel}^2)$ and take the square root of both sides:

$$G = \frac{1}{(1 - v_{rel}^2)^{1/2}}$$

But the right side is just the definition of the time stretch factor γ, so that

$$G = \gamma$$

Substitute this into equation (L-8) to find B:

$$B = v_{rel}\gamma$$

These results plus equations (L-1) and (L-6) yield the Lorentz transformation equations:

The Lorentz transformation

$$
\begin{aligned}
t &= v_{rel}\gamma x' + \gamma t' \\
x &= \gamma x' + v_{rel}\gamma t' \\
y &= y' \\
z &= z'
\end{aligned}
\qquad \text{(L-10a)}
$$

or, substituting for the value of gamma, $\gamma = 1/(1 - v_{rel}^2)^{1/2}$:

$$
\begin{aligned}
t &= \frac{v_{rel}x' + t'}{(1 - v_{rel}^2)^{1/2}} \\[2mm]
x &= \frac{x' + v_{rel}t'}{(1 - v_{rel}^2)^{1/2}} \\[2mm]
y &= y' \quad \text{and} \quad z = z'
\end{aligned}
\qquad \text{(L-10b)}
$$

In summary, the Lorentz transformation equations rest fundamentally on the required linearity of the transformation and on the invariance of the spacetime interval. Invariance of the interval was used twice in the derivation. First, we examined a pair of events both of which occur at a the same fixed location in the rocket, so that rocket time between these events — proper time, wristwatch time — equals the spacetime interval between them (Section L.3). Second, we demanded that the interval also be invariant between every possible event and the reference event (the present section).

L.6 INVERSE LORENTZ TRANSFORMATION

from laboratory event coordinates, reckon rocket coordinates

Equations (L-10) provide laboratory coordinates of an event when one knows the rocket coordinates of the same event. But suppose that one already knows the laboratory coordinates of the event and wishes to predict the coordinates of the event measured by the rocket observer. What equations should be used for this purpose?

An algebraic manipulation of equations (L-10) provides the answer. The first two of these equations can be thought of as two equations in the two unknowns x' and t'. Solve for these unknowns in terms of the now-knowns x and t. To do this, multiply both sides of the second equation by v_{rel} and subtract corresponding sides of the resulting second equation from the first. Terms in x' cancel to yield

$$t - v_{rel} x = \gamma t' - v_{rel}^2 \gamma t' = \gamma(1 - v_{rel}^2)t' = \frac{\gamma}{\gamma^2} t' = \frac{t'}{\gamma}$$

Here we have used the definition $\gamma^2 = 1/(1 - v_{rel}^2)$. The equation for t' can then be written

$$t' = -v_{rel} \gamma x + \gamma t$$

A similar procedure leads to the equation for x'. Multiply the first of equations (L-10) by v_{rel} and subtract corresponding sides of the first equation from the second — try it! The y and z components are respectively equal in both frames, as before. Then the **inverse Lorentz transformation equations** become

$$
\begin{aligned}
t' &= -v_{rel} \gamma x + \gamma t \\
x' &= \gamma x - v_{rel} \gamma t \\
y' &= y \\
z' &= z
\end{aligned}
$$

(L-11a)

Or, substituting again for gamma, $\gamma = 1/(1 - v_{rel}^2)^{1/2}$:

$$t' = \frac{-v_{rel} x + t}{(1 - v_{rel}^2)^{1/2}}$$

$$x' = \frac{x - v_{rel} t}{(1 - v_{rel}^2)^{1/2}}$$

(L-11b)

$$y' = y \quad \text{and} \quad z' = z$$

Equations (L-11) transform coordinates of an event known in the laboratory frame to coordinates in the rocket frame.

> A simple but powerful *argument from symmetry* leads to the same result. The symmetry argument is based on the relative velocity between laboratory and rocket frames. With respect to the laboratory, the rocket by convention moves with known speed in the *positive x*-direction. With respect to the rocket, the laboratory moves with the same speed but in the opposite direction, the *negative x*-direction. This convention about positive and negative directions — not a law of physics! — is the only difference between laboratory and rocket frames that can be observed from either frame. Lorentz transformation equations must reflect this single difference. In consequence, the "inverse" (laboratory-to-rocket) transformation can be obtained from the "direct" (rocket-to-laboratory) transformation by changing the sign of relative velocity, v_{rel}, in the equations and interchanging laboratory and rocket labels (primed and unprimed coordinates). Carrying out this operation on the Lorentz transformation equations (L-10) yields the inverse transformation equations (L-11). ✍

Long derivation of inverse Lorentz transformation

Inverse Lorentz transformation

Short derivation of inverse Lorentz transformation

L.7 ADDITION OF VELOCITIES

add light velocity to light velocity: get light velocity!

The Lorentz transformation permits us to answer decisively the apparent contradiction to special relativity outlined in Section L.2, namely the apparent addition of velocities to yield a resultant velocity greater than that of light.

Return to velocity addition
paradox

I travel in a rocket that you observe to move at 4/5 light speed. Out the front of my rocket I fire a bullet that I observe to fly forward at 4/5 light speed. Then you measure this bullet to streak forward at 4/5 + 4/5 = 8/5 = 1.6 light speed, which is greater than the speed of light. There!

SAMPLE PROBLEM L-1

TRANSFORMING OVER AND BACK

A rocket moves with speed $v_{rel} = 0.866$ (so $\gamma = 2$) along the x-direction in the laboratory. In the rocket frame an event occurs at coordinates $x' =$ 10 meters, $y' = 7$ meters, $z' = 3$ meters, and $t' = 20$ meters of light-travel time with respect to the reference event.

a.   What are the coordinates of the event as observed in the laboratory?

b.   Transform the laboratory coordinates back to the rocket frame to verify that the resulting coordinates are those given above.

SOLUTION

a.   We already know from Section 3.6 — as well as from the Lorentz transformation, equation (L-10) — that coordinates transverse to direction of relative motion are equal in laboratory and in rocket. Therefore we know immediately that

$$y = y' = 7 \text{ meters}$$
$$z = z' = 3 \text{ meters}$$

The x and t coordinates of the event as observed in the laboratory make use of the first two equations (L-10):

$$t = v_{rel}\gamma x' + \gamma t' = (0.866)(2)(10 \text{ meters}) + (2)(20 \text{ meters})$$
$$= 17.32 + 40 = 57.32 \text{ meters}$$

and

$$x = \gamma x' + v_{rel}\gamma t' = 2(10 \text{ meters}) + (0.866)(2)(20 \text{ meters})$$
$$= 20 + 34.64 = 54.64 \text{ meters}$$

So the coordinates of the event in the laboratory are $t = 57.32$ meters, $x = 54.64$ meters, $y = 7$ meters, and $z = 3$ meters.

b.   Use equation (L-11) to transform back from laboratory to rocket coordinates.

$$t' = -v_{rel}\gamma x + \gamma t = -(0.866)(2)(54.64 \text{ meters}) + (2)(57.32 \text{ meters})$$
$$= -94.64 + 114.64 = 20.00 \text{ meters}$$

and

$$x' = \gamma x - v_{rel}\gamma t = 2(54.64 \text{ meters}) - (0.866)(2)(57.32 \text{ meters})$$
$$= 109.28 - 99.28 = 10.00 \text{ meters}$$

as given in the original statement of the problem.

To analyze this experiment, convert statements about the bullet to statements about events, since event coordinates are what the Lorentz transformation transforms. Event 1 is the firing of the gun, event 2 the arrival of the bullet at the target. The Lorentz transformation equations can give locations x_1, t_1 and x_2, t_2 of these events in the laboratory frame from their known locations x'_1, t'_1 and x'_2, t'_2 in the rocket frame. In particular:

$$x_2 = \gamma x_2' + v_{rel} \gamma t'_2$$
$$x_1 = \gamma x_1' + v_{rel} \gamma t_1'$$

Subtract corresponding sides of these two equations:

$$(x_2 - x_1) = \gamma (x'_2 - x'_1) + v_{rel} \gamma (t'_2 - t'_1)$$

We are interested in the *differences* between the coordinates of the two emissions. Indicate these differences with the Greek uppercase delta, Δ, for example Δx. Then this x-equation and the corresponding t-equation become

$$\Delta x = \gamma \Delta x' + v_{rel} \gamma \Delta t'$$
$$\Delta t = v_{rel} \gamma \Delta x' + \gamma \Delta t' \qquad \text{(L-12)}$$

Incremental event separations define velocities

The subscript "rel" distinguishes *relative* speed between laboratory and rocket frames from other speeds, such as particle speeds in one frame or the other.

Bullet speed in any frame is simply space separation between two events on its trajectory measured in that frame divided by time between them, observed in the same frame. In the special case chosen, only the x-coordinate needs to be considered, since the bullet moves along the direction of relative motion. Divide the two sides of the first equation (L-12) by the corresponding sides of the second equation to obtain laboratory speed:

$$\frac{\Delta x}{\Delta t} = \frac{\gamma \Delta x' + v_{rel} \gamma \Delta t'}{v_{rel} \gamma \Delta x' + \gamma \Delta t'}$$

Then the time stretch factor γ cancels from the numerator and denominator on the right. Divide every term in numerator and denominator on the right by $\Delta t'$.

$$\frac{\Delta x}{\Delta t} = \frac{(\Delta x'/\Delta t') + v_{rel}}{v_{rel}(\Delta x'/\Delta t') + 1}$$

Now, $\Delta x'/\Delta t'$ is just distance covered per unit time by the particle as observed in the rocket, its speed — call it v', with a prime. And $\Delta x/\Delta t$ is particle speed in the laboratory — call it simply v. Then (reversing order of terms in the denominator to give the result its usual form) the equation becomes

$$v = \frac{v' + v_{rel}}{1 + v' v_{rel}} \qquad \text{(L-13)}$$

Law of Addition of Velocities

This is called the **Law of Addition of Velocities** in one dimension. A better name is the **Law of Combination of Velocities,** since velocities do not "add" in the usual sense. Using the Law of Combination of Velocities, we can predict bullet speed in the laboratory. The bullet travels at $v' = 4/5$ with respect to the rocket and the rocket moves at $v_{rel} = 4/5$ with respect to the laboratory. Therefore, speed v of the bullet

(continued on page 110)

SAMPLE PROBLEM L-2

"ET TU, SPACETIME!"

Julius Caesar was murdered on March 15 in the year 44 B.C. at the age of 55 approximately 2000 years ago. Is there some way we can use the laws of relativity to save his life?

Let Caesar's death be the reference event, labeled 0: $x_o = 0$, $t_o = 0$. Event A is you reading this exercise. In the Earth frame the coordinates of event A are $x_A = 0$ light-years, $t_A = 2000$ years. Simultaneous with event A in your frame, Starship Enterprise cruising the Andromeda galaxy sets off

a firecracker: event B. The Enterprise moves along a straight line in space that connects it with Earth. Andromeda is 2 million light-years distant in our frame. Compared with this distance, you can neglect the orbit of Earth around Sun. Therefore, in our frame, event B has the coordinates $x_B = 2 \times 10^6$ light-years, $t_B = 2000$ years. Take Caesar's murder to be the reference event for the Enterprise too ($x_o' = 0$, $t_o' = 0$).

a. How fast must the Enterprise be going in the Earth frame in order that Caesar's murder is happening NOW (that is, $t_B' = 0$) in the Enterprise rest frame? Under these circumstance is the Enterprise moving toward or away from Earth?

b. If you are acquainted with the spacetime diagram (Chapter 5), draw a spacetime diagram for the Earth frame that displays event 0 (Caesar's death), event A (you reading this exercise), event B (firecracker exploding in Andromeda), your line of NOW simultaneity, the position of the Enterprise, the worldline of the Enterprise, and the Enterprise NOW line of simultaneity. The spacetime diagram need not be drawn to scale.

c. In the Enterprise frame, what are the x and t coordinates of the firecracker explosion?

d. Can the Enterprise firecracker explosion warn Caesar, thus changing the course of Earth history? Justify your answer.

SOLUTION

a. From the statement of the problem,

$$x_o = x_o' = 0 \qquad x_A = 0 \qquad x_B = 2 \times 10^6 \text{ light-years}$$
$$t_o = t_o' = 0 \qquad t_A = 2000 \text{ years} \qquad t_B = 2000 \text{ years}$$

We want the speed v_{rel} of the Enterprise such that $t_B' = 0$. The first two Lorentz transformation equations (L-10) with $t_B' = 0$ become

$$t_B = v_{rel} \gamma x_B'$$
$$x_B = \gamma x_B'$$

We do not yet know the value of x_B'. Solve for v_{rel} by dividing the two sides of the first equation by the respective sides of the second equation. The unknown x_B' drops out (along with γ), and we are left with v_{rel} in terms of the known quantities t_B and x_B:

$$v_{rel} = \frac{t_B}{x_B} = \frac{2 \times 10^3 \text{ years}}{2 \times 10^6 \text{ years}} = 10^{-3} = 0.001$$

This is the desired speed v_{rel} between Earth and Enterprise frames. This velocity is a positive quantity, so the Enterprise moves in the positive x-direction, namely away from Earth.

Surprised to see a speed given as the ratio of a time separation to a space separation: t_B/x_B? Then realize that x_B and t_B are not displacements of any particle. Nothing can travel the distance x_B in the time t_B, as discussed in **d**. The goal here is to find a frame in which Caesar's death and the firecracker explosion are simultaneous. For this limited purpose the rocket speed $v_{rel} = t_B/x_B$ is correct.

Why is the relative velocity v_{rel} so small compared with the speed of light? Because of the large denominator x_B in the equation that leads to this value. Consider the string of Earth clocks stretching toward Andromeda when all Earth clocks read zero time (Caesar's death). Enterprise clocks read (from equations L-11 with $t = 0$) as follows: $t' = -v_{rel}\gamma x$. This is an example of the relativity of simultaneity (Section 3-4). The farther the x-distance from Earth, the earlier will Enterprise clock read. With $x = 2$ million light-years, the relative speed v_{rel} does not have to be large to carry Enterprise time back 2000 years for Earth.

b.

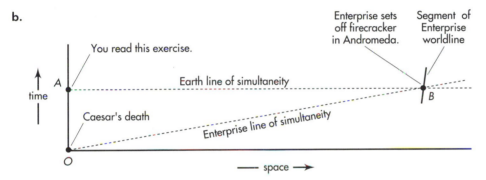

Earth spacetime diagram, showing events O, A, and B. Not to scale.

c.   We need the value of gamma, γ, for the inverse Lorentz transformation equation (L-11). This value is very close to unity, and from it come $t_B{'}$ and $x_B{'}$.

$$\gamma = \frac{1}{[1 - v_{rel}^2]^{1/2}} = \frac{1}{[1 - (10^{-3})^2]^{1/2}} = \frac{1}{[1 - 10^{-6}]^{1/2}} \approx 1 + \frac{10^{-6}}{2}$$

$$t_B{'} = -v_{rel}\gamma x_B + \gamma t_B = \gamma(-10^{-3} \times 2 \times 10^6 + 2 \times 10^3)$$
$$= \gamma(-2 \times 10^3 + 2 \times 10^3) = 0 \text{ years}$$

$$x_B{'} = \gamma x_B - v_{rel}\gamma t_B = \gamma(2 \times 10^6 - 10^{-3} \times 2 \times 10^3) = 2\gamma(1 - 10^{-6})\, 10^6$$

$$= 2\left(1 + \frac{10^{-6}}{2}\right)(1 - 10^{-6})10^6 = 2\left(1 - \frac{10^{-6}}{2} - \frac{10^{-12}}{2}\right)10^6$$

$$\approx 1.999999 \times 10^6 \text{ light-years.}$$

We *chose* the relative velocity so that the time of the firecracker explosion as observed in the rocket is the same as the time of Caesar's death, namely $t_B{'} = 0$. The x-coordinate of this explosion is not much different in the two frames because their relative velocity is so small.

d.   There exists a frame — the rest frame of the Enterprise — in which Caesar's death and the firecracker explosion occur at the same time. In this frame a signal connecting the two events would have to travel at infinite speed. But this is impossible. Therefore the Enterprise cannot warn Caesar; his death is final. Sorry. (Note: In the language of Chapter 6, the relation between the two events is spacelike, and spacelike events cannot have a cause–effect relationship.)

BOX L-1

WHY NO THING TRAVELS FASTER THAN LIGHT

A material object traveling faster than light? No! If one did, we could violate the normal order of cause and effect in a million testable ways, totally contrary to all experience. Here we investigate one example, making use of Lorentz transformation equations.

The Peace Treaty of Shalimar was signed four years before the Great Betrayal. So pivotal an event was the Great Betrayal that it was taken as zero of space and time.

By the Treaty of Shalimar, the murderous Klingons agreed to stop attacking Federation outposts in return for access to the Federation Technical Database. Federation negotiators left immediately after signing the Shalimar Treaty in a ship moving at 0.6 light speed.

Within four years the Klingons used the Federation Technical Database to develop a faster-than-light projectile, the slaughtering Super. On that dark day of Great Betrayal (reference event 0), the Klingons launched the Super at three times light speed toward the retreating Federation ship.

Two Federation space colonies lay between the Klingons and the point of impact of the Super with the Federation ship. A lonely lookout at the first colony witnessed with awe the blinding passage of the Super (event 1). Later many citizens of the second colony gaped as the Super demolished one of their communication structures (event 2) and zoomed on. Both colonies desperately sent warnings toward the Federation ship, but to no avail since the Super outran the radio signals.

Finally, at event 3, the Super overtook and destroyed the Federation ship. All Federation negotiators were lost in a terrible flash of light and scattering of debris. A long dark period of renewed warfare began.

But wait! Look again at events of the Great Betrayal, this time from the point of view of the Federation rocket ship. Where and when does the Great Betrayal occur in this frame? The Great Betrayal is the "hinge of history," the reference event, the zero of space and time coordinates for all laboratory and rocket frames.

Where and when does the Super explode (event 3) in this rocket frame? In the Klingon "laboratory" frame, event 3 has coordinates $x_3 = 3$ light-years and $t_3 = 1$ year. Use the inverse Lorentz transformation equations to find the location of event 3 in the rocket frame of the Federation negotiators. Calculate the time stretch factor γ using speed of the Federation rocket, $v_{rel} = 0.6$, with respect to the Klingon frame:

Klingon ("laboratory") spacetime diagram. The Klingon worldline is the vertical time axis. The Treaty of Shalimar is followed four years later by the Great Betrayal (event 0) at which Klingons launch the Super, which moves at three times light speed. Traveling from left to right, the Super passes one Federation colony (event 1) and then another (event 2). Finally the Super destroys the retreating ship of Federation negotiators (event 3).

$$\gamma = \frac{1}{[1 - v_{rel}^2]^{1/2}} = \frac{1}{[1 - (0.6)^2]^{1/2}} = \frac{1}{[1 - 0.36]^{1/2}}$$

$$= \frac{1}{[0.64]^{1/2}} = \frac{1}{0.8} = 1.25$$

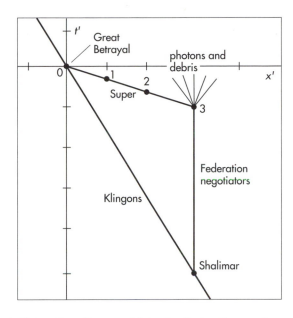

"Rocket" spacetime diagram of departing Federation negotiators. In this frame their destruction comes first (event 3), followed by the passage of the Super from right to left past Federation colonies in reverse order (event 2 followed by event 1). Finally, the Super enters the Klingon launcher without doing further damage (event 0). The Great Betrayal has become the Great Confusion of Cause and Effect.

Substitute these values into equations (L-11) to reckon the rocket coordinates of event 3:

$$t'_3 = -v_{rel}\gamma x_3 + \gamma t_3$$
$$= -(0.6)(1.25)(3 \text{ years}) + (1.25)(1 \text{ year})$$
$$= -2.25 \text{ years} + 1.25 \text{ years} = -1 \text{ year}$$
$$x'_3 = \gamma x_3 - v_{rel}\gamma t_3$$
$$= (1.25)(3 \text{ years}) - (0.6)(1.25)(1 \text{ year})$$
$$= 3.75 \text{ years} - 0.75 \text{ year} = 3 \text{ years}$$

Event 3 is plotted in the rocket diagram and the worldline of the Super drawn by connecting event 3 with the launching of the Super at event 0. Notice that this worldline slopes downward to the right. More about the significance of this in a minute.

In a similar manner find the rocket coordinates of the

treaty signing at Shalimar (subscript Sh), which has laboratory coordinates $x_{Sh} = 0$ and $t_{Sh} = -4$ years:

$$t'_{Sh} = -v_{rel}\gamma x_{Sh} + \gamma t_{Sh}$$
$$= -(0.6)(1.25)(0 \text{ years}) + (1.25)(-4 \text{ years})$$
$$= -5 \text{ years}$$
$$x'_{Sh} = \gamma x_{Sh} - v_{rel}\gamma t_{Sh}$$
$$= (1.25)(0 \text{ years}) - (0.6)(1.25)(-4 \text{ years})$$
$$= +3 \text{ years}$$

In the Federation (rocket) spacetime diagram, the worldline of Federation negotiators extends from treaty signing at Shalimar vertically to explosion of the Super (event 3). The worldline of the Klingons extends from Shalimar diagonally through the launch of the Super at event 0.

In the Federation spacetime diagram, the worldline for the Super tilts downward to the right. In this frame deaths of Federation negotiators (event 3) occur at a time $t'_3 = minus 1 year$, that is, *before* the treacherous Klingons launch the Super at the event of Great Betrayal (reference event 0). From the diagram one would say that the Super moves with three times light speed *from* Federation ship *toward* the Klingons. This seems to be verified by the fact that in this frame the Super passes Federation colonies in reverse order, event 2 followed by event 1, going in the opposite direction. Yet Federation negotiators have created no such terrible weapon and in fact are destroyed by it at the moment they are supposed to launch it, as proved by the flying photons and debris. More: Klingons suffer no damage from the mighty impact of the slaughtering Super (event 0). Rather, in this frame it enters their launching cannon mild as a lamb.

What have we here? A confusion of cause and effect, a confusion that cannot be straightened out as long as we assume that the Super—or any other material object—travels faster than light in a vacuum.

Why does no signal and no object travel faster than light in a vacuum? Because if either signal or object did so, the entire network of cause and effect would be destroyed, and science as we know it would not be possible.

relative to the laboratory comes from the expression

Velocity addition paradox resolved

$$v = \frac{4/5 + 4/5}{1 + (4/5)(4/5)} = \frac{8/5}{1 + 16/25} = \frac{8/5}{41/25} = \frac{40}{41}$$

Thus the bullet moves in the laboratory at a speed less than light speed.

As a limiting case, suppose that the "bullet" shot out from the front of the rocket is, in fact, a pulse of light. Guess: What is the speed of this light pulse in the laboratory? Here is the calculated answer. Light moves with respect to the rocket at speed $v' = 1$ while the rocket continues along at a speed $v_{rel} = 4/5$ with respect to the laboratory. The light then moves with respect to the laboratory at speed v:

Light speed is invariant, as expected.

$$v = \frac{1 + 4/5}{1 + (1)(4/5)} = \frac{9/5}{9/5} = 1$$

So light moves with the same speed in both frames, as required by the Principle of Relativity. Question: Is this true also when a light pulse is shot out of the *rear* of the rocket? ✐

SAMPLE PROBLEM L-3

THE FIRING MESON

A K^o (pronounced "K-naught") meson at rest in a rocket frame decays into π^+ ("pi plus") meson and a π^- ("pi minus") meson, each having a speed of $v' = 0.85$ with respect to the rocket. Now consider this decay as observed in a laboratory with respect to which the K^o meson travels at a speed of $v_{rel} = 0.9$. What is the greatest speed that one of the π mesons can have with respect to the laboratory? What is the least speed?

SOLUTION

Let the speeding K^o-meson move in the positive x-direction in the laboratory. In the rocket frame, daughter π-mesons come off in opposite directions. Their common line of motion can, however, be oriented arbitrarily in this frame. The maximum speed of a daughter π-meson in the laboratory results when it is emitted in the forward x-direction. For such a meson, the law of addition of velocities gives

$$v_{max} = \frac{v' + v_{rel}}{1 + v'v_{rel}} = \frac{0.85 + 0.9}{1 + (0.85)(0.9)} = \frac{1.75}{1.765} = 0.9915$$

Thus adding a speed of 0.85 to a speed of 0.9 does not yield a resulting speed greater than 1, light speed.

The slowest laboratory speed for a daughter meson occurs when it is emitted in the negative x-direction in the rocket frame. In this case the velocity of the daughter meson is negative and the law of addition of velocities becomes a law of subtraction of velocities:

$$v_{min} = \frac{-v' + v_{rel}}{1 - v'v_{rel}} = \frac{-0.85 + 0.9}{1 - (0.85)(0.9)} = \frac{0.05}{0.235} = 0.2128$$

Although the minimum-speed meson moves to the left in the rocket, it moves to the right in the laboratory because of the very great speed of the original K^o-meson in the laboratory.

L.8 SUMMARY

Lorentz transformation deals with coordinates, not invariant quantities

Given the space and time coordinates of an event with respect to the reference event in one free-float frame, the **Lorentz coordinate transformation equations** tell us the coordinates of the same event in an overlapping free-float frame in relative motion with respect to the first. The equations that transform rocket coordinates (primed coordinates) to laboratory coordinates (unprimed coordinates) have the form

$$t = \frac{v_{rel}x' + t'}{(1 - v_{rel}^2)^{1/2}}$$

$$x = \frac{x' + v_{rel}t'}{(1 - v_{rel}^2)^{1/2}}$$

(L-10b)

$$y = y' \quad \text{and} \quad z = z'$$

where v_{rel} stands for relative speed of the two frames (rocket moving in the positive x-direction in the laboratory). The **inverse Lorentz transformation equations** transform laboratory coordinates to rocket coordinates:

$$t' = \frac{-v_{rel}x + t}{(1 - v_{rel}^2)^{1/2}}$$

$$x' = \frac{x - v_{rel}t}{(1 - v_{rel}^2)^{1/2}}$$

(L-11b)

$$y' = y \quad \text{and} \quad z' = z$$

in which v_{rel} is treated as a positive quantity. In both these sets of equations, coordinates of events are measured with respect to a reference event. It is really only the *difference* in coordinates between events that matter, for example $x_2 - x_1 = \Delta x$ for any two events 1 and 2, not the coordinates themselves. This is important in deriving the Law of Addition of Velocities.

The **Law of Addition of Velocities** or **Law of Combination of Velocities** in one dimension follows from the Lorentz transformation equations. This law tells us the velocity v of a particle in the laboratory frame if we know its velocity v' with respect to the rocket and relative speed v_{rel} between rocket and laboratory,

$$v = \frac{v' + v_{rel}}{1 + v'v_{rel}}$$

(L-13)

REFERENCE

Sample Problem L-3, The Firing Meson, was adapted from A. P. French, *Special Relativity* (W.W. Norton, New York, 1968), page 159.

SPECIAL TOPIC EXERCISES

PRACTICE

L-1   a super-speed super?

Take two more steps in the parable of the Great Betrayal (Box L-1).

a   Find the speed of a new rocket frame moving relative to the Klingon frame such that the Super travels at 6 times the speed of light in this new frame. Hint: Examine the coordinates x' and t' of event 3 in the new frame. The ratio of these two, x'/t', is the speed of the Super in this frame. We know the coordinates of event 3 in the Klingon frame. Therefore . . .

b   Find the speed of yet another rocket frame, relative to the Klingon frame, such that the Super travels with infinite speed in this frame. Hint: What does infinite speed imply about the time t' between events 0 and 3 in this new frame?

L-2   a bad clock

Note: This exercise uses spacetime diagrams, introduced in Chapter 5.

A pulse of light is reflected back and forth between mirrors A and B separated by 2 meters of distance in the x-direction in the Earth frame, as shown in the figure (left). A swindler tells us that this device constitutes a clock that "ticks" every time the pulse arrives at either mirror.

The swindler claims that events 1 through 6 are sequential "ticks" of this clock (center). However, we notice that the ticking of the clock is uneven in a rocket frame moving with speed v_{rel} in the Earth frame (right). For example, there is less time between events 0 and 1 than between events 1 and 2 as measured in the rocket frame.

a   What is the physical basis for the "bad" behavior of this clock? Use the Lorentz transformation

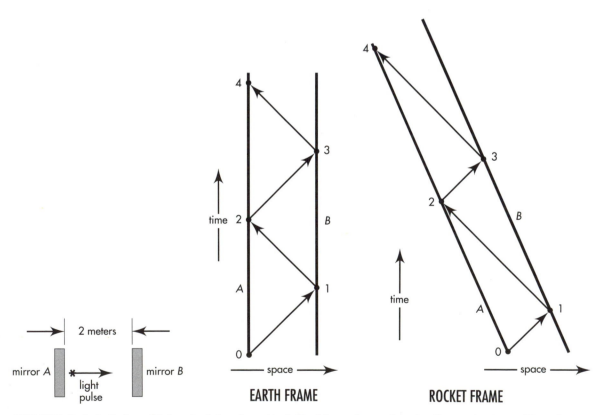

EXERCISE L-2. *Left: Horizontal light-pulse clock as observed in the Earth frame. Center: Spacetime diagram showing worldlines of mirrors A and B and the "uniformly ticking" light pulse as observed in the Earth frame. Right: Time lapses between sequential ticks of the light-pulse clock are not uniform as observed in the rocket frame.*

equations to account for the uneven ticking of this clock in the rocket frame.

b Use some of the same events 0 through 4 to define a "good" clock that ticks evenly in both the laboratory frame and the rocket frame. From the spacetime diagrams, show qualitatively that your good clock "runs slow" as observed from the rocket frame—as it must, since the clock is in motion with respect to the rocket frame.

c Explain why the clock of Figure 1-3 in the text is a "good" clock.

L-3 the Galilean transformation

a Use everyday, nonrelativistic Newtonian arguments to derive transformation equations between reference frames moving at low relative velocities. Show that the result is

$$x' = x - v_{conv}\, t_{sec} \qquad \text{(Newtonian: } v_{conv} \ll c) \qquad (1)$$

$$t'_{sec} = t_{sec} \qquad \text{(Newtonian: } v_{conv} \ll c) \qquad (2)$$

where t_{sec} is time measured in seconds and v_{conv} is speed in conventional units (meters/second for example). List the assumptions you make in your derivation.

b Convert equations (1) and (2) to measure time t in meters and unitless measure of relative velocity, $v_{rel} = v_{con}/c$. Show the results are:

$$x' = x - v_{rel}\, t \qquad \text{(Newtonian: } v \ll 1) \qquad (3)$$

$$t' = t \qquad \text{(Newtonian: } v \ll 1) \qquad (4)$$

Do the new units make these equations correct at high relative velocity between frames?

c Use the first two terms in the binomial expansion to find a low-velocity approximation for γ in the Lorentz transformation.

$$\gamma = \frac{1}{(1 - v_{rel}^2)^{1/2}} = (1 - v_{rel}^2)^{-1/2} \approx 1 + \frac{v_{rel}^2}{2}$$

Show that this expression differs from unity by less than one percent provided v is less than $1/7$. A sports car can accelerate uniformly from rest to 60 miles/hour (about 27 meters/second) in 7 seconds. Roughly how many days would it take for the sports car to reach $v = 1/7$ at the same constant acceleration?

d Set $\gamma = 1$ in the Lorentz transformation equations. Show that the resulting "low-velocity Lorentz transformation" is

$$x' = x - v_{rel}\, t \qquad \text{(Lorentz: } v \ll 1) \qquad (5)$$

$$t' = -v_{rel}\, x + t \qquad \text{(Lorentz: } v \ll 1) \qquad (6)$$

What is the difference between the time transformations for the "Newtonian low-velocity limit" of equation (4) and the "Lorentz low-velocity limit" of equation (6)? How can they both be correct? The term $-v_{rel}x$ does not depend on any time lapse, but only on the separation x of the event from the laboratory origin. This term is due to the difference of synchronization of clocks in the two frames.

e In each of the following cases a laboratory clock (measuring t) at a distance x from the origin as measured in the laboratory frame is compared with a passing rocket clock (measuring t'). Say whether or not the time difference $t - t' = v_{rel}x$ can be detected using wristwatches (accuracy of 10^{-1} second $= 3 \times 10^7$ meters of light-travel time) and using modern electronic clocks (accuracy of 10^{-9} second $= 0.3$ meter of time).

(1) Sports car traveling at 100 kilometers/hour (roughly 30 meters/second) located 1000 kilometers down the road from the origin as measured in the Earth frame.

(2) Moon probe traveling at 30,000 kilometers/hour passing Moon, 3.8×10^5 kilometers from the origin on Earth as measured in the Earth frame.

(3) Distance from origin on Earth at which space probe traveling at 30,000 kilometers/hour leads to detectable time difference between rocket wristwatch and adjacent Earth-linked latticework clock. Compare with Earth–Sun distance of 1.5×10^{11} meters.

f Summarize in a sentence or two the conditions under which the regular Galilean transformation equations (3) and (4) will lead to correct predictions.

L-4 limits of Newtonian mechanics

Use the particle speed $v_{crit} = 1/7$ (Exercise L-3) as an approximate maximum limit for the validity of Newtonian mechanics. Determine whether or not Newtonian mechanics is adequate to analyze motion in each of the following cases, following the example.

Example: Satellite circling Earth at 30,000 kilometers/hour $= 18,000$ miles/hour. **Answer:** Light moves at a speed $v_{conv} = (3 \times 10^5$ kilometers/second) $\times$ (3600 seconds/hour) $= 1.08 \times 10^9$ kilometers/hour. Therefore the speed of the satellite in meters/meter is $v = v_{conv}/c = 2.8 \times 10^{-5}$. This is much less than $v_{crit} = 1/7$, so the Newtonian description of satellite motion is adequate.

a   Earth circling Sun at an orbital speed of 30 kilometers/second.

b   Electron circling a proton in the orbit of smallest radius in a hydrogen atom. **Discussion:** The classical speed of the electron in the inner orbit of an atom of atomic number Z, where Z is the number of protons in the nucleus, is given, for low velocities, by the expression $v = Z/137$. For hydrogen, $Z = 1$.

c   Electron in the inner orbit of the gold atom, for which $Z = 79$.

d   Electron after acceleration from rest through a voltage of 5000 volts in a black-and-white television picture tube. **Discussion:** We say that this electron has a kinetic energy of 5000 electron-volts. One electron-volt is equal to 1.6×10^{-19} joule. Try using the Newtonian expression for kinetic energy.

e   Electron after acceleration from rest through a voltage of 25,000 volts in a color television picture tube.

f   A proton or neutron moving with a kinetic energy of 10 MeV (million electron-volts) in a nucleus.

PROBLEMS

L-5   Doppler shift

A sparkplug at rest in the rocket emits light with a frequency f' pulses or waves per second. What is the frequency f of this light as observed in the laboratory? Let this train of waves (or pulses) of light travel in the positive x-direction with speed c, so that in the course of one meter of light-travel time, f/c of these pulses pass the origin of the laboratory frame. It is understood that the *zeroth* or "fiducial" crest or pulse passes the origin at the zero of time — and that the origin of the rocket frame passes the origin of the laboratory frame at this same time.

a   Show that the x-coordinate of the nth pulse or wave crest is related to the time of observation t (in meters) by the equation

$$n = (f/c)(t - x)$$

b   The same argument, applied in the rocket frame, leads to the relation

$$n = (f'/c)(t' - x')$$

Express this rocket formula in laboratory coordinates x and t using the Lorentz transformation. Equate the resulting expression for f' to the labora-

tory formula for f in terms of x and t to derive the simple formula for f in terms of f' and v_{rel}, the relative speed of laboratory and rocket frames.

$$f = \left(\frac{1 + v_{rel}}{1 - v_{rel}}\right)^{1/2} f' \qquad \text{[wave moves in positive x-direction]}$$

c   Now observe a wave moving along the negative x-direction from the same source at rest in the rocket frame. Show that the frequency of the wave observed in the laboratory frame is

$$f = \left(\frac{1 - v_{rel}}{1 + v_{rel}}\right)^{1/2} f' \qquad \text{[wave moves in negative x-direction]}$$

d   Astronomers define the **redshift** z of light from a receding astronomical object by the formula

$$z = \frac{f_{emit} - f_{obs}}{f_{obs}}$$

Here f_{emit} is the frequency of the light measured in the frame in which the emitter is at rest and v_{obs} the frequency observed in another frame in which the emitter moves directly away from the observer.

The most distant quasar reported as of 1991 has a redshift $z = 4.897$. With what fraction of the speed of light is this quasar receding from us?

Reference: D. P. Schneider, M. Schmidt, and J. E. Gunn, *Astronomical Journal,* Volume 102, pages 837–840 (1991).

L-6   transformation of angles

a   A meter stick lies at rest in the rocket frame and makes an angle ϕ' with the x'-axis. Laboratory observers measure the x- and y-projections of the stick as it streaks past. What values do they measure for these projections, compared with the x'- and y'-projections measured by rocket observers? Therefore what angle ϕ does the same meter stick make with the x-axis of the laboratory frame? What is the length of the "meter stick" as observed in the laboratory frame?

b   Make the courageous assumption that the directions of electric-field lines around a point charge transform in the same way as the directions of meter sticks that lie along these lines. (Electric field lines around a point charge are assumed to be infinite in length, so the length transformation of part **a** does not apply.) Draw qualitatively the electric-field lines due to an isolated positive point charge at rest in the rocket frame as observed in (1) the rocket frame and (2) the laboratory frame. What conclusions follow concerning the time variation of electric forces on nearby charges at rest in the laboratory frame?

L-7 transformation of y-velocity

A particle moves with uniform speed $v'_y = \Delta y'/\Delta t'$ along the y'-axis of the rocket frame. Transform $\Delta y'$ and $\Delta t'$ to laboratory displacements Δx, Δy, and Δt using the Lorentz transformation equations. Show that the x-component and the y-component of the velocity of this particle in the laboratory frame are given by the expressions

$$v_x = v_{rel}$$

$$v_y = v'_y(1 - v_{rel}^2)^{1/2}$$

L-8 transformation of velocity direction

A particle moves with velocity v' in the $x'y'$ plane of the rocket frame in a direction that makes an angle ϕ' with the x'-axis. Find the angle ϕ that that velocity vector of this particle makes with the x-axis of the laboratory frame. (Hint: Transform space and time displacements rather than velocities.) Why does this angle differ from that found in Exercise L-6 on transformation of angles? Contrast the two results when the relative velocity between the rocket and laboratory frames is very great.

L-9 the headlight effect

A flash of light is emitted at an angle ϕ' with respect to the x'-axis of the rocket frame.

a Show that the angle ϕ the direction of motion of this flash makes with respect to the x-axis of the laboratory frame is given by the equation

$$\cos \phi = \frac{\cos \phi' + v_{rel}}{1 + v_{rel} \cos \phi'}$$

b Show that your answer to Exercise L-8 gives the same result when the velocity v' is given the value unity.

c A particle at rest in the rocket frame emits light uniformly in all directions. Consider the 50 percent of this light that goes into the forward hemisphere in the rocket frame. Show that in the laboratory frame this light is concentrated in a narrow forward cone of half-angle ϕ_0 whose axis lies along the direction of motion of the particle. The half-angle ϕ_0 is the solution to the following equation:

$$\cos \phi_0 = v_{rel}$$

This result is called the **headlight effect**.

L-10 the tilted meter stick

Note: This exercise uses the results of Exercise L-7.

A meter stick lying parallel to the x-axis moves in the y-direction in the laboratory frame with speed v_y as shown in the figure (left).

a In the rocket frame the stick is tilted upward in the positive x'-direction as shown in the figure (right). Explain why this is, first without using equations.

b Let the center of the meter stick pass the point $x = y = x' = y' = 0$ at time $t = t' = 0$. Calculate the angle ϕ' at which the meter stick is inclined to the x'-axis as observed in the rocket frame. **Discussion:** Where and when does the right end of the meter stick cross the x-axis as observed in the laboratory frame? Where and when does this event of right-end crossing occur as measured in the rocket frame? What is the direction and magnitude of the velocity of the meter stick in the rocket frame (Exercise L-7)? Therefore where is the right end of the meter stick at $t' = 0$, when the center is at the origin? Therefore . . .

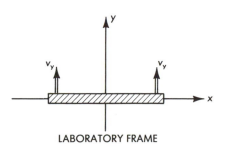

LABORATORY FRAME

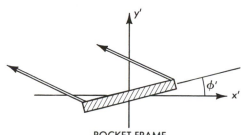

ROCKET FRAME

EXERCISE L-10. *Left: Meter stick moving transverse to its length as observed in the laboratory frame.* *Right: Meter stick as observed in rocket frame.*

L-11   the rising manhole

Note: This exercise uses the results of Exercise L-10.

A meter stick lies along the *x*-axis of the laboratory frame and approaches the origin with velocity v_{rel}. A very thin plate parallel to the *xz* laboratory plane moves upward in the *y*-direction with speed v_y as shown in the figure. The plate has a circular hole with a diameter of one meter centered on the *y*-axis. The center of the meter stick arrives at the laboratory origin at the same time in the laboratory frame as the rising plate arrives at the plane $y = 0$. Since the meter stick is Lorentz-contracted in the laboratory frame it will easily pass through the hole in the rising plate. Therefore there will be no collision between meter stick and plate as each continues its motion. However, someone who objects to this conclusion can make the following argument: "In the rocket frame in which the meter stick is at rest the meter stick is not contracted, while in this frame the hole in the plate is Lorentz-contracted. Hence the full-length meter stick cannot possibly pass through the contracted hole in the plate. Therefore there must be a collision between the meter stick and the plate." Resolve this paradox using your answer to Exercise L-10. Answer unequivocally the question, Will there be a collision between the meter stick and the plate?

Reference: R. Shaw, *American Journal of Physics,* Volume 30, page 72 (1962).

L-12   paradox of the skateboard and the grid

A girl on a skateboard moves very fast, so fast that the relativistic length contraction makes the skateboard very short. On the sidewalk she has to pass over a grid. A man standing at the grid fully expects the fast short skateboard to fall through the holes in the grid. Yet to the fast girl her skateboard has its usual length and it is the grid that has the relativistic contraction. To her

the holes in the grid are much narrower than to the stationary man, and she certainly does not expect her skateboard to fall through them. Which person is correct? The answer hinges on the relativity of rigidity.

Idealize the problem as a one-meter rod sliding lengthwise over a flat table. In its path is a hole one meter wide. If the Lorentz contraction factor is ten, then in the table (laboratory) frame the rod is 10 centimeters long and will easily drop into the one-meter-wide hole. Assume that in the laboratory frame the meter stick moves fast enough so that it remains essentially horizontal as it descends into the hole (no "tipping" in the laboratory frame). Write an equation in the laboratory frame for the motion of the bottom edge of the meter stick assuming that $t = t' = 0$ at the instant that the back end of the meter stick leaves the edge of the hole. For small vertical velocities the rod will fall with the usual acceleration *g*. Note that in the laboratory frame we have assumed that every point along the length of the meter stick begins to fall simultaneously.

In the meter stick (rocket) frame the rod is one meter long whereas the hole is Lorentz-contracted to a 10-centimeter width so that the rod cannot possibly fit into the hole. Moreover, in the rocket frame different parts along the length of the meter stick begin to drop at different times, due to the relativity of simultaneity. Transform the laboratory equations into the rocket frame. Show that the front and back of the rod will begin to descend at different times in this frame. The rod will "droop" over the edge of the hole in the rocket frame—that is, it will not be rigid. Will the rod ultimately descend into the hole in both frames? Is the rod *really* rigid or nonrigid during the experiment? Is it possible to derive any physical characteristics of the rod (for example its flexibility or compressibility) from the description of its motion provided by relativity?

Reference: W. Rindler, *American Journal of Physics,* Volume 29, page 365–366 (1961).

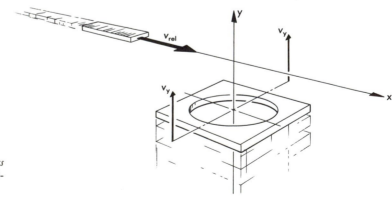

EXERCISE L-11. *Will the "meter stick" pass through the "one-meter-diameter" hole without collision?*

L-13 paradox of the identically accelerated twins

Note: This exercise uses spacetime diagrams, introduced in Chapter 5.

Two identical twins, Dick and Jane, own identical spaceships each containing the same amount of fuel. Jane's ship is initially positioned a distance to the right of Dick's in the Earth frame. On their twentieth birthday they blast off at the same instant in the Earth frame and undergo identical accelerations to the right as measured by Mom and Dad, who remain at home on Earth. Mom and Dad further observe that the twins run out of fuel at the same time and move thereafter at the same speed v. Mom and Dad also measure the distance between Dick and Jane to be the same at the end of the trip as at the beginning.

Dick and Jane compare the ships' logs of their accelerations and find the entries to be identical. However when both have stopped accelerating, Dick and Jane, in their new rest frame, discover that Jane is older than Dick! How can this be, since they have an identical history of accelerations?

a   Analyze a simpler trip, in which each spaceship increases speed not continuously but by impulses, as shown in the first spacetime diagram and the event table. How far apart are Dick and Jane at the beginning of their trip, as observed in the Earth frame? How far apart are they at the end of their accelerations? What is the final speed v (not the average speed) of the two spaceships? How much does each astronaut age along the worldline shown in the diagram? (The answer is not the Earth time of 12 years.)

b   The second spacetime diagram shows the two worldlines as recorded in a rocket frame moving with the final velocity of the two astronauts. Copy the figure. On your copy extend the worldlines of Dick and Jane after each has stopped accelerating. Label your figure to show that Jane stopped accelerating before Dick as observed in this frame. Will Dick age the same between events 0 and 3 in this frame as he aged in the Earth frame? Will Jane age the same between events 4 and 7 in this frame as she aged in the Earth frame?

c   Now use the Lorentz transformation to find the space and time coordinates of one or two critical events in this final rest frame of the twins in order to answer the following questions

(1)   How many years earlier than Dick did Jane stop accelerating?

(2)   What is Dick's age at event 3? (not the rocket time t' of this event!)

(3)   What is Jane's age at event 7?

(4)   What is Jane's age at the same time (in this frame) as event 3?

(5)   What are the ages of Dick and Jane 20 years after event 3, assuming that neither moves again with respect to this frame?

(6)   How far apart in space are Dick and Jane when both have stopped accelerating?

(7)   Compare this separation with their initial (and final!) separation measured by Mom and Dad in the Earth frame.

d   Extend your results to the general case in which Mom and Dad on Earth observe a period of identical *continuous* accelerations of the two twins.

(1)   At the two start-acceleration events (the two events at which the twins start their rockets), the twins are the same age as observed in the Earth frame. Are they the same age at these events as observed in every rocket frame?

(2)   At the two stop-acceleration events (the two events at which the rockets run out of fuel), are the twins the same age as observed in the Earth frame? Are they the same age at these events as observed in every rocket frame?

(3)   The two stop-acceleration events are simultaneous in the Earth frame. Are they simultaneous as observed in every rocket frame? (No!) Whose stop-acceleration event occurs first as observed in the final frame in which both twins come to rest? (Recall the Train Paradox, Section 3.4.)

(4)   "When Dick stops accelerating, Jane is older than Dick." Is this statement true according to the astronauts in their final rest frame? Is the statement true according to Mom and Dad in the Earth frame?

(5)   Criticize the lack of clarity (swindle?) of the word *when* in the statement of the problem: "However when both have stopped accelerating, Dick and Jane, in their new rest frame, discover that Jane is older than Dick!"

e   Suppose that Dick and Jane both accelerate to the left, so that Dick is in front of Jane, but their history is otherwise the same. Describe the outcome of this trip and compare it with the outcome of the original trip.

f   Suppose that Dick and Jane both accelerate in a direction perpendicular to the direction of their separation. Describe the outcome of this trip and compare it with the outcome of the original trip.

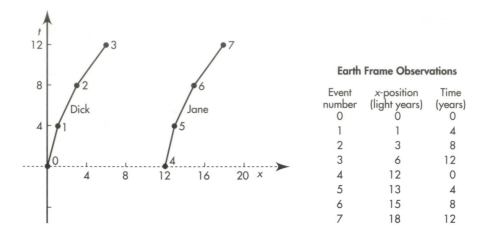

Event number	x-position (light years)	Time (years)
0	0	0
1	1	4
2	3	8
3	6	12
4	12	0
5	13	4
6	15	8
7	18	12

Earth Frame Observations

EARTH FRAME

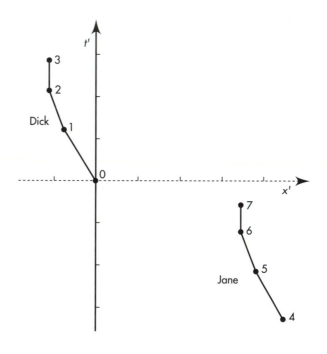

ROCKET FRAME

EXERCISE L-13. *Top: Worldlines of Dick and Jane as observed in the Earth frame of Mom and Dad. Bottom: Worldlines of Dick and Jane as observed in the "final" rocket frame in which both Dick and Jane come to rest after burnout.*

Discussion: Einstein postulated that physics in a uniform gravitational field is, locally and for small particle speeds, the same as physics in an accelerated frame of reference. In this exercise we have found that two accelerated clocks separated along the direction of acceleration do not remain in synchronism as observed simultaneously in their common frame. Rather, the forward clock reads a later time ("runs faster") than the rearward clock as so observed. Conclusion from Einstein's postulate: Two clocks one above the other in a uniform gravitational field do not remain in synchronism; rather the higher clock reads a later time ("runs faster") than the lower clock. General relativity also predicts this result, and experiment verifies it. (Read about the patrol plane experiment in Section 4.10.)

Reference: S. P. Boughn, *American Journal of Physics,* Volume 57, pages 791–793 (September 1989). Reference to general relativity result: Wolfgang Rindler, *Essential Relativity* (Springer, New York, 1977), pages 17 and 117.

L-14   how do rods Lorentz-contract?

Note: Calculus is used in the solution to this exercise; so is the formula for Lorentz contraction from Section 5.8.

Laboratory observers measure the length of a moving rod lying along its direction of motion in the laboratory frame. Then the rod speeds up a little. Again laboratory observers measure its length, which they find to be a little shorter than before. They call this shortening of length Lorentz contraction. How did this shortening of length come about? As happens so often in relativity, the answer lies in the relativity of simultaneity.

First, how much shortening takes place when the rod changes from speed v to speed $v + dv$? Let L_o be the proper length of the rod when measured at rest. At speed v its laboratory-measured length L will be shorter than this by the Lorentz contraction factor (Section 5.8):

$$L = (1 - v^2)^{1/2} L_o$$

a   Using calculus, show that when the rod speeds up from v to a slightly greater speed $v + dv$, the change in length dL is given by the expression

$$dL = -\frac{L_o v dv}{(1 - v^2)^{1/2}}$$

The negative sign means that the change is a shortening of the rod. We want to explain this change in length.

How is the rod to be accelerated from v to $v + dv$? Fire a rocket attached to the rear of the rod? No. Why not? Because the rocket pushes only against the rear of the rod; this push is transmitted along the rod to the front at the speed of a compression wave — very slow! We want the front and back to change speed "at the same time" (exact meaning of this phrase to be determined later). How can this be done? Only by prearrangement! Saw the rod into a thousand equal pieces and tap each piece in the forward direction with a mallet "at exactly 12 noon" as read off a set of synchronized clocks. To simplify things for now, set aside all but the front and back pieces of the rod. Now tap the front and back pieces "at the same time." The change in length of the rod dL is then the change in distance between these two pieces as a result of the tapping. So much for how to accelerate the "rod."

Now the central question: What does it mean to tap the front and back pieces of the rod "at the same time"? To answer this question, ask another: What is our final goal? Answer: To account for the Lorentz

contraction of a fast-moving rod of proper length L_o. More: We want a careful inspector riding on the fast-moving rod to certify that it has the same proper length L_o as it did when it was at rest in the laboratory frame. To achieve this goal, the inspector insists that the pair of accelerating taps be applied to the front and back rod pieces at the same time *in the current rest frame of the rod*. Otherwise the distance between these pieces would not remain the same in the frame of the rod; the rod would change proper length. [Notice that in Exercise L-13 the taps occur at the same time in the laboratory (Earth) frame. This leads to results different from those of the present exercise.]

b   You are the inspector riding along with the front and back pieces of the rod. Consider the two events of tapping the front and back pieces. How far apart $\Delta x'$ are these events along the x-axis in your (rocket) frame? How far apart $\Delta t'$ in time are these events in your frame? Predict how far apart in time Δt these events are as measured in the laboratory frame. Use the Lorentz transformation equation (L-10):

$$\Delta t = v \gamma \Delta x' + \gamma \Delta t'$$

The relative velocity v_{rel} in equation (L-10) is just v, the current speed of the rod. In the laboratory frame is the tap on the rear piece earlier or later than the tap on the front piece?

Your answer to part **b** predicts how much earlier the laboratory observer measures the tap to occur on the back piece than on the front piece of the rod. Let the tap increase the speed of the back end by dv as measured in the laboratory frame. Then during laboratory time Δt the back end is moving at a speed dv faster than the front end. This relative motion will shorten the distance between the back and front ends. After time interval Δt the front end receives the identical tap, also speeds up by dv, and once again moves at the same speed as the back end.

c   Show that the shortening dL predicted by this analysis is

$$dL = -dv\Delta t = -\gamma \Delta x' v dv = -v \gamma L_o dv$$
$$= -\frac{L_o v dv}{(1 - v^2)^{1/2}}$$

which is identical to the result of part **a**, which we wanted to explain. QED.

d   Now start with the front and back pieces of the rod at rest in the laboratory frame and a distance L_o apart. Tap them repeatedly and identically. As they speed up, be sure these taps take place simultaneously in the rocket frame in which the two ends are currently at rest. (This requires you, the ride-along inspector, to resynchronize your rod-rest-frame clocks after each set of front-and-back taps.) Make a logically rigorous argument that after many taps, when the rod is moving at high speed relative to the laboratory, the length of the rod measured in the laboratory can be reckoned using the first equation given in this exercise.

e   Now, by stages, put the rod back together. The full thousand pieces of the rod, lined up but not touching, are all tapped identically and at the same time in the current rest frame of the rod. One set of taps increases the rod's speed from v to $v + dv$ in the laboratory frame. Describe the time sequence of these thousand taps as observed in the laboratory frame. If you have studied Chapter 6 or the equivalent, answer the following questions: What kind of interval — timelike, lightlike, or spacelike — separates any pair of the thousand taps in this set? Can this pair of taps be connected by a light flash? by a compression wave moving along the rod when the pieces are glued back together? Regarding the "logic of acceleration," is there any reason why we should *not* glue these pieces back together? Done!

f   During the acceleration process is the reglued rod *rigid* — unchanging in dimensions — as observed in the rod frame? As observed in the laboratory frame? Is the *rigidity* property of an object an invariant, the same for all observers in uniform relative motion? Show how an ideal rigid rod could be used to transmit signals instantaneously from one place to another. What do you conclude about the idea of a "rigid body" when applied to high-speed phenomena?

Reference: Edwin F. Taylor and A. P. French, *American Journal of Physics*, Volume 51, pages 889–893, especially the Appendix (1983).

L-15   the place where both agree

At any instant there is just one plane in which both the laboratory and the rocket clocks agree.

a   By a symmetry argument, show that this plane lies perpendicular to the direction of relative motion. Using the Lorentz transformation equations, show that velocity of this plane in the laboratory frame is equal to

$$v_{t=t'} = \frac{1}{v_{rel}} [1 - (1 - v_{rel}^2)^{1/2}]$$

b   Does the expression for $v_{t=t'}$ seem strange? From our everyday experience we might expect that by symmetry the "plane of equal time" would move in the laboratory at half the speed of the rocket. Verify that indeed this is correct for the low relative velocities of our everyday experience. Use the first two terms of

the binomial expansion

$$(1 + z)^n \approx 1 + nz \text{ for } |z| \ll 1$$

to show that for low relative velocity, $v_{t=t'} \rightarrow v_{rel}/2$.

c   What is $v_{t=t'}$ for the extreme relativistic case in which $v_{rel} \rightarrow 1$? Show that in this case $v_{t=t'}$ is completely different from $v_{rel}/2$.

d   Suppose we want to go from the laboratory frame to the rocket frame in two equal velocity jumps. Try a first jump to the plane of equal laboratory and rocket times. Now symmetry does work: Viewed from this plane the laboratory and rocket frames move apart with equal and opposite velocities, whose magnitude is given by the equation in part **a**. A second and equal velocity jump should then carry us to the rocket frame at speed v_{rel} with respect to the laboratory. Verify this directly by using the Law of Addition of Velocities (Section L.7) to show that

$$v_{rel} = \frac{v_{t=t'} + v_{t=t'}}{1 + v_{t=t'}v_{t=t'}}$$

L-16   Fizeau experiment

Light moves more slowly through a transparent material medium than through a vacuum. Let v_{medium} represent the reduced speed of light measured in the frame of the medium. Idealize to a case in which this reduced velocity is independent of the wavelength of the light. Place the medium at rest in a rocket moving at velocity v_{rel}, to the right relative to the laboratory frame, and let light travel through the medium, also to the right. Use the Law of Addition of Velocities (Section L.7) to find an expression for the velocity v of the light in the laboratory frame. Use the first two terms of the binomial expansion

$$(1 + z)^n \approx 1 + nz \text{ for } |z| \ll 1$$

to show that for small relative velocity v_{rel} between the rocket and laboratory frames, the velocity v of the light with respect to the laboratory frame is given approximately by the expression

$$v \approx v_{medium} + v_{rel}(1 - v_{medium}^2)$$

This expression has been tested by Fizeau using water flowing in opposite directions in the two arms of an interferometer similar (but not identical) to the interferometer used later by Michelson and Morley (Exercise 3-12).

Reference: H. Fizeau, *Comptes rendus*, Volume 33, pages 349–355 (1851). A fascinating discussion (in French) of some central themes in relativity theory — delivered more than fifty years before Einstein's first relativity paper.

CHAPTER 4

TRIP TO CANOPUS

4.1 INVITATION TO CANOPUS

is one lifetime enough?

Approximately ninety-nine light-years from Earth lies the star Canopus. The Space Agency asks us to visit it, photograph it, and return home with our records.

"But that's impossible," we object. "We have only a little over forty more years to live. We can spare at most twenty years for the outward trip, and twenty years for the return trip. Even if we could travel at the speed of light, we would need ninety-nine years merely to get there."

We are greeted with a smile and a cheery, "Think about our request a little longer, won't you?" ❦

4.2 STRIPPED-DOWN FREE-FLOAT FRAME

throw away most clocks and rods

Troubled thoughts fill us tonight. We dream about invariance of the spacetime interval (Chapter 3). In our dream we find ourselves aboard the rocket used to establish that result (Section 3.7). However, the numbers somehow have changed from meters of distance and meters of light-travel time to light-years of distance and years of time. Suddenly we see things in a new perspective. Three revelations crowd in on us.

The flash of light that got reflected did its work—revelation number one—in establishing the identity of the spacetime interval as measured in either of the two frames. We can remember invariance of the interval and forget about the reflected flash. Eliminating it, we eliminate mirror, photodetector and, most of all, those upward-extended arrays of printout clocks in rocket and laboratory frames whose only purpose was to track the light flash.

The economy goes further. For us aboard the rocket, one reliable calendar clock is enough. As we start our trip from Earth in our dream, that clock by a happy coincidence shows noon on the Fourth of July, 2000 A.D.—and so do clocks at the Space Agency Center on Earth. We celebrate our start by setting off a firecracker.

Later by 6 years—for us—and with a long shipboard program of research and study already completed, our rocket clock—still in our dreams—tells us it is again noon on the Fourth of July and we set off a second firecracker. At that very instant, thanks to the particular speed we had chosen for our rocket relative to Earth, we are passing Lookout Station Number 8. Lonely lighthouse, it has in it little more than a sentry person and a printout clock, one of a series that we have been passing on our trip. They have been stationed out in space, fixed one light-year apart according to Earth measurements. Each clock is calibrated and synchronized to the reference clock on Earth using a reference flash as described in Section 2.6. The laboratory latticework of Figure 2-6 has been reduced to a single rightward-stretching string of lookout stations and their clocks. That we can thus simplify our vision of what is going on from three space dimensions to one is our first revelation. ✒

Retain a single string of Earth-linked clocks

4.3 FASTER THAN LIGHT?

choose your frame. then measure velocity!

Revelation number two strikes us as—still dreaming—we pass Lookout Station Number 8, 8 light-years from Earth: What speed! We glance out of our window and see the lookout station clock print out "Fourth of July 2010 A.D."—10 years later than the Earth date of our departure. Our rocket clock reads 6 years. We are not shocked by the discrepancy in times for, apart from the change in scale from meters of light-travel time to years, the numbers are numbers we have seen before. Nor are we astonished at the identity of the spacetime interval as evaluated in the two very different frames. What amazes us is our speed. Have we actually covered a distance of 8 light-years from Earth in a time of 6 years? Can this mean we have traveled faster than light?

We have often been told that no one and no object can go faster than light. Yet here we are—in our dream—doing exactly that. Speed, yes, we suddenly say to ourselves, but speed in which frame? Ha! What inconsistency! We took the distance covered, 8 light-years, in the Earth-linked laboratory frame, but the time to cover it, 6 years, in the rocket frame!

At this point we recognize that we can talk about our speed in one reference frame or our speed in the other frame, but we get nonsense when we mix together numbers from two distinct reference frames. So we reform. First we pick for reference frame the rocket. But then we get nothing very interesting, because we did not go anywhere with respect to the rocket—we just stayed inside.

Speed: Measure distance and time in same frame

$$\begin{pmatrix} \text{our speed} \\ \text{relative to} \\ \text{rocket frame} \end{pmatrix} = \frac{\begin{pmatrix} \text{distance we cover} \\ \text{with respect to rocket} \end{pmatrix}}{\begin{pmatrix} \text{time we take to cover} \\ \text{it in rocket frame} \end{pmatrix}} = \frac{(0 \text{ light-years})}{(6 \text{ years})} = 0$$

In contrast, our speed relative to the Earth-linked reference frame, the extended laboratory, equals

$$\begin{pmatrix} \text{our speed} \\ \text{relative to} \\ \text{Earth frame} \end{pmatrix} = \frac{\begin{pmatrix} \text{distance we cover} \\ \text{with respect to Earth} \end{pmatrix}}{\begin{pmatrix} \text{time we take to cover} \\ \text{it in Earth frame} \end{pmatrix}} = \frac{(8 \text{ light-years})}{(10 \text{ years})} = 0.8 \text{ light-speed}$$

In other words we—and the rocket—travel, relative to Earth, at 80 percent of the maximum possible speed, the speed of light. Revelation number two is our discovery that speed in the abstract makes no sense, that speed has meaning only when referred to a clearly stated frame of reference. Relative to such a frame we can approach arbitrarily close to light speed but never reach it.

4.4 ALL OF SPACE IS OURS!

in one lifetime: go anywhere in the cosmos

Revelation number three strikes us as—dreaming on—we think more about passing Earth-linked lookout stations. Moving at 80 percent of light speed, we travel 8 light-years in the Earth-linked frame in 6 years of our rocket time. Continuing at the same rate will get us to Canopus in 74 years of our rocket time. Better than 99 years, but not good enough.

Let's use—in imagination—a faster rocket! We suddenly remember the super-rocket discussed in demonstrating the invariance of the spacetime interval (Section 3.8). Converting meters of distance and time to years, we realize that traveling in the super-rocket would bring us to Earth-linked Lookout Station Number 20, *20* Earth-frame light-years from Earth, in 6 years of our rocket time. When passing that station, we can see that station clock reads 20.88 years. Therefore in the Earth-linked frame our super-rocket speed amounts to $20/20.88 = 0.958$ light speed. Continuing at the same speed would bring us to Canopus in 29.7 years of our rocket time. This is nearly short enough to meet our goal of 20 years.

Revelation number three gives us a dizzying new sense of freedom. By going fast enough we can get to Canopus in five minutes of our rocket time if we want! In fact, no matter how far away an object lies, and no matter how short the time allotted to us, nothing in principle stops us from covering the required distance in that time. We only have to be quite careful in explaining this new-found freedom to our Space Agency friends. Yes, we can go any distance the agency requires, however great, provided they specify the distance in the *Earth-linked* reference frame. Yes, we can make it in any nonzero time the agency specifies, however short, provided they agree to measure time on the *rocket* clock we carry along with us.

To be sure, the Earth-linked system of lookout stations and printout clocks will record us as traveling at less than the speed of light. Lookouts will ultimately complain to the Space Agency how infernally long we take to make the trip. But when our Space Agency friends quiz the lookouts a bit more, they will have to confess the truth: When they look through our window as we shoot by station after station, they can see that our clock reads much less than theirs, and in terms of our own rocket clock we are meeting the promised time for the trip.

Our dream ends with sunlight streaming through the window. We lie there savoring the three revelations: economy of description of two events in a reference frame stripped down to one space dimension, speed defined always with respect to a

Five minutes to Canopus— or to any star!

specified reference frame and thus never exceeding light speed, and freedom to go arbitrarily far in a lifetime. ➤

4.5 FLIGHT PLAN

out and back in 40 years to meet our remote descendents

Wide awake now, we face yesterday's question: Shall we go to Canopus, 99 light-years distant, as the Space Agency asks? Yes. And yes, we shall live to return and report.

We take paper and pencil and sketch our plan. The numbers have to be different from those we dreamed about. Trial and error gives us the following plan: After a preliminary run to get up to speed, we will zoom past Earth at $99/101 = 0.9802$ light speed. We will continue at that speed all the 99 light-years to Canopus. We will make a loop around it and record in those few minutes, by high-speed camera, the features of that strange star. We will then return at unaltered speed, flashing by our finish line without any letup, and as we do so, we will toss out our bundle of records to colleagues on Earth. Then we will slow down, turn, and descend quietly to Earth, our mission completed.

The first long run takes 101 Earth years. We have already decided to travel at a speed of 99/101, or 99 light-years of distance in 101 years of time. Going at that speed for 101 Earth years, we will just cover the 99 light-years to Canopus. The return trip will likewise take 101 Earth years. Thus we will deliver our records to Earth 202 Earth-clock years after the start of our trip.

Even briefer will be the account of our trip as it will be perceived in the free-float rocket frame. Relative to the ship we will not go anywhere, either on the outbound or on the return trip. But time will go on ticking away on our shipboard clock. Moreover our biological clock, by which we age, and all other good clocks carried along will tick away in concord with it. How much time will that rocket clock rack up on the outbound trip? Twenty years. How do we know? We reach this answer in three steps. First, we already know from records in the Earth-linked laboratory frame that the spacetime interval — the proper time — between departure from Earth and arrival at Canopus will equal 20 years:

Round trip: 202 Earth years

Round trip: 40 astronaut years

$$\text{Laboratory} \qquad \text{Laboratory}$$
$$(\text{interval})^2 = (\text{time separation})^2 - (\text{space separation})^2$$
$$= (101 \text{ years})^2 - (99 \text{ years})^2$$
$$= 10{,}201 \text{ years}^2 - 9801 \text{ years}^2$$
$$= 400 \text{ years}^2 = (20 \text{ years})^2$$

Second, as the saying goes, "interval is interval is interval": The spacetime interval is invariant between frames. The interval as registered in the rocket frame must therefore also have this 20-year value. Third, in the rocket frame, separation between the two events (departure from Earth and arrival at Canopus) lies all in the time dimension, zero in the space dimension, since we do not leave the rocket. Therefore separation in rocket time itself between these two events is the proper time and must likewise be 20 years:

$$\text{Rocket} \qquad \text{Rocket}$$
$$(\text{interval})^2 = (\text{time separation})^2 - (\text{space separation})^2$$
$$= (\text{time separation})^2 - (\text{zero})^2$$
$$= (\text{rocket time})^2 = (\text{proper time})^2$$
$$= (20 \text{ years})^2$$

We boil down our flight plan to bare bones and take it to the Space Agency for approval: Speed $99/101 = 0.9802$ light speed; distance 99 light-years out, 99 light-years back; time of return to Earth 202 years after start; astronaut's aging during trip, 40 years. The responsible people greet the plan with enthusiasm. They thank us for volunteering for a mission so unprecedented. They ask us to take our proposal before the Board of Directors for final approval. We agree, not realizing what a hornets' nest we are walking into.

The Board of Directors consists of people from various walks of life, set up by Congress to assure that major projects have support of the public at large. The media have reported widely on our proposal in the weeks before we meet with the board, and many people with strong objections to relativity have written to voice their opinions. A few have met with board members and talked to them at length. We are unaware of this as we enter the paneled board room.

At the request of the chairman we summarize our plan. The majority appear to welcome it. Several of their colleagues, however, object. ✐

4.6 **TWIN PARADOX**

a kink in the path explains the difference

"Your whole plan depends on relativity," stresses James Fastlane, "but relativity is a swindle. You can see for yourself that it is self-contradictory. It says that the laws of physics are identical in all free-float frames. Very well, here's your rocket frame and here's Earth frame. You tell me that identical clocks, started near Earth at identical times, each in one of these free-float frames, will read very different time lapses. You go away and return only 40 years older, while we and our descendants age 202 years. But if there's any justice, if relativity makes any sense at all, it should be equally possible to regard *you* as the stay-at-home. Relative to you, *we* speed away in the opposite direction and return. Hence we should be younger than you when we meet again. In contrast, you say you will be younger than we are. This is a flat contradiction. Nothing could show more conclusively that neither result can be right. Aging is aging. It is impossible to live long enough to cover a distance of 99 light-years twice — going and coming. Forget the whole idea."

"Jim," we reply, "your description is the basis for the famous Twin Paradox, in which one twin stays on Earth while the other takes the kind of round trip we have been describing. Which twin is older when they come together again? I would like to leave this question for a minute and consider a similar trip across the United States.

"We all know, Jim, that every July you drive straight north on Interstate Highway 35 from Laredo, Texas, on the Mexican border, to Duluth, Minnesota, near the Canadian border. Your tires roll along a length of roadway equal to 2000 kilometers and the odometer on your car shows it.

"I too drive from Laredo to Duluth, but last year I had to make a stop in Cincinnati, Ohio, on the way. I drove northeast as straight as I could from Laredo to Cincinnati, 1400 kilometers, and northwest as straight as I could from Cincinnati to Duluth, another 1400 kilometers. Altogether, my tires rolled out 2800 kilometers. When we left Laredo you could have said that my route was deviating from yours, and I could have said with equal justice that yours was deviating from mine. The great difference between our travels is this, that my course has a sharp turn in it. That's why my kilometerage is greater than yours in the ratio of 2800 to 2000."

Fastlane interrupts: "Are you telling me that the turn in the rocket trajectory at Canopus explains the *smaller* aging of the rocket traveler? The turn in your trip to Duluth made your travel distance *longer*, not shorter."

Which twin travels?

Curved path in space is a longer path

"That is the difference between path length in Euclidean space geometry and wristwatch time in Lorentz spacetime geometry," we reply. "In Euclidean geometry the *shortest path length* between two points is achieved by the traveler who does not change direction. All indirect paths are longer than this minimum. In spacetime the *greatest aging* between two events is experienced by the traveler who does not change direction. For all travelers who change direction, the total proper time, the total wristwatch time, the total aging is *less* than this maximum.

"The distinction between distance in Euclidean geometry and aging in spacetime comes directly from the contrast between *plus* sign in the expression for distance between two locations and *minus* sign in the expression for interval between two events. In going to Duluth by way of Cincinnati I use the *plus* sign:

$$\begin{pmatrix} \text{distance:} \\ \text{Laredo to} \\ \text{Cincinnati} \end{pmatrix}^2 = \begin{pmatrix} \text{northward} \\ \text{separation:} \\ \text{Laredo to} \\ \text{Cincinnati} \end{pmatrix}^2 + \begin{pmatrix} \text{eastward} \\ \text{separation:} \\ \text{Laredo to} \\ \text{Cincinnati} \end{pmatrix}^2$$

"Contrast this with motion in spacetime. In analyzing my trip to Canopus, I use the minus sign:

$$\begin{pmatrix} \text{proper time:} \\ \text{Earth to} \\ \text{Canopus} \end{pmatrix}^2 = \begin{pmatrix} \text{rocket time:} \\ \text{Earth to} \\ \text{Canopus} \end{pmatrix}^2 = \begin{pmatrix} \text{Earth time:} \\ \text{Earth to} \\ \text{Canopus} \end{pmatrix}^2 - \begin{pmatrix} \text{Earth distance:} \\ \text{Earth to} \\ \text{Canopus} \end{pmatrix}^2$$

"The contrast between a plus sign and a minus sign: This is the distinction between distance covered during travel in space and time elapsed—aging—during travel in spacetime."

4.7 LORENTZ CONTRACTION

go a shorter distance in a shorter time

As James Fastlane ponders this response, Dr. Joanne Short breaks in. "The Twin Paradox is not the only one you have to explain in order to convince us of the correctness of your analysis. Look at the outward trip as observed by you yourself, the rocket traveler. You reach Canopus after just 20 years of your time. Yet we know that Canopus lies 99 light-years distant. How can you possibly cover 99 light-years in 20 years?"

"That is exactly what I dreamed about, Joanne!" we reply. "First of all, it is confusing to combine distances measured in one reference frame with time measured in another reference frame. The 99-light-year distance to Canopus is measured with respect to the Earth-linked frame, while the 20 years recorded on the outward traveler's clock refers to the rocket frame. No wonder the result appears to imply a rate of travel faster than light. Why not take what I paid for fuel for *my* car last week and divide it by the number of gallons you bought today for *your* car, to figure the cost of a gallon of fuel? A crazy, mixed-up, wrong way to work out cost—but no crazier than that way to figure speed!

"But your question about time brings up a similar question about distance: distance between Earth and Canopus measured in the frame in which they are at rest does not agree with the distance between them measured from a rocket that moves along the line connecting them.

Astronaut who turns around ages less . . .

. . . because of a minus sign!

Canopus much closer for astronaut

"Any free-float frame is as good as any other for analyzing motion—that is the Principle of Relativity! So think of the entire outward trip in terms of rocket measurements. At the starting gun (or firecracker) Earth is rushing past the rocket at speed 99/101. Twenty years later Canopus arrives at the rocket, Canopus also traveling at that speed, 99/101 in that rocket frame. This means that for the rocket traveler the Earth-Canopus distance is only about 20 light-years. In fact it is just the fraction (99/101) of 20 light-years, so that at speed 99/101 this distance is covered in exactly 20 years."

"Of course. We are dealing with **Lorentz contraction**," huffs Professor Bright, who thinks any objection to relativity is a waste of time. He has no head for politics, so does not appreciate how important it is for the public to accept the expenditures proposed for this project.

Lorentz contraction

He continues, "Think of a very long stick lying with one end at Earth, the other end at Canopus. Each observer, with the help of colleagues, measures the position of the two ends of this stick *at the same time* in his or her frame. By this means the outward rocket traveler measures a shorter length of the stick—a smaller Earth–Canopus distance—than does an observer in the Earth-linked frame in which the stick lies at rest.

"The factor by which the stick appears contracted in the rocket frame is just the same as the ratio of rocket time to Earth time for the outward trip. This ratio is (20 years)/(101 years). Hence the rocket observer measures the Earth–Canopus distance to be (99 light-years)(20/101) = 19.6 light-years—just a bit less than 20 light-years, as you said.

"Everybody has a satisfactory picture: The astronaut can get to Canopus in 20 years of rocket time because the astronaut's measurements show Canopus to be slightly less than 20 light-years distant. We on Earth agree that the time lapse on the rocket clock is 20 years, but our 'explanation' rests on the invariance of the interval between the events of departure from Earth and arrival at Canopus." Professor Bright pounds the table: "Why are you giving this poor astronaut such a hard time, when relativity is so utterly simple?" He is surprised by the outburst of laughter from other board members and the audience in the room. ➥

4.8 TIME TRAVELER

visit the future. don't come back.

Laura Long has been thoughtfully following the argument. She comments, "You know, we have been discussing you as a space traveler. But you are a *time* traveler as well. Do you realize that by traveling to Canopus and back at 99/101 of light speed, you journey six generations forward in time: 202 years at 33 years per generation? So you will be able to visit your great-great-great-great-great-grandchildren at a cost of only 40 years of your life."

"Yes, I did think of that," we reply. "Time and space are not so different in this respect. Just as we can travel to as great an Earth-linked distance as we want in as short a rocket time as we want, so we can also travel as far forward into Earth's future as we wish.

Travel to Earth's future

"While I was trying various numbers in making up the proposed plan, I realized that if we traveled not at 99/101 light speed but at 9999/10,001 light speed, then a round trip would take not 40 rocket years but only 3.96 rocket years and 198 Earth years. Ten such round trips will age us 39.6 years and bring us back finally at an Earth time about two thousand years in the future, or some year in the fortieth century. That is not six generations ahead, but sixty generations, an additional time equal to one third of recorded history on Earth.''

''Why stop there?'' pursues Laura Long excitedly. ''Why not go even faster, make more round trips, and learn the ultimate fate of Earth and its solar system — or even the still more remote future of the Universe as a whole? Then you could report back to us whether the Universe expands forever or ends in a crunch.''

''Sorry, but no report back to our century is possible,'' smiles Professor Bright. ''There are differences between travel in time and travel in space. To begin with, we can stand still on Earth if we choose and go nowhere in space with respect to that frame. Concerning travel through time, however, we have no such choice! Even when we stand stock still on Earth, we nevertheless travel gently but inevitably forward in time. Time proceeds inexorably!

Time travel is one way

''Second, time travel is one way. You may be able to buy a round-trip ticket to Canopus, but you can get only a one-way ticket to the fortieth century. You can't go backward in time. Time won't reverse.''

Turning to us he adds, ''As for the fate of the solar system and the end of the Universe, our descendants may meet you there as fellow observers, but we ourselves will have to bid you a firm and final 'good-bye' as you leave us on any of the trips we have been discussing. The French *au revoir* — until we meet again — will not do.''

4.9 RELATIVITY OF SIMULTANEITY

we turn around; our changing colleagues say Earth's clock flies forward

By this time James Fastlane has gotten his second wind. ''I am still stuck in this Twin Paradox thing. The time for the outward trip is less as measured in the rocket frame than as measured in the Earth frame. But if relativity is correct, every free-float frame is equivalent. As you sit on the rocket, you feel yourself to be at rest, stationary, motionless; you measure our Earth watch-station clocks to be zipping by you at high velocity. Who cares about labels? For you these Earth clocks are in motion! Therefore the time for the outward trip should be less as measured on the ('moving') Earth clock than as measured on your ('stationary') rocket clock.''

Rocket observer: Fewer Earth-clock ticks on outward trip . . .

We nod assent and he continues. ''Nothing prevents us from supposing the existence of a series of rocket lookout stations moving along in step with your rocket and strung out at separations of one light-year as measured in your rocket frame, all with clocks synchronized in your rocket frame and running at the same rate as your rocket clock. Now, as Earth passes each of these rocket lookout stations in turn, won't those stations read and record the times on the passing Earth clock to be less than their own times? Otherwise how can relativity be correct?''

''Yes, your prediction is reasonable,'' we reply.

. . . also fewer Earth-clock ticks on return trip

''And on the return trip will not the same be true: Returning-rocket lookout stations will measure and record time lapses on the passing Earth clock to be less than on their own clocks?''

''That conclusion is inevitable if relativity is consistent.''

''Aha!'' exclaims Mr. Fastlane, ''Now I've got you! If Earth clock is measured by rocket lookout stations to show smaller time lapses during the outward trip — and also during the return trip — then obviously total Earth time must be less than rocket round-trip time. But you claim just the opposite: that total rocket time is less than Earth time. This is a fundamental contradiction. Your relativity is wrong!'' Folding his arms he glowers at us.

There is a long silence. Everyone looks at us except Professor Bright, who has his head down. It is hard to think with all this attention. Yet our mind runs over the trip again. Going out . . . coming back . . . turning around . . . that's it!

"All of us have been thinking the wrong way!" we exclaim. "We have been talking as if there is only a single rocket frame. True, the same vehicle, with its traveler, goes out and returns. True, a single clock makes the round trip with the traveler. But this vehicle *turns around*—reverses its direction of travel—and that changes everything.

"Maybe it's simpler to think of two rockets, each moving without change of velocity. We ride on the first rocket going out and on the second rocket coming back. Each of these two is really a rocket *frame:* each has its own long train of lookout stations with recording clocks synchronized to its reference clock (Figure 4-1). The traveler can be thought of as 'jumping trains' at Canopus—from outward-bound rocket frame to inward-bound rocket frame—carrying the calendar clock.

"Now follow Mr. Fastlane's prescription to analyze the trip in the rocket frame, but with this change: make this analysis using *two* rocket frames—one outward bound, the other inward bound.

"It is 20 years by outward-rocket time when the traveler arrives at Canopus. That is the reading on all lookout station clocks in that outward-rocket frame. One of those lookout stations is passing Earth when this rocket time arrives. Its clock, synchronized to the clock of the outward traveler at Canopus, also reads 20 years. What time does that rocket lookout-station guard read on the passing Earth clock? For the rocket observer Earth clock reads less time by the same factor that rocket clocks read less time (20 years at arrival at Canopus) for Earth observers (who read 101 years on their own clocks). This factor is 20/101. Hence for the outward-rocket observer the Earth clock must read 20/101 times 20 years, or 3.96 years."

"What!" explodes Fastlane. "According to your plan, the turnaround at Canopus occurs at 101 years of Earth time. Now you say this time equals less than 4 years on Earth clock."

"No sir, I do *not* say that," we reply, feeling confident at last. "I did say that *at the same time* as the outgoing rocket arrives at Canopus, Earth clock reads 3.96 years *as measured in that outgoing rocket frame*. An equally true statement is that *at the same time* as the outgoing rocket arrives at Canopus, Earth clock reads 101 years *as measured in the Earthbound frame*. Apparently observers in different reference frames in relative motion do not agree on what events occur *at the same time* when these events occur far apart along the line of relative motion."

Once again Professor Bright supplies the label. "Yes, that is called **relativity of simultaneity.** Events that occur at the same time—simultaneously—judged from

Astronaut jumps from outgoing frame to returning frame

Outgoing rocket:
As it arrives at Canopus,
Earth clock reads 3.96 years

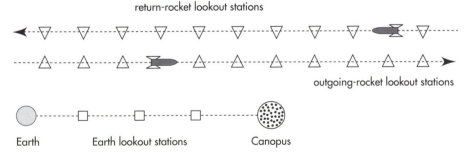

return-rocket lookout stations

outgoing-rocket lookout stations

Earth   Earth lookout stations   Canopus

FIGURE 4-1. *Schematic plot in the Earth-linked frame showing the outgoing rocket and the return rocket used in the round trip between Earth and Canopus. The two rockets meet at Canopus, where the traveler jumps from outgoing rocket to return rocket. Each reference frame has its own string of lookout stations, at rest and synchronized in that frame, shown by small squares, triangles, and inverted triangles. In this figure the outgoing and return rocket lines of motion are displaced vertically for purposes of analysis; in reality, all motion lies along the single line between Earth and Canopus. The figure is* not *to scale!*

one free-float frame but far apart along the line of relative motion do not occur simultaneously as judged from another free-float frame.

"As an example of relativity of simultaneity, consider either chain of lookout stations strung along the line of relative motion. If all clocks in the lookout stations of one frame strike exactly at noon in that frame, these strikes are not simultaneous as measured in another frame in relative motion with respect to the first. This is called **relative synchronization of clocks.**

"Incidentally, most of the so-called 'paradoxes' of relativity, one of which we are considering now, turn on misconceptions about relativity of simultaneity."

Dr. Short breaks in. "What about the returning rocket? What time on the Earth clock will the returning rocket lookout station measure as the traveler starts back?"

Returning rocket: As it leaves Canopus, Earth clock reads 198.04 years

"That shouldn't be too difficult to figure out," we reply. "We know that the clock on the returning rocket reads 40 years when we arrive home on Earth. And the Earth clock reads 202 years on that return. Both of these readings occur at the same place (Earth), so we do not need to worry about relativity of simultaneity of that reading. And during the return trip Earth clock records less elapsed time than rocket clocks' 20 years by the same factor, 20/101, or a total elapsed time of $20 \times 20/101 = 3.96$ years according to return rocket observations. Therefore at the earlier turnaround, return rocket observers will see Earth clock reading $202 - 3.96 = 198.04$ years."

"Wait a minute!" bellows Fastlane. "First you say that the rocket observer sees the Earth clock reading 3.96 years at turnaround in the outward-bound frame. Now you say that the rocket observer sees the Earth clock read 198.04 years at turnaround in the inward-bound frame. Which one is right?"

"Both are right," we reply. "The two observations are made from two different frames. Each of these frames has a duly synchronized system of lookout-station clocks, as does the Earth-linked frame (Figure 4-1). The so-called Twin Paradox is resolved by noticing that between the Earth-clock reading of 3.96 years, taken from the outward rocket lookout station at turnaround and the Earth-clock reading of 198.04 years, taken by the returning-rocket lookout station at turnaround, there is a difference of 194.08 years.

Forward "jump" in Earth clock results from frame change

"This 'jump' appears on no single clock but is the result of the traveler changing frames at Canopus. Yet this jump, or difference, resolves the paradox: For the traveler, the Earth clock reads small time lapses on the outward leg—and also small time lapses on the return leg—but it jumps *way* ahead at turnaround. This jump accounts for the large value of Earth-aging during the trip: 202 years. In contrast the traveler ages only 40 years during the trip (Table 4-1).

"And notice that the traveler is unique in the experience of changing frames; only the traveler suffers the terrible jolt of reversing direction of motion. In contrast, the

---

◁ **TABLE 4-1** ▷

OBSERVATIONS OF EVENTS ON CANOPUS TRIP

| Event | Time measured in Earth-linked frame | Time measured by traveler | Earth-clock reading observed by | |
			outgoing-rocket lookout stations passing Earth	return-rocket lookout stations passing Earth
Depart Earth	0 years	0 years	0 years	
Arrive Canopus	101 years	20 years	20 years $\times$ 20/101 $= 3.96$ years	
Depart Canopus	101 years	20 years	3.96 years	$202 - 3.96$ $= 198.04$ years
Arrive Earth	202 years	40 years		202 years

Earth observer stays relaxed and comfortable in the same frame during the astronaut's entire trip. Therefore there is no symmetry between rocket traveler and Earth dweller, so no genuine contradiction in their differing time lapses, and the story of the twins is not a paradox.

"In fact, the observer in each of the three frames — Earth-linked, outward-rocket, and inward-rocket — has a perfectly consistent and nonparadoxical interpretation of the sequence of events. However, in accounting for disagreements between his or her readings and those of observers in other free-float frames, each observer infers some misbehavior of measuring devices in these other frames. Each observes less elapsed time on clocks in the other frame than on his or her own clocks (time stretching or time dilation). Each thinks that an object lying along the line of relative motion and at rest in another frame is contracted (Lorentz contraction). Each thinks that lookout-station clocks in other frames are not synchronized with one another (relative synchronization of clocks). As a result, each cannot agree with other observers as to which events far apart along the line of relative motion occur at the same time (relativity of simultaneity)."

All observers agree on result, disagree on reason

"Boy," growls Fastlane, "all these different reference frames sure do complicate the story!"

"Exactly!" we exclaim. "These complications arise because observations from any one frame are limited and parochial. All disagreements can be bypassed by talking only in the invariant language of spacetime interval, proper time, wristwatch time. The proper time from takeoff from Earth to arrival at Canopus equals 20 years, period. The proper time from turnaround at Canopus to rearrival at Earth equals 20 years, period. The sum equals 40 years as experienced by the astronaut, period. On the Earth clock, the proper time between departure and return is 202 years, period. End of story. Observers in all free-float frames reckon proper times — spacetime intervals between these events — using their differing space and time measurements. However, once the data are translated into the common language of proper time, every observer agrees. Proper times provide a universal language independent of reference frame."

Spacetime interval is universal language

4.10 EXPERIMENTAL EVIDENCE

objects large and small, slow and fast: many witnesses for the Canopus trip

Alfred Missouri has remained silent up to this point. Now he declares, "All this theory is too much for me. I won't believe a word you say unless you can show me an experimental demonstration."

We reply, "Atomic clocks have been placed on commercial airliners and carried around Earth, some in an eastward direction, others in a westward direction. In each case the airliner clocks were compared with reference clocks at the U.S. Naval Observatory before and after their trips. These clocks disagreed. Results were consistent with the velocity-related predictions of special relativity.

"Airliner" test of twin effect

"This verification of special relativity has two minor difficulties and a major one. Minor difficulties: (1) Each leg of a commercial airliner's trip may be at a different speed, not always accurately known and for which the time-stretching effect must be separately calculated. Also, temperature and pressure effects on airborne clocks are hard to control in a commercial airliner. (2) More fundamentally, Earth rotates, carrying the reference Naval Observatory clocks eastward around the center of Earth. Earth center can be regarded as the inertial point in free-float around Sun. With

BOX 4-1

DO WE NEED GENERAL RELATIVITY? NO!

The group takes a break and mills around the conference room, chatting and eating refreshments. Joanne Short approaches us juggling coffee, a donut, and her notes.

"I didn't want to embarrass you in public," she says, "but isn't your plan faulty because of the turnaround? You can't be serious about leaping from one high-speed rocket to another rocket going in the opposite direction. That means certain death! Be realistic: You and your rocket will have to slow down over some time period, come to rest at Canopus, then speed up again, this time headed back toward Earth. During this change of velocity you will be thrown against the front of the rocket ship, as I'm thrown when I slam on my car brakes. Release a test particle from rest and it will hurtle forward! Surely you are not in an inertial (free-float) frame. Therefore you cannot use special relativity in your analysis of this time period. What does that do to your description of the 'jump ahead' of Earth clocks as you slow down and speed up again? Don't you need general relativity to analyze events in accelerated reference frames?"

"Oh yes, general relativity can describe events in the accelerated frame," we reply, "but so can special relativity if we take it in easy steps! I like to think of a freight yard with trains moving at different speeds along parallel tracks. Each train has its own string of recording clocks along its length, each string synchronized in that particular train frame. Each adjacent train is moving at a slightly different speed from the one next to it. Now we can change frames by walking *across* the trains, stepping from the top of one freight car to the top of the freight car rolling next to it at a slightly different speed.

"Let these trains become rocket trains in space. Each train then has an observer passing Earth as we step on that train. Each observer, by prearrangement, reads the Earth clock *at the same time* that we step onto his train ('at the same time' as recorded in *that* frame). When you assemble all this data later on, you find that the set of observers on the sequence of trains see the Earth clock jumping forward in time much faster than would be expected. The net result is similar to the single horrible jerk as you jump from the outgoing rocket to the incoming rocket.

"Notice that it takes a whole set of clocks in different frames, all reading the single Earth clock, to establish this result. So there is never any contradiction between a single clock in one frame and a single clock in any other frame. In this case special relativity can do the job just fine."

The directors reassemble and Joanne Short, smiling, takes her place with them.

---

respect to this center, one airborne clock moves even faster eastward than Earth's surface, while the other one — heading west with respect to the surface — with respect to Earth's center also moves eastward, but more slowly. Taking account of these various relative velocities adds further complication to analysis of results.

"We overcome these two minor difficulties by having an airplane fly round and round in circles in the vicinity of a single ground-based reference atomic clock.

Then—to a high accuracy—only *relative* motion of these two clocks enters into the special-relativity analysis.

"On November 22, 1975, a U.S. Navy P3C antisubmarine patrol plane flew back and forth for 15 hours at an altitude of 25,000 to 35,000 feet (7600 to 10,700 meters) over Chesapeake Bay in an experiment arranged by Carroll Alley and collaborators. The plane carried atomic clocks that were compared by laser pulse with identical clocks on the ground. Traveling at an average speed of 270 knots (140 meters per second), the airborne clocks lost an average of 5.6 nanoseconds $= 5.6 \times 10^{-9}$ seconds due to velocity-related effects in the 15-hour flight. The expected special-relativity difference in clock readings for this relative speed is 5.7 nanoseconds. This result is remarkably accurate, considering the low relative velocity of the two clocks: 4.7×10^{-7} light speed.

"Circling airplane" test of twin effect

"The *major* difficulty with all of these experiments is this: A high-flying airplane is significantly farther from Earth's center than is the ground-based clock. Think of an observer in a helicopter reading the clocks of passing airplanes and signaling these readings for comparison to a ground-based clock directly below. These two clocks—the helicopter clock and the Earthbound clock—are at rest with respect to one another. Are they in the same inertial (free-float) frame? The answer is No.

Trouble: Large frame is not inertial

"We know that a single inertial reference frame near Earth cannot extend far in a vertical direction (Section 2.3). Even if the two clocks—helicopter and Earthbound—were dropped in free fall, they could not both be in the same inertial frame. Released from rest 30,000 feet one above the other, they would increase this relative distance by 1 millimeter in only 0.3 second of free fall—too rapid a change to be ignored. But the experiment went on not for 0.3 second but for 15 hours!

"Since the helicopter clock and Earthbound clock are not in the same inertial frame, their behavior cannot be analyzed by special relativity. Instead we must use general relativity—the theory of gravitation. General relativity predicts that during the 15-hour flight the higher-altitude clock in the Chesapeake Bay experiment will record *greater* elapsed time by 52.8 nanoseconds due to the slightly reduced gravitational field at altitudes at which the plane flew. From this must be subtracted the 5.7 nanoseconds by which the airborne clock is predicted to record *less* elapsed time due to effects of relative velocity. These velocity effects are predicted by both special relativity and general relativity and were the only results quoted above. The overall predicted result equals $52.8 - 5.7 = 47.1$ nanoseconds net gain by the high-altitude clock compared with the clock on the ground. Contrast this with the measured value of 47.2 nanoseconds.

Solution: Use general relativity

"Hence for airplanes flying at conventional speeds and conventional altitudes, tidal-gravitational effects on clocks can be greater than velocity-dependent effects to which special relativity is limited. In fact, the Chesapeake Bay experiment was conducted to verify the results of general relativity: The airplane pilot was instructed to fly as slowly as possible to reduce velocity effects! The P3C patrol plane is likely to stall below 200 knots, so a speed of 270 knots was chosen.

"In all these experiments the time-stretching effect is small because the speed of an airplane is small compared to the speed of light, but atomic clocks are now so accurate that these speed effects are routinely taken into account when such clocks are brought together for direct comparison."

Professor Bright chimes in. "What the astronaut says is correct: We do not have large clocks moving fast on Earth. On the other hand, we have a great many small clocks moving very fast indeed. When particles collide in high-speed accelerators, radioactive fragments emerge that decay into other particles after an average lifetime that is well known when measured in the rest frame of the particle. When the radioactive particle moves at high speed in the laboratory, its average lifetime is significantly longer as measured on laboratory clocks than when the particle is at rest. The amount of lengthening of this lifetime is easily calculated from the particle speed in the same way the astronaut calculates time stretching on the way to and from

"High-speed radioactive particle" test of twin effect

Canopus. The time-stretch factor can be as great as 10 for some of these particles: the fast-moving particles are measured to live 10 times longer, on average, than their measured lifetime when at rest! The experimental results agree with these calculations in all cases we have tried. Such time stretching is part of the everyday experience of high-energy particle physicists.

"And for these increased-lifetime experiments there is no problem of principle in making observations in an inertial, free-float frame. While they are decaying, particles cover at most a few tens of meters of space. Think of the flight of each particle as a separate experiment. An individual experiment lasts as long as it takes one high-speed particle to move through the apparatus — a few tens of meters of light-travel time. Ten meters of light-travel time equals about 33 nanoseconds, or 33×10^{-9} seconds.

"Can we construct an inertial frame for such happenings? Two ball bearings released from rest say 20 meters apart do not move together very far in 33 nanoseconds! Therefore these increased-lifetime experiments could be done, in principle, in free-float frames. It follows that special relativity suffices to describe the behavior of the 'radioactive-decay clocks' employed in these experiments. We do not need the theory of gravitation provided by general relativity.

"Of course, in none of these high-speed particle experiments do particles move back and forth the way our astronaut friend proposes to do between Earth and Canopus. Even that back-and-forth result has been verified for certain radioactive iron nuclei vibrating with thermal agitation in a solid sample of iron. Atoms in a hotter sample vibrate back and forth faster, on average, and thus stay younger, on average, than atoms in a cooler sample. In this case the 'tick of the clock' carried by an iron atom is the period of electromagnetic radiation ('gamma ray') given off when its nucleus makes the transition from a radioactive state to one that is not radioactive. For detailed reasons that we need not go into here, this particular 'clock' can be read with very high accuracy. Beyond all such details, the experimental outcome is simply stated: Clocks that take one or many round trips at higher speed record a smaller elapsed time than clocks that take one or many round trips at lower speed.

"These various results — plus many others we have not described — combine to give overwhelming experimental support for the predictions of the astronaut concerning the proposed trip to Canopus."

Dr. Bright sits back in his chair with a smile, obviously believing that he has disposed of all objections single-handedly.

"Yes," we conclude, "about the reality of the effect there is no question. Therefore if you all approve, and the Space Agency provides that new and very fast rocket, we can be on our way."

The meeting votes approval and our little story ends. 🐟

Earth frame: Free-float for particle experiments

"Oscillating iron atom" test of twin effect

Twin effect verified!

REFERENCES

The "airliner check" of time stretching (Section 4.10) is reported in J. C. Hafele and Richard E. Keating, *Science,* Volume **177,** pages 166–167 and 168–170 (14 July 1972).

The "patrol plane" check of general relativity (Section 4.10) is reported by Carroll O. Alley in *Quantum Optics, Experimental Gravity, and Measurement Theory,* edited by Pierre Meystre and Marlan O. Scully (Plenum, New York, 1983). See also 1976 physics Ph.D. theses by Robert A. Reisse and Ralph E. Williams, University of Maryland.

The "radioactive nuclei" check of time stretching (Section 4.10) is reported in R. V. Pound and G. A. Rebka, Jr., *Physical Review Letters,* Volume **4,** pages 274–275 (1960).

CHAPTER 4 EXERCISES

Note: The following exercises are related to the story line of this chapter. Additional exercises may be selected from Chapter 3 or the Special Topic on the Lorentz Transformation following Chapter 3.

4-1   practical space travel

In 2200 A.D. the fastest available interstellar rocket moves at $v = 0.75$ of the speed of light. James Abbott is sent in this rocket at full speed to Sirius, the Dog Star (the brightest star in the heavens as seen from Earth), a distance $D = 8.7$ light-years as measured in the Earth frame. James stays there for a time $T = 7$ years as recorded on his clock and then returns to Earth with the same speed $v = 0.75$. Assume Sirius is at rest relative to Earth. Let the departure from Earth be the reference event (the zero of time and space for all observers).

According to Earth-linked observers:

a   At what time does the rocket arrive at Sirius?

b   At what time does the rocket leave Sirius?

c   At what time does the rocket arrive back at Earth?

According to James's observations:

d   At what time does he arrive at Sirius?

e   At what time does he leave Sirius?

f   At what time does he arrive back at Earth?

g   As he moves toward Sirius, James is accompanied by a string of *outgoing* lookout stations along his direction of motion, each one with a clock synchronized to his own. What is the spatial distance between Earth and Sirius, according to observations made with this outgoing string of lookout stations?

h   One of James's outgoing lookout stations, call it Q, passes Earth at the same time (in James's outgoing frame) that James reaches Sirius. What time does Q's clock read at this event of passing? What time does the clock on Earth read at this same event?

i   As he moves back toward Earth, James is accompanied by a string of *incoming* lookout stations along his direction of motion, each one with a clock synchronized to his own. One of these incoming lookout stations, call it Z, passes Earth at the same time (in James's incoming frame) that James leaves Sirius to return home. What time does Z's clock read at this event of passing? What time does the clock on Earth read at this same event?

To *really* understand the contents of Chapter 4, repeat this exercise many times with new values of v, D, and T that you choose yourself.

4-2   one-way twin paradox?

A worried student writes, "I still cannot believe your solution to the Twin Paradox. During the outward trip to Canopus, each twin can regard the other as moving away from him; so how can we say which twin is younger? The answer is that the twin in the rocket makes a turn, and in Lorentz spacetime geometry, the greatest aging is experienced by the person who does not turn. This argument is extremely unsatisfying. It forces me me to ask: What if the rocket breaks down when I get to Canopus, so that I stop there but cannot turn around? Does this mean that it is no longer possible to say that I have aged less than my Earthbound twin? But if not, then I would never have gotten to Canopus alive." Write a half-page response to this student, answering the questions politely and decisively.

4-3   a relativistic oscillator

In order to test the laws of relativity, an engineer decides to construct an oscillator with a very light oscillating bob that can move back and forth very fast. The lightest bob known with a mass greater than zero is the electron. The engineer uses a cubical metal box, whose edge measures one meter, that is warmed slightly so that a few electrons "boil off" from its surfaces (see the figure). A vacuum pump removes air from the box so that electrons may move freely inside without colliding with air molecules. Across the middle of the box — and electrically insulated from it — is a metal screen charged to a high positive voltage by a power supply. A voltage-control knob on the power supply can be turned to change the DC voltage V_o between box and screen. Let an electron boiled off from the inner wall of the box have very small velocity initially (assume that the initial velocity is zero). The electron is attracted to the positive screen, increases speed toward the screen, passes through a hole in the screen, slows down as it moves away from the attracting screen, stops just short of the opposite wall of the box, is pulled back toward the screen; and in this way oscillates back and forth between the walls of the box.

a   In how short a time T can the electron be made to oscillate back and forth on one round trip between the walls? The engineer who designed the equipment claims that by turning the voltage control knob high enough he can obtain as high a frequency of oscillation $f = 1/T$ as desired. Is he right?

b   For sufficiently low voltages the electron will be nonrelativistic — and one can use Newtonian me-

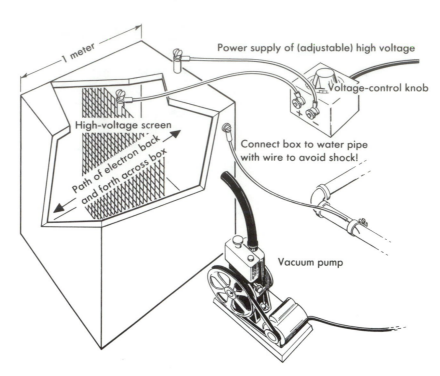

EXERCISE 4-3. *Relativistic oscillator with electron as oscillating bob.*

chanics to analyze its motion. For this case the frequency of oscillation of the electron is increased by what factor when the voltage on the screen is doubled? **Discussion:** At corresponding points of the electron's path before and after voltage doubling, how does the Newtonian kinetic energy of the electron compare in the two cases? How does its velocity compare in the two cases?

c What is a definite formula for frequency f as a function of voltage in the nonrelativistic case? Wait as late as possible to substitute numbers for mass of electron, charge of electron, and so forth.

d What is the frequency in the extreme relativistic case in which over most of its course the electron is moving . . . (rest of sentence suppressed!) . . . ? Call this frequency $f_{\max}$.

e On the same graph, plot two curves of the dimensionless quantity $f/f_{\max}$ as functions of the dimensionless quantity $q\mathrm{V_o}/(2mc^2)$, where q is the charge on the electron and m is its mass. First curve: the nonrelativistic curve from part **c** to be drawn heavily in the region where it is reliable and indicated by dashes elsewhere. Second curve: the extreme relativistic value from part **d**, also with dashed lines where not reliable. From the resulting graph estimate quantitatively the voltage of transition from the nonrelativistic to the relativistic region. If possible give a simple argument explaining why your result does or does not make sense as regards order of magnitude (that is, overlooking factors of 2, π, etc.).

f Now think of the round-trip "proper period" of oscillation τ experienced by the electron and logged by its recording wristwatch as it moves back and forth across the box. At low electron speeds how does this proper period compare with the laboratory period recorded by the engineer? What happens at higher electron speeds? At extreme relativistic speeds? How is this reflected in the "proper frequency" of oscillation f_{proper} experienced by the electron? On the graph of part **e** draw a rough curve in a different color or shading showing qualitatively the dimensionless quantity $f_{\mathrm{proper}}/f_{\max}$ as a function of $q\mathrm{V_o}/(2mc^2)$.

TREKKING THROUGH SPACETIME

5.1 TIME? NO. SPACETIME MAP? YES.

no such thing as the unique time of an event!

Events are the sparkling grains of history. They define spacetime. Spacetime, yes. Time, no.

"Time, no"? How come? Time here in Tokyo, at this enthronement of the successor of the Emperor Hirohito? Where is any meter to be seen that shows any such quality of location as time? Meter to measure the temperature here and now? Yes, this thermometer. Meter to measure atmospheric pressure here and now? Yes, this barometer. But look as we will, nowhere can we see any meter that we can poke into the space hereabouts to measure its "time." The time of an event? Impossible! No such thing. Time is not "meterable."

Anything with which to compare time? Yes. Odometer reading, whether miles or kilometers, on the dashboard of our car. There's no such thing as the odometer reading of Tokyo. Try every gadget one can, thrust it out into this Tokyo air, not one will register anything with the slightest claim to be called the odometer reading of these hereabouts.

What about looking at the dashboards of the cars in this neighborhood? Not all of them; that would be nonsense. Only the cars that were new, with odometer reading zero, at the time of Hirohito's own enthronement.

Now at last we are getting into a line of questioning that shows some prospect of clearing up what we mean by "time." We ask our companion, "What do all those day-and-year-counting wristwatches now read that were set to zero at the time of that earlier ceremony?"

"Sixty-two years, two days," is her first reply. But then we ask, "What about that team that zoomed out to the nearest eye-catching star, Alpha Centauri, and back with almost the speed of light? Didn't they get back ten years younger than we stay-at-homes?"

"Time" of an event has no unique meaning

Car mileage depends on car's path between places

137

Wristwatch reading depends on its history of travel between events

"Yes," she agrees, "surely their wristwatches now read fifty-two years, not sixty-two. So let me draw the lesson. There is no such thing as time. There is only totalized *interval* of time, time as that interval is racked up between the enthronement of Hirohito and the enthronement of of the new Emperor Akihito, between event *A* and event *B*, on a wristwatch that has undergone its own individual history of travel from *A* to *B*."

"I agree. The concept of time does not apply to location in spacetime. It applies to individual history of travel through spacetime."

"How apt the comparison with odometer reading. Each dashboard shows, not the kilometerage of Akihito, but the kilometers traveled by that particular car between the one imperial ceremony and the other."

Geographic map assigns kilometer coordinates to places

Yes, it is nonsense to attribute a kilometer reading to Tokyo. However, it is not at all nonsense to make a map showing where Tokyo lies relative to all the towns roundabout, a map in which kilometers do appear, kilometers north and south, kilometers east and west. Likewise the term "the time" of an event is totally without meaning. However, that event — and every event near it — lends itself to display on a spacetime diagram (Figure 5-1), with distance (the locator of latticework clock) running in one direction, and in another direction time (the reading printed out by that clock on the occasion of that event). Time as employed in this sense acquires meaning only because it serves as a measure on a latticework-defined map. A different lattice-work? A different set of clocks, different readings on those clocks, a different map — but same events, same spacetime, same tools to measure the history-dependent interval between event and event.

Spacetime map assigns space and time coordinates to events

Only on such a spacetime plot does one see at a glance the layout of all nearby events, and how one history of travel from event *A* to event *B* differs from another.

One problem in making our map: Spacetime has four dimensions — three space dimensions plus time. We picture our event points most readily when they occupy a two-dimensional domain and let themselves be dotted in on a two-dimensional page. Therefore for the present we limit attention to time and one space dimension; to events, whatever their timing, that occur on one line in space. All events that do not occur on this line we ignore for now. The space location of each event on this line we plot along a **horizontal axis** on the page. The lattice-clock time at which an event occurs we plot along a **vertical axis,** from bottom to top of the page. Space and time we measure in the same unit, for example meters of distance and meters of time — or light-years of distance and years of time. We call the result a **spacetime map** or a **spacetime diagram.** Each spacetime map represents data from a particular reference frame, for example "the laboratory frame." Figure 5-1 shows such a spacetime map.

Limit attention to one space dimension plus one time dimension

Five sample event points appear on the laboratory spacetime map of Figure 5-1, events labeled *0, A, B, C,* and *D.*

- **Event *0*** is the **reference event,** the firing of the starting gun, which we take to locate zero position in space and the zero of time. For our own convenience, we place point *0* at the origin of the spacetime map and measure space and time locations of all other events with respect to it.

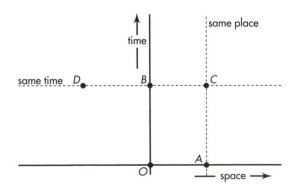

FIGURE 5-1. *Laboratory spacetime map, showing the reference event* O, *other events* A, B, C, *and* D, *a horizontal dashed line of simultaneity in time, and a vertical dashed line of equal position in space.*

- **Event B** stands on the vertical time axis, directly above reference event O. Therefore event B occurs at a later time than event O. Event B lies neither to the right of the reference event nor to the left; its horizontal (space) location is zero. Therefore it occurs at the same place as the reference event O in the laboratory but later in time.

- **Event A** lies on the horizontal space axis, directly to the right of reference event O. Therefore event A occurs at a different space location than event O. It is neither above nor below event O; its vertical (time) location is zero. Therefore it occurs at the same time as reference event O as observed in the laboratory.

- **Event C** rests above and to the right of the reference event. Standing higher than the reference event on the map, event C occurs later in time than O in this frame. Since it lies to the right, event C occurs at a positive space location with respect to event O in this frame.

- **Event D** reposes above and to the left of the reference event. It also occurs later in time than reference event O but at a negative space location with respect to event O as observed in the laboratory.

Scatter other event points on the spacetime map. Each event point can represent an important happening. Then a single glance at the spacetime map gives us, in principle, a global picture of *all* significant events that have occurred along *one* line in space and as far back in time as we wish to look. The spacetime map puts all this history at our fingertips!

In exploring history, we may want to know which events occurred at the same time as others in the laboratory free-float frame. Two events that occur at the same time have the same vertical (time) location on the spacetime map. A horizontal line drawn through one event point passes through all events simultaneous with that event in the given frame. In Figure 5-1, the dashed horizontal line shows that events B and D are simultaneous as observed in the laboratory frame, although they occur at different locations in space. Similarly, events O and A are simultaneous as observed in this frame.

When we wish to "retell history," we draw a sequence of horizontal lines above one another on the spacetime map. We mimic the advance of time by stepping in imagination from one horizontal line to the next horizontal line above it, noting which events occur at each time.

Vertical lines on the spacetime map indicate which events occur at the same place along the single line in space. Events A and C in Figure 5-1 occur at the same space location as measured in the laboratory, but at different times as measured in this frame. Similarly, events O and B occur at the same place as one another in the laboratory. 🐟

Spacetime map assigns space and time coordinates to events

5.2 SAME EVENTS; DIFFERENT FREE-FLOAT FRAMES

different frames: different points for an event on their spacetime maps, but same spacetime interval between two events

Figure 5-1 demonstrates two great payoffs of the spacetime map: (1) It places space and time on an equal footing, thus recognizing a basic symmetry of nature. (2) It allows us to review at a single glance the whole history of events and motions that have occurred along the given line in space.

Same events, different frames:
Different spacetime maps

We want to take advantage of a third payoff of the spacetime map: Plot the same events on two, three, or more spacetime maps based on two, three, or more different free-float frames in uniform relative motion. Compare. In this way analyze the various space and time relations among these events as measured in different frames. Why do this? In order to find out what is different in the different frames and what remains the same.

Figure 5-3 shows three spacetime maps — for laboratory, rocket, and super-rocket free-float frames. The super-rocket moves faster than the rocket with respect to the laboratory (but not faster than light!). On each of the three spacetime maps we plot the same two events: the events of emission E and reception R of a light flash. These are the two events analyzed in Chapter 3 to derive the expression for the spacetime interval. As a reminder of the physical phenomena behind events E and R, refer to Figure 5-2.

The light flash is emitted (event E) from a sparkplug attached to the reference clock of the first rocket. Take event E as the reference event, called event O in Figure 5-1. By prearrangement the sparkplug fires at the instant when both the rocket reference clock and the super-rocket reference clock pass the laboratory reference clock. All three

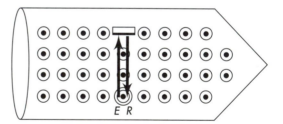

LABORATORY PLOT

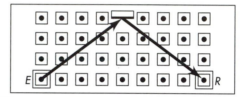

ROCKET PLOT

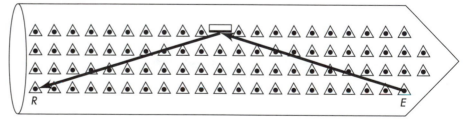

SUPER-ROCKET PLOT

FIGURE 5-2 (Figure 3-5 repeated). *The flash path as recorded in three different frames, showing event* **E**, *emission of the flash, and event* **R**, *its reception after reflection. Squares, circles, and triangles represent the latticework of recording clocks in laboratory, rocket, and super-rocket frames, respectively. The super-rocket frame moves to the right with respect to the rocket, so that the event of reception,* **R**, *occurs to the left of the event of emission,* **E**, *as measured in the super-rocket frame. The reflecting mirror is fixed in the rocket, hence appears to move from left to right in the laboratory and from right to left in the super-rocket.*

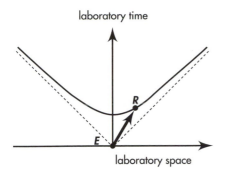

LABORATORY
SPACETIME MAP

FIGURE 5-3. *Spacetime maps for three frames, showing emission of the reference flash and its reception after reflection.* The hyperbola drawn in each map satisfies the equation for the invariant interval (or proper time), which has the same value in all three frames: $(interval)^2 = (time)^2 - (space)^2$.

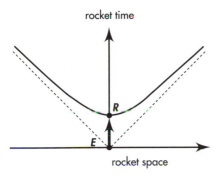

ROCKET
SPACETIME MAP

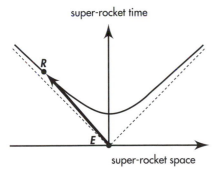

SUPER-ROCKET
SPACETIME MAP

reference clocks are set to read zero at this reference event, whose event point is placed at the origin of all three spacetime maps.

Now use the latticework of meter sticks and clocks in each free-float frame (clocks pictured in Figure 5-2) to measure the position and time of every other event with respect to the reference event. In particular, record the position and time of the reception (event R) of the flash in each of the three frames.

The reception of the light ray (event R) occurs at different locations and at different times as measured in the three frames. In the rocket the reception of the reflected flash occurs back at the reference clock (the zero of position) and 6 meters of time later, as seen in Figure 5-2 and more directly in Figure 5-3 (center):

Same events, different frames:
Different space and time
coordinates

$$\textbf{Rocket: (position of reception, event } R) = 0$$
$$\textbf{Rocket: (time of reception, event } R) = 6 \text{ meters}$$

Emission and reception occur at the same place in the rocket frame. Therefore the rocket time, 6 meters, is just equal to the interval, or proper time, between these two events:

$$(\text{proper time})^2 = \left(\begin{array}{c}\textbf{Rocket}\\ \text{time of}\\ \text{reception}\end{array}\right)^2 - \left(\begin{array}{c}\textbf{Rocket}\\ \text{position of}\\ \text{reception}\end{array}\right)^2$$

$$= \left(\begin{array}{c}\textbf{Rocket}\\ \text{time of}\\ \text{reception}\end{array}\right)^2 - (\text{zero})^2 = (6 \text{ meters})^2$$

In the laboratory the reception event R occurs at a time greater than 6 meters, as can be seen from the expression for interval:

$$\left(\begin{array}{c}\textbf{Laboratory}\\ \text{time of}\\ \text{reception}\end{array}\right)^2 - \left(\begin{array}{c}\textbf{Laboratory}\\ \text{position of}\\ \text{reception}\end{array}\right)^2 = (6 \text{ meters})^2$$

In this equation the square of 6 meters results from subtracting a positive quantity from the square of the laboratory time of reception. Therefore the laboratory time of reception itself must be greater than 6 meters:

$$\textbf{Laboratory: (position of reception, event } R) = 8 \text{ meters}$$
$$\textbf{Laboratory: (time of reception, event } R) = 10 \text{ meters}$$

In the laboratory frame, reception appears to the right of the emission, as seen in Figure 5-2. Hence it is plotted to the right of the origin in the laboratory map (Figure 5-3, top).

In the super-rocket frame, moving faster than the rocket with respect to the laboratory, the event of reception appears to the left of the emission (Figure 5-2). Therefore the space separation is called negative and plotted to the left of the origin in the super-rocket map (Figure 5-3, bottom). The time separation in the super-rocket is greater than 6 meters, by the same argument used for the time of reception in the laboratory frame:

Same events, different frames:
Same spacetime interval

$$\left(\begin{array}{c}\textbf{Super-rocket}\\ \text{time of}\\ \text{reception}\end{array}\right)^2 - \left(\begin{array}{c}\textbf{Super-rocket}\\ \text{position of}\\ \text{reception}\end{array}\right)^2 = (6 \text{ meters})^2$$

In this equation, the space separation is a negative quantity. Nevertheless its square is a positive quantity. So the equation says that the square of 6 meters results from subtracting a positive quantity from the square of the super-rocket time of reception. Therefore the super-rocket time separation must also be greater than 6 meters:

$$\textbf{Super-rocket: (position of reception, event } R) = -20 \text{ meters}$$
$$\textbf{Super-rocket: (time of reception, event } R) = 20.88 \text{ meters}$$

5.3 INVARIANT HYPERBOLA

all observers agree: "event point lies somewhere on this hyperbola"

Different reception points marked *R* in different spacetime maps all refer to the same event. What do these different separations of the same event from the reference event have in common? They all satisfy invariance of the interval, reflected in the equation

$$(\text{time separation})^2 - (\text{space separation})^2 = (\text{interval})^2 = \text{constant}$$

Constant? Constant with respect to what?

With respect to free-float frame. Record different space and time measurements in different frames, but figure out from them always the same interval.

Curves drawn on the three maps conform to this equation. This kind of curve, in which the difference of two squares equals a constant, is called a **hyperbola.** Somewhere on this hyperbola is recorded the time and position of one and the same reception event as measured in every possible rocket and super-rocket frame. Same reception event, different frames, all summarized in one hyperbola, the **invariant hyperbola.**

Spacetime arrows in all three maps connect the same pair of events. They imply the identical invariant interval. They embody the same spacetime reality. In a deep sense these three arrows on the page represent the same arrow in spacetime. Spacetime maps of different observers show different projections — different perspectives — of the same arrow in spacetime.

Invariant hyperbola: Locus of same event in all rocket frames

The same arrow? The same magnitude for the spacetime arrow pictured in all three maps of Figure 5-3? Then why do the three arrows have obviously different lengths in the three maps?

Because the paper picture of spacetime is a lie! The length of an arrow on a piece of paper is Euclidean, related to the sum of squares of the space separations of the endpoints in two perpendicular directions. Euclidean geometry works fine if what is being represented is flat space, for example the map of a township. But Euclidean geometry is the wrong geometry and betrays us when we try to lay out time along one direction on the page. Instead we need to use Lorentz geometry of spacetime. In Lorentz geometry, time must be combined with space through a *difference* of squares to find the correct magnitude of the resulting spacetime vector — the interval. That is why the arrows in the different spacetime maps of Figure 5-3 seem to be of different lengths. The reality that these lengths represent, however — the value of the interval between two events — is the same in all three spacetime maps. ▰

5.4 WORLDLINE

the moving particle traces out a line — its *worldline* — on the spacetime diagram

We describe the world by listing events and showing how they relate to one another. Until now we have focused on pairs of events and spacetime intervals between them. Now we turn to a whole chain of events, events that track the passage of a particle through spacetime. Think of a speeding sparkplug that emits a spark every meter of time read on its own wristwatch. Each spark is an event; the collection of spark events forms a chain that threads through spacetime, like pearls. String the pearls together. The thread connecting the pearl events, tracing out the path of a particle through spacetime, has a wonderfully evocative name: **worldline.** The sparkplug travels through spacetime trailing its worldline behind it.

String of event pearls: Worldline!

The speeding sparkplug is only an example. Every particle has a worldline that connects events along its spacetime path, events such as collisions or near-collisions (close calls) with other particles.

Events — pearls in spacetime — exist independent of any reference frame we may choose to describe them. A worldline strings these event pearls together. The worldline, too, exists independent of any reference frame. A particle traverses spacetime — follows a worldline — totally oblivious to our poor efforts to describe its motion using one or another free-float frame. Yet we are accustomed to using a free-float frame and its associated latticework of rods and clocks. One clock after another records its encounter with the particle. The worldline of the particle connects this chain of encounter events.

Wordline versus line on spacetime map

We can draw this worldline of a particle on the spacetime map for this reference frame. Such worldlines are shown in Figure 5-5 and in later figures of this chapter. Strictly speaking, the line drawn on the spacetime map is not the worldline itself. It is an image of the worldline — a strand of ink printed on a piece of paper. When we use a highway map, we often refer to a line drawn on the paper as "the highway." Yet is not the highway itself, but an image. Ordinarily this causes no confusion; no one tries to drive a car across a highway map! Similarly, we loosely refer to the line drawn on the spacetime map as the worldline, even though the worldline in spacetime stands above and beyond all our images of it.

Examples of worldlines

The worldline is seen in no way more clearly than through example. Particle 1 starts at the laboratory reference clock at zero time and moves to the right with constant speed (Figure 5-4). As particle 1 zooms along a line of laboratory latticework of clocks, each clock it encounters records the time at which the particle passes. Each clock record shows where the clock is located and the time at which particle 1 coincides with the clock. "Where and when" determines an event, the **event of coincidence** of particle and recording clock. Afterwards the chief observer travels throughout the lattice of clocks, collecting the records of these coincidence events. She plots these events as points on her spacetime map. She then draws a line through event points in sequence — the worldline of particle 1 (Figure 5-5).

Particle 1 moves with constant speed along a single direction in space. The distance it covers is equal for each tick of the laboratory clocks. The worldline of particle 1 shows equal changes in space during equal lapses of time by being straight on the spacetime map.

Particle 2 moves to the right faster than particle 1 and so covers a greater distance in the same time lapse (Figure 5-4). Lattice clocks record their events of coincidence with particle 2, and the observer collects these records and plots the worldline of particle 2 on the same spacetime map (worldline shown in Figure 5-5).

And so it goes: Particle 3 is a light flash and moves to the right in space (Figure 5-4) with maximum speed: one meter of distance per meter of time. With horizontal and vertical axes calibrated in meters, the light-flash worldline rises at an angle of 45 degrees (Figure 5-5).

Particle 4 does not move at all in laboratory space; it rests quietly next to the laboratory reference clock. Like you sitting in your chair, it moves only along the time dimension; in the laboratory spacetime map its worldline is vertical (Figure 5-5).

Particle 5 moves not to the right but to the left in space according to the laboratory observer (Figure 5-4), so its worldline angles up and leftward in the laboratory spacetime map (Figure 5-5).

Each of these particles moves with constant speed, so each traces out a straight worldline. After 3 meters of time as measured in the laboratory frame, different

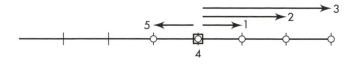

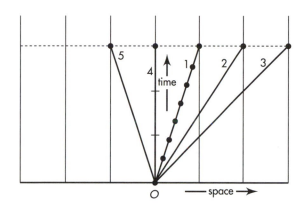

FIGURE 5-4. *Trajectories in space (not in spacetime!) of particles 1 through 5 during 3 meters of time. Each particle starts at the reference clock (the square) at zero of time and moves with a constant velocity.*

FIGURE 5-5. *Worldlines in spacetime of the particles shown in Figure 5-4, plotted for the laboratory frame. Only the worldline for particle 1 includes a sample set of event points that are connected to make up the worldline.*

particles have moved different distances from the starting point (Figure 5-4). In the laboratory spacetime map their space positions after 3 meters of time lie along the upper horizontal **line of simultaneity,** shown dashed in Figure 5-5.

Particle 4 is not the only object stationary in space. Every laboratory clock lies at rest in the laboratory frame; it moves neither right nor left as time passes. Nevertheless each laboratory clock moves forward in time, tracing out its own vertical worldline in the laboratory spacetime map. The background vertical lines in Figure 5-5 are worldlines of the row of laboratory clocks.

What is the difference between a "path in space" and a "worldline in spacetime"?

The transcontinental airplane leaves a jet trail in still air. That trail is the plane's path in space. Take a picture of that trail and you have a *space map* of the motion. From that space map alone you cannot tell how fast the jet is moving at this or that different point on its path. The space map is an incomplete record of the motion.

The plane moves not only in space but also in time. Its beacon flashes. Plot those emissions as events on a spacetime map. This spacetime map has not only a horizontal space axis but also a vertical time axis. Now connect those event points with a worldline. The worldline gives a *complete* description of the motion of the jet as recorded in that frame. For example, from the worldline we can reckon the speed of the plane at every event along its path.

Worldline gives spacetime map of the journey of the jet. Likewise a worldline drawn on a spacetime map images the journey of any particle through spacetime. A worldline is not a physical path, not a trajectory, not a line in *space*. An object at rest in your frame has, for you, no path at all through space; it stays always at one space point. Yet this stationary particle traces out a "vertical" worldline in your spacetime map (such as line 4 in Figure 5-5). A particle *always* has a worldline in *spacetime*. As you sit quietly in your chair reading this book, you glide through spacetime on your own unique worldline. Every stationary object lying near you also traces out a worldline, parallel to your own on your spacetime map.

Path in space versus worldline in spacetime

Not all particles move with constant speed. When a particle changes speed with respect to a free-float frame, we know why: A force acts on it. Think of a train moving on a straight stretch of track. A force applied by the locomotive speeds up all the cars. *Small speed:* small distance covered in a given time lapse; worldline inclined slightly to the vertical in the spacetime map. *Great speed:* great distance covered in an the same stretch of time; worldline inclined at a greater angle to the vertical in the spacetime map. *Changing speed:* changing distances covered in equal time periods; worldline that changes inclination as it ascends on the spacetime map—a curved worldline!

Changing speed means curving worldline

Wait a minute! The train moves along a straight track. Yet you say its worldline is curved. Straight or curved? Make up your mind!

Straight in space does not necessarily mean straight in spacetime. Place your finger on the straight edge of a table near you. Now move your finger rapidly back and forth along this edge. Clearly this motion lies along a straight line. As your fingertip changes speed and direction, however, it travels different spans of distance in equal time periods. During a spell in which it is at rest on the table edge, your fingertip traces out a vertical portion of its worldline on the spacetime map. When it moves slowly to the right on the table, it traces out a worldline inclined slightly to the right of vertical on the map. When it moves rapidly to the left, your fingertip leaves a spacetime trail inclined significantly to the left on the map. Changing inclination of the worldline from point to point results in a curved worldline. Your finger moves straight in space but follows a curved worldline in spacetime!

Figure 5-6 shows a curved worldline, not for a locomotive, but for a particle constrained to travel down the straight track of a linear accelerator. The particle starts at the reference clock at the time of the reference event (*O* on the map). Initially the particle moves slowly to the right along the track. As time passes—advancing upward on the spacetime map—the particle speed increases to a large fraction of the speed of light. Then the particle slows down again, comes to rest at event *Z*, with a vertical tangent to its worldline at that event. Thereafter the particle accelerates to the left in space until it arrives at event *P*.

What possible worldlines are available to the particle that has arrived at event *P*? A material particle must move at less than the speed of light. In other words, it travels less than one meter of distance in one meter of time. Its future worldline makes an "angle with the vertical" somewhere between plus 45 degrees and minus 45 degrees when space and time are measured in the same units and plotted to the same scale along horizontal and vertical axes on the graph. These limits of slope—which apply to every point on a particle worldline—are shown as dashed lines emerging from event *P* in Figure 5-6 (and also from event *O*).

Limit on worldline slope: speed of light

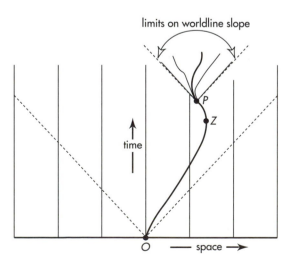

FIGURE 5-6. *Curved laboratory worldline of a particle that changes speed as it moves along a straight line in space.* Some possible worldlines available to the particle after event P.

The worldline gives a *complete* description of particle motion in spacetime. As drawn in the spacetime map for any frame, the worldline tells position and velocity of the particle at every event along its trail. In contrast, the trajectory or orbit or path shape of a particle in space does not give a complete description of the motion. To complete the description we need to know when the particle occupies each location on that trajectory. A worldline in a spacetime map automatically displays all of this information.

The spacetime map provides a tool for retrospective study of events that have already taken place and have been reported to the free-float observer who plots them. Once she plots these event points, this analyst can trace already plotted worldlines backward in time. She can examine at a single glance event points that may have occurred light-years apart in space. These features of the spacetime map do not violate our experience that time moves only forward or that nothing moves faster than light. Everything plotted on a spacetime map is history; it can be scanned rapidly back and forth in the space dimension or the time dimension or both. The spacetime map supplies a comprehensive tool for recognizing patterns of events and teasing out laws of nature, but it is useless for influencing the events it represents. ➤

Spacetime map displays only already detected events

5.5 LENGTH ALONG A PATH

straight line has *shortest* length between two given *points* in *space*

Distance is a central idea in all applications of Euclidean geometry. For instance, using a flexible tape measure it is easy to quantify the total distance along a winding path that starts at one point (point O in Figure 5-7) and ends at another point (point B). Another way to measure distance along the curved path is to lay a series of short straight sticks end to end along the path. Provided the straight sticks are short enough to conform to the gently curving path, total distance along the path equals the sum of lengths of the sticks.

Measure length of curved path with tape measure . . .

The length of a short stick laid between any two nearby points on the path—for instance, points 3 and 4 in Figure 5-7—can also be calculated using the northward separation and the eastward separation between the two ends of the stick as measured by a surveyor.

. . . or with short straight sticks laid end to end along path

$$(\text{length})^2 = (\text{northward separation})^2 + (\text{eastward separation})^2$$

Distance is invariant for surveyors. Therefore the length of this stick is the same when calculated by any surveyor, even though the northward and eastward separations between two ends of the stick have different values, respectively, for different surveyors. The length of another stick laid elsewhere along the path is also agreed on by all surveyors despite their use of different northward directions. Therefore the sum of the lengths of all short sticks laid along the path has the same value for all surveyors. This sum equals the value of the total length of the path, on which all surveyors agree. And this total length is just the length measured using the flexible tape.

All surveyors agree on length of path

It is possible to proceed from O to B along quite another path—for example along straight line OB in Figure 5-7. The length of this alternative path is evidently different from that of the original curved path. This feature of Euclidean geometry is so well known as to occasion hardly any comment and certainly no surprise: In Euclidean geometry a curved path between two specified points is longer than a straight path between them. The existence of this difference of length between two paths violates no law. No one would claim that a tape measure fails to perform properly when laid along a curved path.

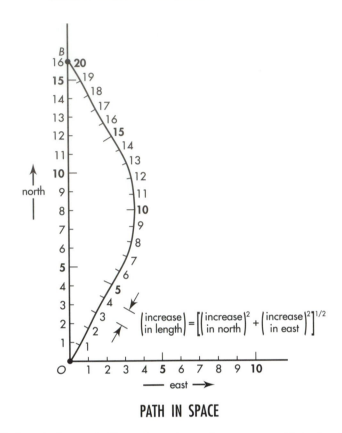

$$\begin{pmatrix} \text{increase} \\ \text{in length} \end{pmatrix} = \left[\begin{pmatrix} \text{increase} \\ \text{in north} \end{pmatrix}^2 + \begin{pmatrix} \text{increase} \\ \text{in east} \end{pmatrix}^2 \right]^{1/2}$$

PATH IN SPACE

FIGURE 5-7. *Length along a winding path starting at the town square.* Notice that the total length along the winding path from point O to point B is greater *than the length along the straight northward axis from O to B.*

Straight path in space has *shortest* length

Among all possible paths between two points in space, the straight-line path is unique. All surveyors agree that this path has the shortest length. When we speak of "the distance between two points," we ordinarily mean the length of this straight path. 🪶

5.6 WRISTWATCH TIME ALONG A WORLDLINE

straight worldline has *longest proper time* between two given *events* in *spacetime*

Measure proper time along curved worldline with wristwatch . . .

A curved path in Euclidean space is determined by laying down a flexible tape measure and recording distance along the path's length. A curved worldline in Lorentz space-time is measured by carrying a wristwatch along the worldline and recording what it shows for the elapsed time. The summed spacetime interval—the proper time read directly on the wristwatch—measures the worldline in Lorentz geometry in the same way that distance measures path length in Euclidean geometry.

A particle moves along the worldline in Figure 5-8. This particle carries a wrist-watch and a sparkplug; the sparkplug fires every meter of time $(1, 2, 3, 4, \ldots)$ as read off the particle's wristwatch. The laboratory observer notes which of his clocks the

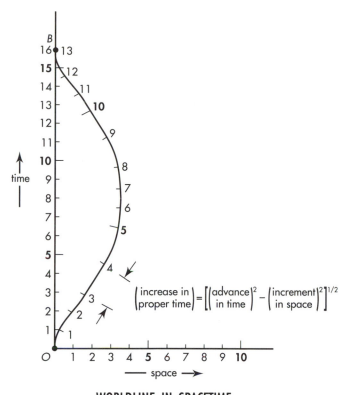

$$\binom{\text{increase in}}{\text{proper time}} = \left[\binom{\text{advance}}{\text{in time}}^2 - \binom{\text{increment}}{\text{in space}}^2 \right]^{1/2}$$

WORLDLINE IN SPACETIME

FIGURE 5-8. *Proper time along a curved worldline.* Notice that the total proper time along the curved worldline from event O to event B is smaller *than the proper time along the straight line from* O *to* B.

traveling particle is near every time the sparkplug fires. He plots that location and that lattice clock time on his spacetime map, tracing out the worldline of the particle. He numbers spark points sequentially on the resulting worldline, as shown in Figure 5-8, knowing that these numbers register meters of time recorded on the moving wrist-watch.

Consider the spacetime interval between two sequential numbered flashes of the sparkplug, for instance those marked 3 and 4 in the figure. In the laboratory frame these two sparks are separated by a difference in position and also by a difference in time (the time between them). The squared interval — the proper time — between the sparks is given by the familiar spacetime relation:

$$(\text{proper time})^2 = (\text{difference in time})^2 - (\text{difference in position})^2$$

What about the proper time between sparks 3 and 4 calculated from measurements made in the sparkplug frame? In this frame, both sparks occur at the same place, namely at the position of the sparkplug. The difference in position between the sparks equals zero in this frame. As a result, the time difference in the sparkplug frame — the "wristwatch time" — is equal to the proper time between these two events:

$$(\text{proper time})^2 = (1 \text{ meter})^2 - (\text{zero})^2 = (1 \text{ meter})^2 \quad \text{[recorded on traveling wristwatch]}$$

This analysis assumes that sparks are close together in both space and time. For sparks close enough together, the velocity of the emitting particle does not change much from one spark to the next; the particle velocity is effectively constant between sparks; the piece of curved worldline can be replaced with a short straight segment. Along this straight segment the particle acts like a free-float rocket. The proper time is

... or as sum of intervals between adjacent events

invariant in free-float rocket and free-float laboratory frames. Thus the laboratory observer can compute the value of the proper time between events 3 and 4 and predict the time lapse — one meter — on the traveling wristwatch, which measures the proper time directly.

Elsewhere along the worldline the particle moves with a different speed. Nevertheless the proper time between each consecutive pair of sparks must also be independent of the free-float frame in which that interval is reckoned. For sparks close enough together, this proper time equals the time read directly on the wristwatch.

All observers agree on the proper time between every sequential pair of sparks emitted by the sparkplug. Therefore the sum of of all individual proper times has the same value for all observers. This sum equals the value of the total proper time, on which all free-float observers agree. And this total proper time is just the wristwatch time measured by the traveling sparkplug.

In brief, proper time is the time registered in a rocket by its own clock, or by a person through her own wristwatch or her own aging. Like aging, proper time is cumulative. To obtain total proper time racked up along a worldline between some marked starting event and a designated final event, we first divide up the worldline into segments so short that each is essentially "straight" or "free-float." For each segment we determine the interval, that is, the lapse of proper time, the measurement of aging experienced on that segment. Then we add up the aging, the proper time for each segment, to get total aging, total wristwatch time, total lapse of proper time.

An automobile may travel the most complicated route over an entire continent, but the odometer adds it all up and gives a well-understood number. The traveler through the greater world of spacetime, no matter how many changes of speed or direction she undergoes, has the equivalent of the odometer with her on her journey. It is her wristwatch and her body — her aging. Your own wristwatch and your biological clock automatically add up the bits of proper time traced out on all successive segments of your worldline.

It is possible to proceed from event O to event B along quite another worldline — for example, along the straight worldline OB in Figures 5-8 and 5-9 (bottom). The proper time from O to B along this new worldline can be measured directly by a flashing clock that follows this new worldline. It can also be calculated from records of flashes emitted by the clock as recorded in any laboratory or rocket frame.

Total proper time along this alternative worldline has a different value than total proper time along the original worldline. In Lorentz geometry a curved worldline between two specified events is *shorter* than the direct worldline between them — shorter in terms of total proper time, total wristwatch time, total aging.

Total proper time, the aging along any given worldline, straight or curved, is an *invariant:* it has the same value as reckoned by observers in all overlapping free-float frames. This value correctly predicts elapsed time recorded directly on the wristwatch of the particle that travels this worldline. It correctly predicts the aging of a person or a mouse that travels this worldline. A different worldline between the same two events typically leads to a different value of aging — a new value also agreed on by all free-float observers: Aging is maximal along the straight worldline between two events. This uniqueness of the straight worldline is also a matter of complete agreement among all free-float observers. All agree also on this: The straight worldline is the one actually followed by a free particle. Conclusion: Between two fixed events, a free particle follows the worldline of maximal aging. This more general prediction of the worldline of a free particle is called the **Principle of Maximal Aging.** It is true not only for "straight" particle worldlines in the limited regions of spacetime described by special relativity but also, with minor modification, for the motion of free particles in wider spacetime regions in the vicinity of gravitating mass. The Principle of Maximal Aging provides one bridge between special relativity and general relativity.

The stark contrast between Euclidean geometry and Lorentz geometry is shown in Figure 5-9. In Euclidean geometry distance between nearby points along a curved

---

Margin notes:

All observers agree on proper time along worldline

Straight worldline has *longest* proper time

Principle of Maximal Aging predicts motion of free particle

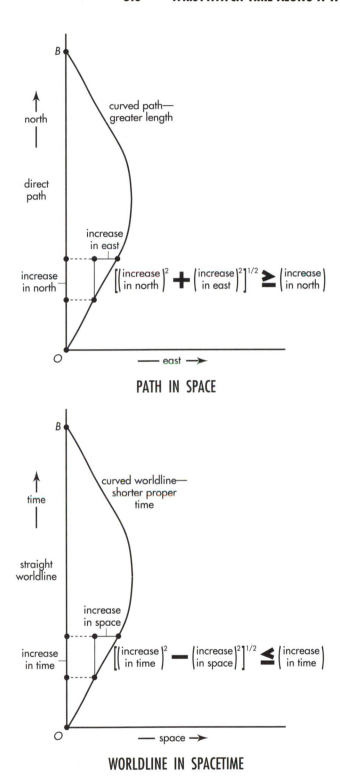

PATH IN SPACE

WORLDLINE IN SPACETIME

FIGURE 5-9. *Path in space: In* Euclidean *geometry the* curved *path has* greater *length.* ***Worldline in spacetime:*** *In* Lorentz *geometry the* curved *worldline is traversed in* shorter *proper time.*

path is always equal to or *greater* than the northward separation between those two points. In contrast, proper time between nearby events along a curved worldline is always equal to or *less* than the corresponding time along the direct worldline as measured in that frame.

Stark contrast between Euclidean and Lorentz geometries

The difference of proper time between two alternative worldlines in spacetime violates no law, just as the difference of length between two alternative paths in space violates no law. There is nothing wrong with a wristwatch that reads different proper times when carried along different worldlines between events O and B in spacetime, just as there is nothing wrong with a tape measure that records different lengths for different paths between points O and B in space. In both cases the measuring device is simply giving evidence of the appropriate geometry: Euclidean geometry for space, Lorentz geometry for spacetime.

Proper times compare worldlines

In brief, the determination of cumulative interval, proper time, wristwatch time, aging along a worldline between two events is a fundamental method of comparing different worldlines that connect the same two events.

Among all possible worldlines between two events, the straight worldline is unique. All observers agree that this worldline is straight and has the longest proper time—greatest aging—of any possible worldline connecting these events.

5.7 KINKED WORLDLINE

kink in the worldline decreases aging along that worldline

The change in slope of the worldline from event to event in Figures 5-8 and 5-9 (bottom) means that the clock being carried along this worldline changes velocity: It accelerates. Different clocks behave differently when accelerated. Typically a clock can withstand a great acceleration only when it is small and compact. A pendulum clock is not an accurate timepiece when carried by car through stop-and-go traffic; a wristwatch is fine. A wristwatch is destroyed by being slammed against a wall; a radioactive nucleus is fine. Typically, the smaller the clock, the more acceleration it can withstand and still register properly, and the sharper can be the curves and kinks on its worldline. In all figures like Figures 5-8 and 5-9 (bottom), we assume the ideal limit of small (acceleration-proof) clocks.

Acceleration-proof clocks

We are now free to analyze a motion in which particle and clock are subject to a great acceleration. In particular, consider the simple special case of the worldline of Figure 5-8. That worldline gradually changes slope as the particle speeds up and slows down. Now make the period of speeding up shorter and shorter (great driving force!); also make the period of slowing down shorter and shorter. In this way come eventually to the limiting case in which episodes of acceleration and deceleration—curved portions of the worldline—are too short even to show up on the scale of the spacetime map (worldline OQB in Figure 5-10). In this simple limiting case the whole history of motion is specified by (1) initial event O, (2) final event B, and (3) turnaround event Q, halfway in time between O and B. In this case it is particularly easy to see how the lapse of proper time between O and B depends on the location of the halfway event—and thus to compare three worldlines, OPB, OQB, and ORB.

Simplify: Worldlines with straight segments

Path OPB is the worldline of a particle that does not move in space; it stays next to the reference-frame clock. Proper time from O to B by way of P is evidently equal to time as measured in the free-float frame of this reference clock:

$$(\text{total proper time along } OPB) = 10 \text{ meters of time}$$

In contrast, on the way from O to B via R, for each segment the space separation equals the time separation, so the proper time has the value zero:

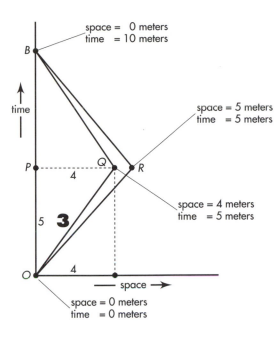

space = 0 meters
time = 10 meters

B

↑
time

space = 5 meters
time = 5 meters

P   Q   R
         4

space = 4 meters
time = 5 meters

5   **3**

4

O

——— space →

space = 0 meters
time = 0 meters

FIGURE 5-10. *Three alternative worldlines connecting events* O *and* B. *The sharp changes of velocity at events* Q *and* R *have been drawn for the ideal limit of small clocks that tolerate great acceleration. The bold-face number 3 is the proper time along the segment* OQ, *reckoned from the difference between the squared time separation and the squared space separation:* $3^2 = 5^2 - 4^2$.

$$(\text{proper time along leg } OR)^2 = (\text{time})^2 - (\text{space})^2$$
$$= (5 \text{ meters})^2 - (5 \text{ meters})^2$$
$$= 0$$
$$(\text{total proper time along } ORB) = 2 \times (\text{proper time along } OR)$$
$$= 0$$

Zero proper time for light

As far as we know, only three things can travel 5 meters of distance in 5 meters of time: light (photons), neutrinos, and gravitons (see Box 8-1). No material clock can travel at light speed. Therefore the worldline *ORB* is not actually attainable by a material particle. However, it can be approached arbitrarily closely. One can find a speed sufficiently close to light speed — and yet less than light speed — so that a trip with this speed first one way then the other will bring an ideal clock back to the reference clock with a lapse of proper time that is as short as one pleases. In the same way we can, in principle, go to the star Canopus and back in as short a round-trip rocket time as we choose (Section 4.8).

As distinguished from the limiting case *ORB*, worldline *OQB* demands an amount of proper time that is greater than zero but still less than the 10 meters of proper time along the direct worldline *OPB:*

$$(\text{proper time along leg } OQ)^2 = (5 \text{ meters})^2 - (4 \text{ meters})^2$$
$$= 25 \text{ (meters)}^2 - 16 \text{ (meters)}^2$$
$$= 9 \text{ (meters)}^2$$
$$= (3 \text{ meters})^2$$

Reduced proper time along kinked worldline

so

$$(\text{proper time along leg } OQ) = 3 \text{ meters}$$

and

$$(\text{total proper time along both legs } OQB) = 2 \times (\text{proper time along } OQ)$$
$$= 6 \text{ meters}$$

This is less proper time than (proper time along *OPB*) = 10 meters that characterized the "direct" worldline *OPB*. Our trip to Canopus and back described in Chapter 4 follows a worldline similar to *OQB*. 🪶

SAMPLE PROBLEM 5-1

MORE IS LESS

In the spacetime map shown, time and space are measured in years. A table shows space and time locations of numbered events in this frame.

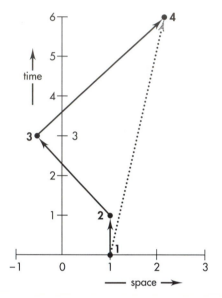

SPACE AND TIME
LOCATIONS OF EVENTS

	Space (years)	Time (years)
Event **1**	1	0
Event **2**	1	1
Event **3**	−0.5	3
Event **4**	2	6

Two alternative worldlines between events 1 and 4

a. One traveler moves along the solid straight worldline segments from event 1 to events 2, 3, and 4. Calculate the time increase on her clock between event 1 and event 2; between event 2 and event 3; between event 3 and event 4. Calculate total proper time — her aging — along worldline 1, 2, 3, 4.

b. Another traveler, her twin brother, moves along the straight dotted worldline from event 1 directly to event 4. Calculate the time increase on his clock along the direct worldline 1, 4.

c. Which twin (solid-line traveler or dotted-line traveler) is younger when they rejoin at event 4?

SOLUTION

a. From the table next to the map, space separation between events 1 and 2 equals 0. Time separation equals 1 year. Therefore the interval is reckoned from (interval)$^2 = 1^2 - 0^2 = 1^2$. Thus the proper time lapse on a clock carried between events 1 and 2 equals 1 year.

Space separation between event 2 and event 3 equals $1 - (-0.5) = 1.5$ light-years. Time separation equals 2 years. Therefore the square of the interval is $2^2 - (1.5)^2 = 4 - 2.25 = 1.75$ (years)^2 and the advance of proper time equals the square root of this, or 1.32 years.

Between event 3 and event 4 space separation equals 2.5 light-years and time separation 3 years. The square of the interval has the value $3^2 - (2.5)^2 = 9 - 6.25 = 2.75$ (years)^2 and proper time between these two events equals the square root of this, or 1.66 years.

Total proper time—aging—along worldline 1, 2, 3, 4 equals the sum of proper times along individual segments: $1 + 1.32 + 1.66 = 3.98$ years.

b. Space separation between events 1 and 4 equals 1 light-year. Time separation is 6 years. The squared interval between them equals $6^2 - 1^2 = 36 - 1 = 35$ (years)^2. A traveler who moves along the direct worldline from event 1 to event 4 records a span of proper time equal to the square root of this value, or 5.92 years.

c. The brother who moves along straight worldline 1, 4 ages 5.92 years during the trip. The sister who moves along segmented worldline 1, 2, 3, 4 ages less: 3.98 years. As always in Lorentz geometry, the direct worldline (shown dotted) is longer—that is, it has more elapsed proper time, greater aging—than the indirect worldline (shown solid).

5.8 STRETCH FACTOR

ratio of frame-clock time to wristwatch time

A speeding beacon emits two flashes, F and S, in quick succession. These two flashes, as recorded in the rocket that carries the beacon, occur with a 6-meter separation in time but a zero separation in space. Zero space separation? Then 6 meters is the value of the interval, the proper time, the wristwatch time between F and S. As registered in the laboratory, in contrast, the second flash S occurs 10 meters of time later than the first flash F. The ratio between this frame time, 10 meters, and the proper time, 6 meters, between the two events we call the time stretch factor, or simply **stretch factor**. Some authors use the lowercase Greek letter gamma, γ, for the stretch factor, as we do occasionally. We will also use the Greek letter tau, τ, for proper time.

The same two events register in the super-rocket frame that overtakes and passes the beacon—register with a separation in time of 20.88 meters. In this frame, the time stretch factor between the two events is $(20.88)/6 = 3.48$. In the beacon frame the stretch factor is unity: $6/6 = 1$. Why? Because in this beacon frame flashes F and S occur at the same place, so beacon-frame clocks record the proper time directly. This proper time is less than the time between the two flashes as measured in either laboratory or super-rocket frame. The larger value of time observed in laboratory and super-rocket frames shows up in Figure 5-11 (center and right). Among all conceivable frames, the separation in time between the two flashes evidently takes on its minimum value in the beacon frame itself, the value of the proper time τ.

Hold it! In Sections 5.6 and 5.7 you insisted that the time along a straight worldline is a MAXIMUM. Now you show us a straight worldline along which the time is—you say—a MINIMUM. Maximum or minimum? Please make up your mind!

The worldline taken by the beacon wristwatch from F to S is straight. It is straight whether mapped in the beacon frame itself or in the rocket or super-rocket frame. The beacon racks up 6 meters of *proper* time regardless of the frame in which we reckon this time. When we turn from this wristwatch time to what different

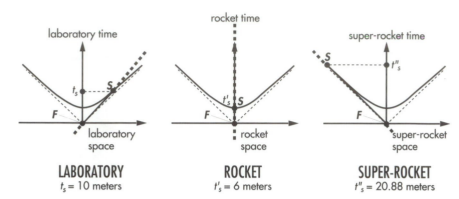

LABORATORY
$t_s = 10$ meters

ROCKET
$t'_s = 6$ meters

SUPER-ROCKET
$t''_s = 20.88$ meters

FIGURE 5-11. *Spacetime maps of Figure 5-3, modified to show the worldline of the speeding beacon (heavy dashed line) and the segment of this line between emission* F *of the first flash and the second flash* S *(solid section of worldline). Emission* F *is taken as the zero of space and time. Time* t_s *of the second emission* S *is different as recorded in different frames. The shortest time is recorded in that frame in which the two events occur at the same place—in this case the rocket frame.*

free-float frames show for the separation in map time (latticework time, frame time) between the two flashes, however, the record displays a minimal value for that separation in time only in the beacon frame itself.

In contrast, Figure 5-12 (Figure 5-10 in simplified form) shows two *different* worldlines that join events *0* and *B* mapped in the same reference frame. In this case we compare two different proper times: a proper time of 10 meters racked up by a wristwatch carried along the direct course from *0* to *B*, and a proper time of 6 meters recorded by the wristwatch carried along on the kinked worldline *0QB*. In every such comparison made in the context of flat spacetime, the direct worldline displays maximum proper time. Caution: Conditions can be different in curved spacetime (Chapter 9).

In summary, two points come to the fore in these comparisons of the time between two events. (1) Are we comparing map time (frame time, latticework time) between those two events, pure and simple, free of any talk about any worldline that might connect those events? Then separation in time between those events is least as mapped in the free-float frame that shows them happening at the same place. (2) Or are we directing our attention to a worldline that connects the two events? More specifically, to the time racked up by a wristwatch toted along that worldline? Then we have to ask, is that worldline straight? Then it registers maximal passage of proper time. Or does it have a kink? Then the proper time racked up is not maximal.

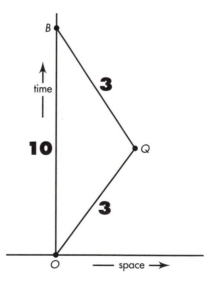

FIGURE 5-12. *Figure 5-10 stripped down to emphasize total proper time (wristwatch time), printed boldface along two different worldlines between the same two events* O *and* B *in a given reference frame. Among all possible worldlines connecting events* O *and* B, *the straight worldline registers maximal lapse of proper time.*

When we find ourselves in a free-float frame and see a beacon zooming past in a straight line with speed v, how much is the factor by which our frame-clock time is stretched relative to the beacon wristwatch time? Answer: The stretch factor is

$$\text{(stretch factor)} = \gamma = \frac{1}{(1 - v^2)^{1/2}} \tag{5-1}$$

How derive this famous formula? If you do not cover up the following lines and derive this answer on your own, here is the reasoning: Start with measurements in the laboratory frame. We know that for this rocket

$$\text{(advance in proper time)}^2 = \text{(advance in lab time)}^2 - \text{(lab distance covered)}^2$$

However, we want to compare lapses in laboratory time and proper time; laboratory distance covered is not of interest. For the laboratory observer the proper clock moving along a straight worldline covers the distance between the two events in the time between the events. Therefore this distance and time are related by particle speed:

Stretch factor
= frame time/proper time

$$\text{(lab distance covered)} = \text{(speed)} \times \text{(advance in lab time)}$$

Substitute this expression into the equation for proper time:

$$\begin{aligned} \text{(proper time)}^2 &= \text{(lab time)}^2 - \text{(speed)}^2 \times \text{(lab time)}^2 \\ &= \text{(lab time)}^2 \, [1 - \text{(speed)}^2] \end{aligned}$$

This leads to an expression for the square of the stretch factor:

$$\frac{\text{(lab time)}^2}{\text{(proper time)}^2} = \text{(stretch factor)}^2 = \frac{1}{1 - \text{(speed)}^2} = \frac{1}{1 - v^2}$$

where we use the symbol $v = v_{conv}/c$ for speed. The equation for the stretch factor becomes

$$\text{(stretch factor)} = \gamma = \frac{1}{(1 - v^2)^{1/2}} \tag{5-1}$$

Stretch factor derived

The stretch factor has the value unity when $v = 0$. For all other values of v the stretch factor is greater than unity. For very high relative speeds, speeds close to that of light ($v \rightarrow 1$), the value of the stretch factor increases without limit.

The value of the stretch factor does not depend on the direction of motion of the rocket that moves from first event to second event: The speed is squared in equation (5-1), so any negative sign is lost.

The stretch factor is the ratio of frame time to proper time between events, where speed ($= v$) is the steady speed necessary for the proper clock to pass along a straight worldline from one event to the other in that frame.

Lorentz-contraction by same "stretch" factor

The stretch factor also describes the Lorentz contraction, the measured shortening of a moving object along its direction of motion when the observer determines the distance between the two ends *at the same time*. For example, suppose you travel at speed v between Earth and a star that lies distance L away as measured in the Earth frame. Your trip takes time $t = L/v$ in the Earth-linked frame. Proper time τ—your wristwatch time—is smaller than this by the stretch factor: $\tau = L/[v \times \text{(stretch factor)}] = (L/v)\,(1 - v^2)^{1/2}$. Now think of a very long rod that reaches from Earth to star and is at rest in the Earth frame. How long is that rod in your rocket frame? In your frame the rod is moving at speed v. One end of the rod, at the position of Earth, passes at speed v. A time τ later in your frame the other end of the rod arrives—along with the star—also moving at speed v according to your rocket measurements. From these data you calculate that the length of the rod in your rocket frame—call it L'—is equal to $L' = v\tau = v(L/v)\,(1 - v^2)^{1/2} = L\,(1 - v^2)^{1/2}$. This is a valid measure of length. By this method the rod is measured to be shorter.

Stretch factor as a measure of speed

Finally, the stretch factor is often used as an alternative measure of particle speed: A particle moves with a speed such that the stretch factor is 10. This statement assumes that the particle is moving with constant speed, so that the separation between any pair of events on the particle worldline has the same stretch factor as the separation between any other pair. This way of describing particle speed can be both convenient and powerful. We will see (Chapter 7) that the total energy of a particle is proportional to the stretch factor.

SAMPLE PROBLEM 5-2

ROUND TRIP OBSERVED IN A DIFFERENT FRAME

Return to the alternative worldlines between events O and B, shown in Figure 5-10 and the spacetime maps in this sample problem. Measure these worldlines from a rocket frame that moves outward with the particle from O to Q and keeps on going forever at the same constant velocity. Show that an observer in this outward-rocket frame predicts the same proper time—wristwatch

time—for worldline OQB as that predicted in the laboratory frame. Similarly show that this outward-rocket-frame observer predicts the same proper time along the *direct* worldline OPB as does the laboratory observer. Finally, show that both observers predict the elapsed wristwatch time along OQB to be less than along OPB.

SOLUTION

Here are laboratory and rocket spacetime maps for these round trips, simplified and drawn to reduced scale.

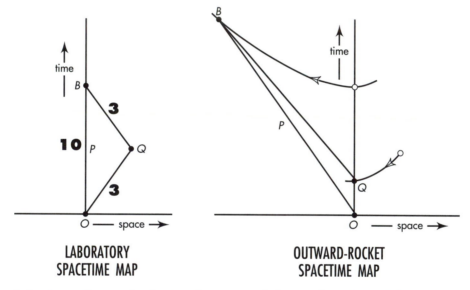

**LABORATORY
SPACETIME MAP**

**OUTWARD-ROCKET
SPACETIME MAP**

Laboratory and outward-rocket spacetime maps, each showing alternative worldlines (direct **OPB** *and indirect* **OQB***) between events* **O** *and* **B***. Laboratory spacetime map:* Figure 5-10, *redrawn to a different scale. Proper times are shown on the laboratory spacetime map.* **Outward-rocket spacetime map:** *The rocket in which the outgoing particle is at rest. Portions of two invariant hyperbolas show how events* **Q** *and* **B** *transform. The direct worldline* **OPB** *has longer total proper time—greater aging—as computed using measurements from either frame.*

Find x'_Q and t'_Q: First compute space and time locations of events Q and B in the outgoing rocket frame—right-hand map. (Event O is the reference event, $x = 0$ and $t = 0$ in all frames by convention.) We choose the rocket frame so that the worldline segment OQ lies vertical and the outbound rocket does not move in this frame. As a result, event Q occurs at rocket space origin: $x'_Q = 0$. (Primes refer to measurements in the outward-rocket frame.) The rocket time t'_Q for this event is just the wristwatch time between O and Q, because the wristwatch is at rest in this frame: $t'_Q = 3$ meters.

In summary, using a prime for rocket measurements:

$$x'_Q = 0$$
$$t'_Q = 3 \text{ meters}$$

Find x'_B and t'_B: In the *laboratory* frame, the particle moves to the *right* from event O to event Q, covering 4 meters of distance in 5 meters of time. Therefore its speed is the fraction $v = 4/5 = 0.8$ of light speed. As measured in the rocket frame, the laboratory frame moves to the *left* with speed $v = 0.8$, by symmetry. Use equation (5-1) with $v = 0.8$ to compute the value of the stretch factor:

$$\frac{1}{[1 - v^2]^{1/2}} = \frac{1}{[1 - (0.8)^2]^{1/2}} = \frac{1}{[1 - 0.64]^{1/2}} = \frac{1}{[0.36]^{1/2}} = \frac{1}{0.6} = \frac{10}{6} = \frac{5}{3}$$

This equals the ratio of rocket time period t'_B to proper time τ_B along the direct path OPB. Hence elapsed rocket time $t'_B = (5/3) \times 10$ meters $= 50/3$ meters of time. In this time, the laboratory moves to the left in the rocket frame by the distance $x'_B = -vt'_B = -(4/5)(50/3) = -200/15 = -40/3$ meters. In summary for outgoing rocket:

$$x'_B = -\frac{40}{3} \text{ meters} = -13\frac{1}{3} \text{ meters}$$

$$t'_B = \frac{50}{3} \text{ meters} = 16\frac{2}{3} \text{ meters of time}$$

Events Q and B are plotted on the rocket spacetime map.

Compare Wristwatch Times: Now compute the total proper time — wristwatch time, aging — along alternative worldlines OPB and OQB using rocket measurements. Direct worldline OB has proper time τ_{OB} given by the regular expression for interval:

$$(\tau_{OB})^2 = (t'_{OB})^2 - (x'_{OB})^2 = \left(\frac{50}{3}\right)^2 - \left(-\frac{40}{3}\right)^2$$

$$= \frac{2500}{9} - \frac{1600}{9} = \frac{900}{9} = 100 \text{ (meters)}^2$$

whence $\tau_{OB} = 10$ meters computed from rocket measurements. This is the same value as computed in the laboratory frame (in which proper time equals laboratory time, since laboratory separation in space is zero).

Worldline OQB has two segments. On the first segment, OQ, proper time lapse is just equal to the rocket time span, 3 meters, since the space separation equals zero in the rocket frame. For the second segment of this worldline, QB, we need to compute elapsed time in this frame:

$$t'_{QB} = t'_B - t'_Q = \frac{50}{3} - 3 = \frac{50}{3} - \frac{9}{3} = \frac{41}{3} \text{ meters}$$

$$x'_{QB} = -\frac{40}{3} \text{ meters}$$

Therefore,

$$(t_{QB})^2 = (t'_{QB})^2 - (x'_{QB})^2 = \left(\frac{41}{3}\right)^2 - \left(\frac{40}{3}\right)^2$$

$$= \frac{1681}{9} - \frac{1600}{9} = \frac{81}{9} = 9 \text{ (meters)}^2$$

whence $\tau_{QB} = 3$ meters. So the total increase in proper time — the total aging — along worldline OQB sums to $3 + 3 = 6$ meters as reckoned from outward-rocket measurements. This is the same as figured from laboratory measurements.

How can these weird results be true? In our everyday lives why don't we have to take account of clocks that record different elapsed times between events, and rods that we measure to be contracted as they speed by us?

In answer, consider two events that occur at the same place in our frame. The proper clock moving in spacetime between these two events has speed zero for us. In this case the stretch factor has the value unity: the frame clock is the proper clock. The same is *approximately* true for events that are much closer together in space (measured in meters) than the time between them (also measured in meters). In these cases the proper clock moving between them has speed v—measured in meters/meter—that is very much less than unity. That is, the proper clock moves very much slower than the speed of light. For such slow speeds, the stretch factor has a value that approaches unity; the proper clock records *very nearly* the same time lapse between two events as frame clocks. This is the situation for all motions on earth that we can follow by eye. For all such "ordinary-speed" motions, moving clocks and stationary clocks record essentially the same time lapses. This is the assumption of Newtonian mechanics: "Absolute, true, and mathematical time, of itself, and from its own nature, flows equably without relation to anything external . . ."

A similar argument leads to the conclusion that Lorentz contraction is negligible for objects moving at everyday speeds. Newton's mechanics—with its unique measured time between events and its unique measured length for an object whether or not it moves—gives correct results for objects moving at everyday speeds. In contrast, for particle speeds approaching light speed (approaching one meter of distance traveled per meter of elapsed time in the laboratory frame), the denominator on the right of equation (5-1) approaches zero and the stretch factor increases without limit. Increased without limit, also, is the laboratory time between ticks of the zooming particle's wristwatch. This is the case for high-speed particles in accelerators and for cosmic rays, very high-energy particles (mostly protons) that continually pour into our atmosphere from space. Newton's mechanics gives results wildly in error when applied to these particles and their interactions; the laws of relativistic mechanics must be used.

More than one cosmic ray has been detected (indirectly by the resulting shower of particles in the atmosphere) moving so fast that it could cross our galaxy in 30 seconds as recorded on its own wristwatch. During this trip a thousand centuries pass as recorded by clocks on Earth! (See Exercise 7-7.) ⬟

5.9 TOURING SPACETIME WITHOUT A REFERENCE FRAME

all you need is worldlines and events

An explosion is an explosion. Your birth was your birth. An event is an event. Every event has a concreteness, an existence, a reality independent of any reference frame. So, too, does a worldline that connects the trail of event points left by a high-speed sparkplug that flashes as it streaks along. Events mark worldlines, independent of any reference frame.

Worldlines also locate events. The intersection of two worldlines locates an event as clearly and sharply as the intersection of two straws specifies the place of a dust speck in a great barn full of hay (Figure 5-13). To say that an event marks a collision between two particles is identification enough. The worldlines of those two particles are rooted in the past and stretch out into the future. They have a rich texture of connections with nearby worldlines. The nearby worldlines in turn are linked in a hundred ways with

Events and worldlines exist independent of any reference frame

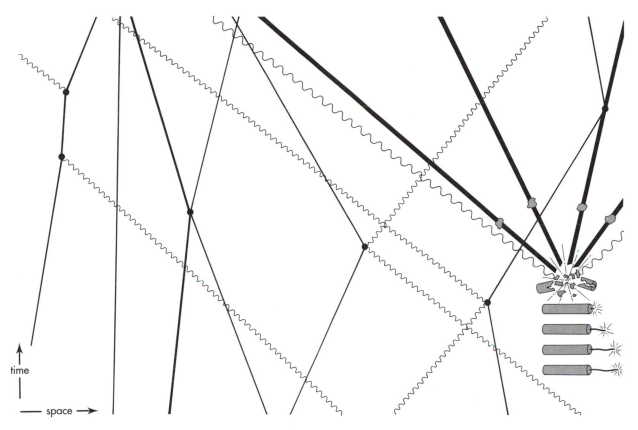

FIGURE 5-13. *The crossing of straws in a barn full of hay is a symbol for the worldlines that fill up spacetime. By their crossings and jogs, these worldlines mark events with a uniqueness beyond all need of reference frames. Straight worldlines track particles with mass; wiggly worldlines trace photons. Typical events symbolized in the map (black dots) from left to right: absorption of a photon; reemission of a photon; collision between a particle and a particle; collision between a photon and another particle; another collision between a photon and a particle; explosion of a firecracker; collision of a particle from outside with one of the fragments of that firecracker.*

worldlines more remote. How then does one tell the location of an event? Tell first what worldlines thread the event. Next follow each of these worldlines. Name additional events that they encounter. These events pick out further worldlines. Eventually the whole barn of hay is cataloged. Each event is named. One can find one's way as surely to a given intersection as the London dweller can pick her path to the meeting of St. James's Street and Piccadilly. No numbers giving space and time location of an event in a given reference frame. No reference frame at all!

Most streets in Japan have no names and most houses no numbers. Yet mail is delivered just the same. Each house is named after its senior occupant, and everyone knows how the streets interconnect these named houses. Now print the map of Japanese streets on a rubber sheet and stretch the sheet this way and that. The postal carrier is not fooled. Each house has its unique name and the same interconnections with neighbor houses as on the unstretched map. So dispense with all maps! Replace them with a catalog or directory that lists each house by name, notes streets passing the house, and tabulates the distance to each neighboring house along the streets.

Similarly, the visual pattern of event dots on a spacetime map (spacetime diagram) and the apparent lengths of worldlines that connect them depend on the reference frame from which they are observed (for example, compare alternative spacetime maps of the same worldline shown in the figure in Sample Problem 5-2). However, each named event is the same for every observer: the event of your birth is unique to you and to everyone connected with you. Moreover, the segment of a worldline that

Locate house at intersection of streets

Locate event at intersection of worldlines

connects one event with the next has a unique magnitude — the interval or proper time — also the same for every observer. Therefore dispense with reference frames altogether! Replace them with a catalog or directory that lists each event by name, notes each worldline that threads the event, and tabulates the interval that connects the event with the next event along each worldline. With this directory in hand we can say precisely how all events are interconnected with each other and which events caused which other events. That is the essence of science; in principle we need no reference frames.

But reference frames are convenient. We are accustomed to them. Most of us prefer to live on named streets with numbered houses. Similarly, most of us speak easily of space separations between events and time separations between the same events as if space and time separations were unconnected. In this way we enjoy the concreteness of using our latticework of rods and clocks while suffering the provinciality of a single reference frame. So be it! Nevertheless, with worldlines Nature gives us power to relate events — to do science — without reference frames at all. ✒

Events and worldlines alone can describe Nature

5.10 SUMMARY

straighter worldline? greater aging!

Events? Yes. Each event endowed with its own location in that great fabric we call spacetime? Yes. But time? No point in all that fabric displays any trace of anything we can identify with any such thing as the "time" of that event. Label that event with a "time" anyway? Sure. No one can stop us. Moreover, such labeling often proves quite useful. But it is *our* labeling! A different reference frame, a different wristwatch brought to that event along a different worldline yields a different time label for that event.

For our own convenience, then, we plot events on a **spacetime map (spacetime diagram)** for a particular free-float frame and its latticework of rods and clocks. This map can be printed on the page of a book if events are limited to one line in space. Distance along this line is plotted horizontally on the spacetime map, with time of the event plotted vertically (Section 5.1). The time and space values of an event are measured with respect to a common **reference event,** plotted at the origin of the spacetime map. The invariance of the interval: $(\text{interval})^2 = (\text{time})^2 - (\text{distance})^2$ between an event and the reference event corresponds to the equation of a **hyperbola,** the same hyperbola as plotted on the spacetime map of every overlapping free-float frame. The event point lies somewhere on the same **invariant hyperbola** as plotted on every one of these spacetime maps (Sections 5.2 and 5.3).

Billions of events sparkle like sand grains scattered over the spacetime map. A given event is unconnected to most other events on the map. Here we pay attention to particular strings of events that are connected. The **worldline** of a particle connects in sequence events that occur at the particle (Section 5.4). The "length" of a worldline between an initial and a final event is the elapsed time measured on a clock carried along the worldline between the two events (Section 5.6). This is called the proper time, wristwatch time, or aging along this worldline. The lapse of proper time is given the symbol τ, in contrast to the symbol t for the frame time read on the latticework clocks in a given free-float frame.

Carry a wristwatch (or grow old!) along a worldline: This is one way to measure the total proper time along it from some initial event (such as the birth of a person or a particle) to some final event (such as death of a person or annihilation of a particle). This method is direct, experimental, simple. A second method? Calculate the interval between each pair of adjacent events that make up the worldline, and then add up all these intervals, assuming that each tiny segment is short enough to be considered straight. This method seems more bothersome and detailed, but it can be carried out by the observer in *any* free-float frame. All such observers will agree with one another — and with the clock-carrier — on the value of the total proper time from the initial event to the final event on the worldline (Section 5.6).

Among all possible worldlines between two given events, the straight line is the worldline of **maximal aging.** This is the actual worldline followed by a free particle that travels from one of these two events to the other (Section 5.6).

As measured in a given free-float frame, the **stretch factor** $= 1/(1 - v^2)^{1/2}$ equals the ratio of elapsed frame time t to elapsed proper time τ along a segment of worldline in which the particle moves with speed v in that frame. The stretch factor is also the Lorentz contraction factor (Section 5.8): Locate, at the same time, the front and back ends of an object moving in a given free-float frame. These end locations will be $(1 - v^2)^{1/2}$ as far apart in that frame as they are in a frame in which the object is at rest.

Worldlines connect events. Like events, they exist independent of any reference frame. In principle, worldlines allow us to relate events to one another — to do science — without using reference frames at all (Section 5.9).

REFERENCES

Newton quotation toward the end of Section 5.8: Sir Isaac Newton, *Mathematical Principles of Natural Philosophy and His System of the World (Philosophiae Naturalis Principia Mathematica),* Joseph Streater, London, July 5, 1686; translated from Latin — the scholarly language of Newton's time — by Andrew Motte in 1729, revised and edited by Florian Cajori and published in two paperback volumes (University of California Press, Berkeley, 1962).

Section 5.9 uses slightly modified passages from Charles W. Misner, Kip S. Thorne, and John Archibald Wheeler, *Gravitation* (W.H. Freeman, New York, 1973), pages 5 – 8. Figure 5-13 is taken directly from this reference, its caption slightly altered from the original.

CHAPTER 5 EXERCISES

PRACTICE

5-1   more is less

The spacetime diagram shows two alternative worldlines from event 1 to event 4. The table shows coordinates of numbered events in this frame. Time and space are measured in years.

a   One traveler moves along the solid segmented worldline from event A to events B, C, and D. Calculate the time increase on his wristwatch (proper clock)

(1) between event A and event B.

(2) between event B and event C.

(3) between event C and event D.

(4) Also calculate the total proper time along worldline A, B, C, D.

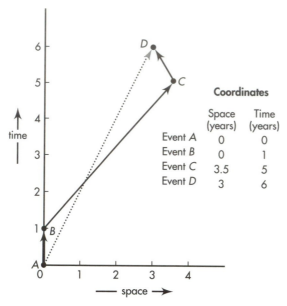

Coordinates

	Space (years)	Time (years)
Event A	0	0
Event B	0	1
Event C	3.5	5
Event D	3	6

EXERCISE 5-1. *Two alternative worldlines between initial event A and final event D.*

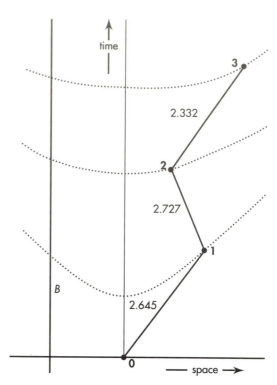

Event Coordinates in the Laboratory Frame

Event 0	$x = 0$	$t = 0$
Event 1	$x = 3.000$	$t = 4.000$
Event 2	$x = 1.750$	$t = 7.000$
Event 3	$x = 5.000$	$t = 11.000$

EXERCISE 5-2. *Two worldlines as recorded in the laboratory frame. Numbers on the segmented worldline are proper times along each straight segment.*

b His twin sister moves along the straight dotted worldline from event A directly to event D. Calculate the time increase on her wristwatch between events A and D.

c Which twin (solid-line or dotted-line traveler) is younger when they rejoin at event D?

5-2   transforming worldlines

The laboratory spacetime diagram in the figure shows two worldlines. One, the vertical line labeled B, is the worldline of an object that is at rest in this frame. The other, the segmented line that connects events 0, 1, 2, and 3, is the worldline of an object that moves at different speeds at different times in this frame. The proper time is written on each segment and invariant hyperbolas are drawn through events 1, 2, and 3. The event table shows the space and time locations in this frame of the four events 0, 1, 2, and 3.

a Trace the axes and hyperbolas onto a blank piece of paper. Sketch a qualitatively correct spacetime diagram for the same pair of worldlines observed in a frame in which the particle on the segmented worldline has zero velocity between event 1 and event 2.

b What is the velocity, in this new frame, of the particle moving along worldline B?

c On each straight portion of the segmented worldline for this new frame write the numerical value of the interval between the two connected events.

5-3   mapmaking in spacetime

Note: Recall Exercise 1-6, the corresponding mapmaking exercise in Chapter 1.

Here is a table of timelike intervals between events, in meters. The events occur in the time sequence ABCD in all frames and along a single line in space in all frames. (They do *not* occur along a single line on the spacetime map.)

INTERVAL to event	A	B	C	D
from event				
A	0	1.0	3.161	5.196
B		0	2.0	4.0
C			0	2.0
D				0

a Use a ruler and the hyperbola graph to construct a spacetime map of these events. Draw this map

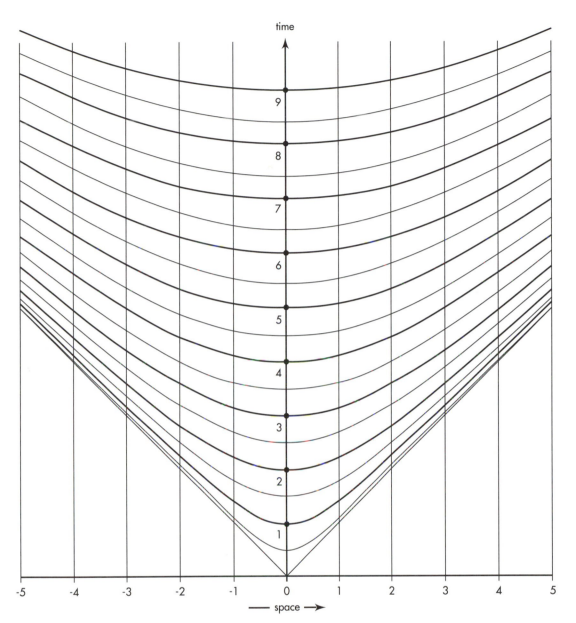

EXERCISE 5-3. *Template of hyperbolas for converting intervals into a spacetime map.*

on thin paper so you can lay it over the hyperbola graph and see the hyperbolas.

Discussion: How to start? With three arbitrary decisions! (1) Choose event *A* to be at the origin of the spacetime map. (2) Choose event *B* to occur at the same place as event *A*. That is, event point *B* is located on the positive time axis with respect to event point *A*. After plotting *B*, use your ruler to draw this straight time axis through event points *A* and *B*. Keep this line parallel to the vertical lines on the hyperbola graph in all later constructions. (3) Even with these choices, there are two spacetime locations (x, t) at which you can locate the event point *C*; choose either of these two

spacetime locations arbitrarily. Then go on to plot event *D*.

Analogy to surveying: In surveying (using Euclidean geometry) you locate all points a given distance from some stake by using that stake as origin and drawing a circle of radius equal to the desired distance. In a spacetime map (using Lorentz geometry) you locate all event points a given interval from some event by using that event point as origin and drawing a hyperbola with nearest point equal to the desired interval.

b Now take a new piece of paper and draw a spacetime map for another reference frame. Choose event *D* to be at the origin of the spacetime map. This means that all other events occur before *D*. Hence turn the hyperbola plot upside down, so that the hyperbolas open downward. Choose event *B* to occur at the same place as *D*. Now find the locations of *A* and *C* using the same strategy as in part **a**.

c Find an approximate value for the relative speed of the two frames for which you have made spacetime plots.

d Hold one of your spacetime maps up to the light with the marks on the side of the paper facing the light. Does the map you see from the back also satisfy the table entries?

PROBLEMS

5-4 the pole and barn paradox

A worried student writes, "Relativity must be *wrong*. Consider a 20-meter pole carried so fast in the direction of its length that it appears to be only 10 meters long in the laboratory frame of reference. Let the runner who carries the pole enter a barn 10 meters long, as shown in the figure. At some instant the farmer can close the front door and the pole will be entirely enclosed in the barn. However, look at the same situation from the frame of reference of the runner. To him the barn appears to be contracted to half its length. How can a 20-meter pole possibly fit into a 5-meter barn? Does not this unbelievable conclusion prove that relativity contains somewhere a fundamental logical inconsistency?"

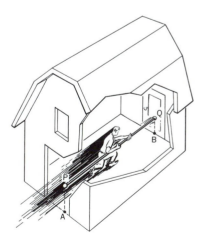

EXERCISE 5-4. *Fast runner with "20-meter" pole enclosed in a "10-meter" barn. In the next instant he will burst through the back door, which is made of paper.*

Write a reply to the worried student explaining clearly and carefully how the pole and barn are treated by relativity without internal contradiction. Use the following outline or some other method.

a Make two carefully labeled spacetime diagrams, one an *xt* diagram for the barn rest frame, the other an *x't'* diagram for the runner rest frame. Referring to the figure, take the event "Q coincides with A" to be at the origin of both diagrams. In both plot the worldlines of A, B, P, and Q. Pay attention to the scale of both diagrams. Label both diagrams with the time (in meters) of the event "Q coincides with B" (derived from Lorentz transformation equations or otherwise). Do the same for the times of events "P coincides with A" and "P coincides with B."

b Discussion question: Suppose the barn has no back door but rather a back wall of steel-reinforced concrete. What happens after the farmer closes the front door on the pole?

c Replace the pole with a line of ten tennis balls the same length as the pole and moving together with the same velocity as the pole. The farmer's ten children line up inside the barn, and each catches and stops one tennis ball at the same time as the farmer closes the front door of the barn. Describe the stopping events as recorded by the observer riding on the last tennis ball. Plot them on your two diagrams.

5-5 radar speed trap

A highway patrolman aims a stationary radar transmitter backward along the highway toward oncoming traffic. A detector mounted next to the transmitter analyzes the radar wave reflected from an approaching car. An internal computer uses the shift in frequency of the reflected wave to reckon and display the car's speed. Analyze this shift in frequency as in parts **a – e** or with some other method. Treat the car as a simple mirror and assume that the radar signals move back and forth along one line on the highway. Radar is an electromagnetic wave that moves with the speed of light.

The figure shows the worldline of the car, worldlines of two adjacent maxima of the radar wave, and the wavelength λ of incident and reflected waves.

a From the 45-degree right triangle *ABC*, show that

$$\Delta t = v\Delta t + \lambda_{\text{reflected}}$$

From the 45-degree right triangle *DEF*, show that

$$\Delta t = \lambda_{\text{incident}} - v\Delta t$$

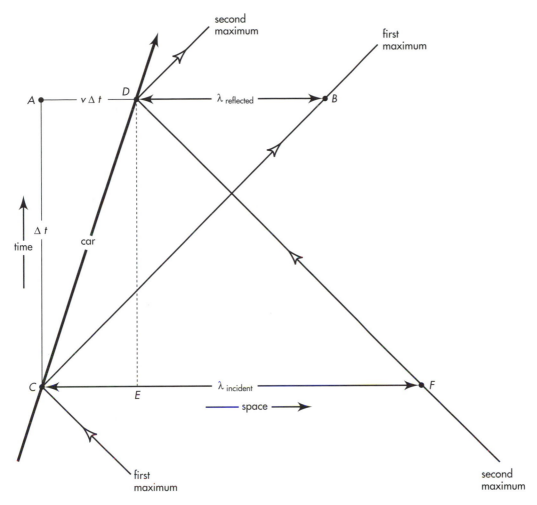

EXERCISE 5-5. *Worldlines of approaching car and two radar wave maxima that reflect from the car. The speed of the car is greatly exaggerated.*

Eliminate Δt from these two equations to find an expression for $\lambda_{\text{reflected}}$ in terms of $\lambda_{\text{incident}}$ and the automobile speed v.

b The frequency f of radar (in cycles/second) is related to its wavelength λ in a vacuum by the formula $f = c/\lambda$, where c is the speed of light ($\approx$ the speed of radar waves in air). Derive an expression or frequency $f_{\text{reflected}}$ of the reflected radar signal in terms of frequency f_{incident} of the incident wave and the speed v of the oncoming automobile. Show that the result is

$$f_{\text{reflected}} = \left(\frac{1+v}{1-v}\right) f_{\text{incident}}$$

c For an automobile moving at a speed $v = v_{\text{conv}}/c$ that is a small fraction of the speed of light, assume that the fractional change in frequency of

reflected radar is small. Under this assumption, use the first two terms of the binomial expansion

$$(1-z)^n \approx 1 - nz \text{ for } |z| \ll 1$$

to show that the fractional change of frequency is given by the approximate expression

$$\frac{\Delta f}{f} \approx 2v$$

Substitute the speed of a car moving at 100 kilometers/hour (= 44.7 meters/second $\approx$ 60 miles/hour) and show that your assumption about the small fractional change is justified.

d One radar gun used by the Massachusetts Highway Patrol operates at a frequency of 10.525×10^9 cycles/second. By how many cycles/

second is the reflected beam shifted in frequency when reflected from a car approaching at 100 kilometers/hour?

e What discrimination between different frequency shifts must the unit have if it can distinguish the speed of a car moving at 100 kilometers/hour from the speed of one moving at 101 kilometers/hour?

Reference: T. M. Kalotas and A. R. Lee, *American Journal of Physics*, Volume 58, pages 187–188 (February 1990).

5-6 a summer evening's fantasy

You are standing alone outdoors at dusk on the first day of summer. You see Sun setting due west and the planet Venus in the same direction. On the opposite horizon the full Moon is rising due east. An alien ship approaches from the east and lands beside you. The occupants inform you that they are from Proxima Centauri, which lies due east beyond the rising Moon. They say they have been traveling straight to Earth and that their reduced approach speed within the solar system was such that the time stretch factor gamma during the approach was 5/3.

At the same instant that the aliens land, you see Sun explode. The aliens admit to you that earlier, on their way to Earth, they shot a laser light pulse at Sun, which caused this explosion. They warn that Sun's explosion emitted an immense pulse of particles moving at half the speed of light that will blow away Earth's atmosphere. In confirmation, shortly after the aliens land you notice that the planet Venus, lying in the direction of Sun, suddenly changes color.

You grab a passing human of the opposite sex and plead with the aliens to take you both away from Earth in order to establish the human gene pool elsewhere. They agree and set the dials to flee in an easterly direction away from Sun at top speed, with time stretch factor gamma of 25/7. The takeoff is to be 7 minutes after the alien landing on Earth.

Do you make it?

Draw a detailed Earth spacetime diagram showing the events and worldlines of this story. Use the following information.

- Sun is 8 light-minutes from Earth.

- Venus is 2 light-minutes from Earth.

- Assume that Sun, Venus, Earth, and Moon all lie along a single direction in space and are relatively at rest during this short story. The incoming and outgoing paths of the alien ship lie along this same line in space.

- All takeoffs and landings involve instantaneous changes from initial to final speed.

- $5^2 - 3^2 = 4^2$ and $(25)^2 - (7)^2 = (24)^2$

a Plot EVENTS labeled with the following NUMBERS.
0. your location when the aliens land (at the origin)
1. Sun explodes
2. light from Sun explosion reaches you
3. Venus's atmosphere blown away
4. light from event 3 reaches you
5. you and aliens depart Earth (you hope!)
6. Earth atmosphere blown away

b Plot WORLDLINES labeled with the following CAPITAL LETTERS.
A. your worldline
B. worldline of Earth
C. aliens' worldline
D. worldline of Sun
E. worldline of Venus
F. worldline of light from Sun's explosion
G. worldline of the "speed-one-half" pulse of particles from Sun's explosion
H. worldline of light emitted when Venus loses atmosphere
J. terminal part of the worldline of the laser cannon pulse fired at Sun by the aliens

c Write numerical values for the speed $v = v_{conv}/c$ on every segment of all worldlines.

5-7 the runner on the train paradox

A letter sent to the Massachusetts Institute of Technology by Hsien-Yen Tsao of Los Angeles poses the following paradox, which he asserts disproves the theory of relativity. The Chairman of the Physics Department sends the inquiry along to you, asking you to respond to Mr. Tsao. You determine to make the answer clear, concise, decisive, and polite—a personal test of your diplomacy and grasp of relativity.

The setting: A train travels at high speed. A runner on the train sprints toward the back of the train with the same speed (with respect to the train) as the train moves forward (with respect to Earth). Therefore the runner is not moving with respect to Earth.

The paradox: We know that, crudely speaking, clocks on the train run "slow" compared to the Earth clock. We also know that the runner's clock runs "slow" compared to the train clocks. Therefore the runner's clock should run "doubly slow" with respect to the Earth clock. *But the runner is not moving with respect to Earth!* Therefore the runner's clock must run at the same rate as the Earth clock. How can it possibly be that the runner's clock runs *"doubly slow"* with respect to the Earth clock and also runs *at the same rate* as the Earth clock?

5-8   the twin paradox put to rest—a worked example

Motto: The swinging line of simultaneity tells all!

Combine the Lorentz transformation with the space-time diagram to clear up—once and for all!—the solution to the Twin Paradox. An astronaut travels from Earth to Canopus (Chapter 4) at speed $v_{rel} = 99/101$, arriving at Canopus $t' = 20$ years later according to her rocket clock, $t = 101$ years later according to Earth-linked clocks—which means that the stretch factor γ has the value $101/20$.

The key idea is "lines of simultaneity" (boxed labels in the figure). A line of simultaneity connects events that occur "at the same time." But events simultaneous in the Earth ("laboratory") frame are typically not simultaneous in the rocket frame (Section 3.4). *Horizontal* is the line of simultaneity on the Earth ("laboratory") spacetime map that connects events occurring at the same time in the Earth frame. *Totally different*—not a horizontal line!—is a line of simultaneity on the Earth spacetime map that connects events simultaneous in the outgoing astronaut frame. To draw this line of outgoing-astronaut simultaneity, start with the inverse Lorentz transformation equation for time:

$$t' = -v_{rel}\gamma x + \gamma t$$

For the outgoing astronaut, $v_{rel} = 99/101$ and $\gamma = 101/20$. We want the line of simultaneity that passes through turnaround event T. So let $t' = 20$ years. Then:

$$20 = -(99/101)(101/20)x + (101/20)\, t$$

Multiply through by $20/101$:

$$400/101 = -(99/101)x + t$$

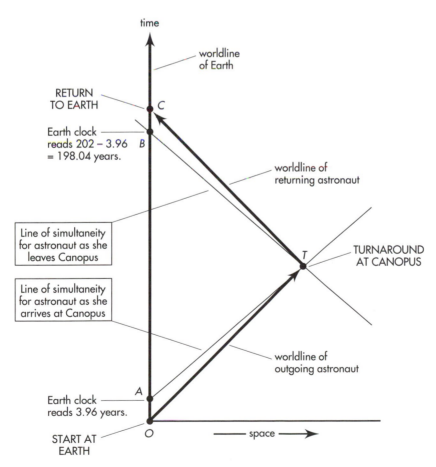

EXERCISE 5-8. *Earth spacetime map of the trip to Canopus and back. As the astronaut arrives at Canopus, her colleagues in her outgoing reference frame record along line AT events simultaneous with this arrival, including Earth-clock reading of 3.96 years at A. At Canopus the astronaut changes frames, thus changing the line of simultaneity, which swings to BT. As she leaves Canopus, her new colleagues take an Earth-clock reading of 198.04 years at B. At turnaround, the ticks on the Earth clock along worldline segment AB go from the outward-moving astronaut's future to the incoming astronaut's past.*

which yields

$$t = 0.980\,x + 3.96$$

This is the equation for a straight line passing through event points A and T in the spacetime diagram. It is the line of simultaneity for the outgoing astronaut, connecting all events simultaneous with the arrival of the rocket at Canopus (simultaneous in that frame). Among these events is event A, the Earth clock reading of 3.96 years, which occurs at Earth position $x = 0$. In brief, at the moment the rocket arrives at Canopus, the Earth clock reads 3.96 years as observed in the outgoing rocket frame.

Now the astronaut jumps to the incoming rocket frame. This reverses the velocity of the astronaut with respect to the Earth-linked frame — *and so reverses the slope of the line of astronaut simultaneity.* This new line of astronaut simultaneity passes through event points B and T in the figure. Event B is the Earth clock reading of $202 - 3.96 = 198.04$ years.

To go back over the astronaut trip while looking at the spacetime map is (finally!) to solve the Twin Paradox. As the astronaut travels outward toward Canopus, many colleagues follow her at the same speed, with clocks synchronized in her frame. As they whiz past Earth, each records the reading on the Earth clock. Later analysis leads them to agree that the time between ticks of Earth's clock is longer than the time between ticks of their own outward-moving clocks. (They say, "The Earth clock runs slow.") At any event point on her outward worldline, the astronaut's line of simultaneity slopes upward to the right in the Earth spacetime diagram, as shown in the figure. Simultaneous with astronaut arrival at Canopus (event T, when *all* outward-moving clocks read 20 years), one of her colleagues reads a time 3.96 years on the Earth clock (event A).

Now the astronaut jumps from the outward-moving rocket to a returning rocket. She inherits a *completely new set of colleagues,* with a new set of synchronized clocks. The astronaut's new line of simultaneity slopes upward to the left in the Earth spacetime diagram. Simultaneous with her departure from Canopus (event T, when *all* inward-moving clocks read 20 years), one of her new colleagues reads a time $202 - 3.96 = 198.04$ years on the passing Earth clock (event B). Thereafter new colleague after new colleague streaks past Earth, recording the fact that Earth clock ticks are farther apart in time than the ticks on their own clocks. (They say, "The Earth clock runs slow.").

The analysis so far accounts for the short time segments OA and BC recorded by the Earth clock on its vertical worldline AC. What about the omitted time lapse AB? This is recorded, sure enough, by the Earth clock plowing forward along worldline OC in its comfortable single free-float frame. However, the story of time AB is quite different for the turn-around astronaut. Before she reaches turnaround at T, events on line AB are *in her future.* All those Earth clock ticks are yet to be recorded by her outgoing colleagues. These events lie *above* her line of simultaneity BT as she arrives at Canopus at T. However, as she turns around, her line of simultaneity also slews forward, swinging from line AT to line BT. Suddenly the events on line AB — all those intermediate ticks of the Earth clock — are in the astronaut's *past*. These events lie *below* the line of simultaneity BT as she starts back at T. Her outward-moving colleague reads 3.96 years on the Earth clock as she reaches Canopus; an instant later on her clock, her new inward-moving colleague reads 198.04 on the Earth clock.

Shall we say that the Earth clock "jumps ahead" as the astronaut turns around? No! Utterly ridiculous! For what single observer does it jump ahead? Not for the Earth observer. Not for the outgoing set of clock-readers. Not for the returning set of clock readers. For whom then? Nobody! *At the same time as she reaches Canopus* — old meaning of simultaneous! — the astronaut's outgoing colleague records 3.96 years for the Earth clock. *At the same time as she leaves Canopus* — new meaning of simultaneous! — her new ingoing colleague records 198.04 years on the Earth clock. The astronaut has nobody but herself to blame for her misperception of a "jump" in the Earth clock reading.

The "lost Earth time" AB in the figure makes consistent the story each observer tells about the clocks. Simple is the story told by the Earth observer: "My clock ticked along steadily at the 'proper' rate from astronaut departure to astronaut return. In contrast, ticks on the astronaut clock were far apart in time on both the outgoing and incoming legs of her trip. We agree that her total ticks are less than my total ticks: she is younger than I when we meet again." More complicated is the astronaut account of clock behavior: "Ticks on the Earth clock were far apart in time as I traveled to Canopus; ticks on the Earth clock were also far apart as I traveled home again. But as I turned around, a whole bunch of Earth clock ticks went from my future to my past. This accounts for the larger number of total ticks on the Earth clock than on my clock during the trip. We agree that I am younger when we meet again."

So saying, the astronaut renounces her profession and becomes a stand-up comedian.

Reference: E. Lowry, *American Journal of Physics,* Volume 31, page 59 (1963).

CHAPTER 6

REGIONS OF SPACETIME

6.1 LIGHT SPEED: LIMIT ON CAUSALITY

no signal reaches us faster than light

Nine-year-old Meredith waves her toy magician's wand and shouts, "Sun is exploding right now!" Is she right? We have no way on Earth of knowing — at least not for a while. Sun lies 150,000 million meters from Earth. Therefore it will take 150,000 million meters of light-travel time for the first light flash from the explosion to reach us. This equals 500 seconds — 8 minutes and 20 seconds. We will just have to wait and see if Meredith is correct . . .

Signal Sun with super speed?

When 8 minutes and 20 seconds pass, we have evidence that Meredith was mistaken: Looking through our special dark glasses, we see no exploding Sun.

But Meredith's wand has started us thinking. What in the laws of nature prohibits the wave of her wand from being the signal for Sun to explode at that same instant? Or — more reasonably, given the awesome event — what prevents Meredith from having instantaneous warning, so that she raises her wand simultaneously with Sun's explosion in order to give us (in light of later developments) a false impression of her power?

Both questions have the same answer: "The speed of light." Whatever her powers, Meredith cannot affect Sun in less than 500 seconds; neither can a warning signal reach us from Sun in less time than that. All during that intervening 500 seconds we would see the accustomed round shape of Sun, apparently healthy as ever.

No, just speed of light

More generally, one event cannot cause another when their spatial separation is greater than the distance light can travel in the time between these events. Light speed sets a limit on causality. No known physical process can overcome this limit: not gravity, not some other field, not a zooming particle of any kind. "Spacetime interval" quantifies this limit on causality. Interval between far-away events — unlike distance between far-away points — can be zero. In this and other ways the spacetime geometry of the real world differs fundamentally from the space geometry of Euclid's 2300-year-old textbook. ✒

171

6.2 RELATION BETWEEN EVENTS: TIMELIKE, SPACELIKE, OR LIGHTLIKE

minus sign yields three possible relations between pairs of events

Using Euclidean geometry, a surveyor reckons the distance between two steel stakes from the sum of the squares of the northward and eastward separations of these stakes:

Squared distance: Positive or zero

$$(\text{distance})^2 = (\text{northward separation})^2 + (\text{eastward separation})^2$$

In consequence, in Euclidean geometry a distance—or its square—always has a positive value or zero.

In contrast, the spacetime interval between events in Lorentz geometry arises from the *difference* of squares of time and space separations:

Squared interval: Positive, zero, or negative

$$(\text{interval})^2 = (\text{separation in time})^2 - (\text{separation in space})^2$$

In consequence of the minus sign, this equation yields a number that may be positive, negative, or zero, depending on whether the time or the space separation predominates. Moreover, whichever of these three descriptions characterizes the interval in one free-float frame also characterizes the interval in any other free-float frame. Why? Because the spacetime interval between two events has the same value in all overlapping free-float frames. In the threefold possibilities for an interval, nature reveals the causal relation between events.

An interval between two events earns the name **timelike** or **spacelike** or **lightlike** depending on whether the time part predominates, the space part predominates, or the time and space parts are equal, respectively, as shown in Table 6-1. For convenience, the minus sign is placed so that the resulting squared interval is greater than or equal to zero.

Timelike Interval: We picture the sequence of sparks emitted by a moving sparkplug. Points representing these sparks on the spacetime map trace out the worldline of the particle (Chapter 5). No material particle has ever been measured to travel faster than light. Every material particle always travels *less* than one meter of distance in one meter of light-travel time. The sparks emitted by the particle have a greater time separation than their separation in space. In other words, the worldline of a particle consists of events that have a timelike relation with one another and with the initial event. We say that a material particle follows a **timelike worldline**.

Timelike interval: Time part dominates

The interval τ between two timelike events reveals itself to the observer in any free-float frame:

$$(\text{timelike interval})^2 = \tau^2 = (\text{time separation})^2 - (\text{space separation})^2 \qquad \text{(6-1)}$$

$\langle\!\langle$ **TABLE 6-1** $\rangle\!\rangle$

CLASSIFICATION OF THE RELATION BETWEEN TWO EVENTS

Description	Squared interval is named and reckoned
Time part of interval dominates space part	$(\text{timelike interval})^2 = \tau^2 = (\text{time})^2 - (\text{distance})^2$
Space part of interval dominates time part	$(\text{spacelike interval})^2 = s^2 = (\text{distance})^2 - (\text{time})^2$
Time part of interval equals space part	$(\text{lightlike interval})^2 = 0 = (\text{time})^2 - (\text{distance})^2$

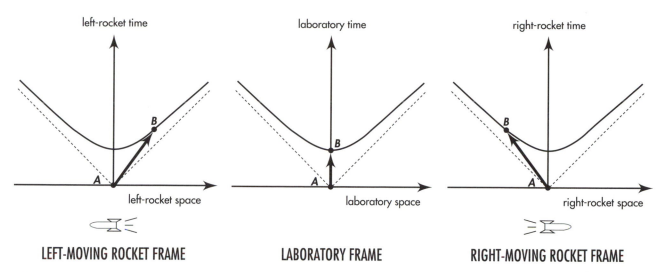

LEFT-MOVING ROCKET FRAME   LABORATORY FRAME   RIGHT-MOVING ROCKET FRAME

FIGURE 6-1. *Events* A *and* B *form a timelike pair (with event* A *arbitrarily chosen as reference event), here recorded in the spacetime maps of three free-float frames. Point* B *lies on a hyperbola opening along the time axis in each frame. The shortest* time *between events* A *and* B *is recorded in the laboratory frame, the frame in which the two events occur at the same place.*

Same two sparks registered in different frames? Different records for the separation in time between those sparks. Different records for the separation in space. *Same* figure for the *timelike interval* between them!

Nobody can keep us from tracing out on one and the same diagram (Figure 6-1) the very different records for the separation *AB* that observers get in different free-float frames. One frame? One point on the diagram. Another frame? Another point on the diagram. And so on. These many records for the same pair of events *AB* trace out a hyperbola. This hyperbola opens out in the time direction.

The two sparks, *A* and *B*—definite locations though they occupy in spacetime—nevertheless register in different frames of reference as having different separations in reference-frame time. Among the many conceivable frames, which one records this separation in time as smallest? Answer: The frame in which spark *B* occurs at the same *place* as spark *A*. In other words, the frame that happens to move along in sync with the sparkplug, even if only briefly. In that frame the clock records a separation in time between *A* and *B* identical with the timelike interval *AB*.

As seen in the left-moving rocket frame in Figure 6-1, spark *B* lies to the right of spark *A*. In contrast, spark *B* occurs to the left of spark *A* in the right-moving rocket. The position of *B* relative to *A* depends on the reference frame from which it is measured. For a pair of events separated by a timelike interval, labels "right" and "left" have no invariant meaning: they are frame-dependent.

Spacelike Interval: The interval between two events *A* and *D* is spacelike when the space part predominates over the time part. Such was the case for a possible explosion of Sun (event *A*) and Meredith's wand waving (event *D*), simultaneous with *A* as recorded in the Earth frame (Section 6.1). Events *A* and *D*, if they occurred, would be separated in the Earth–Sun frame by a distance of 150,000 million meters and separated by a time of zero meters. Clearly the space part predominates over the time part! Whenever the space part predominates, we call the relation between the two events *spacelike*.

The interval *s* (sometimes called by the Greek letter sigma, σ) between two spacelike events reveals itself to the observer in any free-float frame:

$$(\text{spacelike interval})^2 = s^2 = (\text{space separation})^2 - (\text{time separation})^2 \qquad (6\text{-}2)$$

Timelike interval:
Invariant hyperbola opens
along time axis

Spacelike interval:
Space part dominates

Spacelike interval:
Invariant hyperbola opens
along space axis

Events A and D registered in different frames? Then different records for the separation in time between those events. Also different records for the separation in space. *Same* numerical value for the *spacelike interval* between them!

We plot on another spacetime diagram (Figure 6-2) all of the very different records for the separation AD that observers get in different free-float frames. One frame? One point on the diagram. Another frame? Another point on the diagram. And so on. These many records for the same pair of events AD trace out a hyperbola. This hyperbola opens out in the space direction.

The two events, A and D—definite locations though they occupy in spacetime—nevertheless register in different frames of reference as having different separations in reference-frame space. Among the many conceivable frames, which one records this separation in space as smallest? Answer: The frame in which spark D occurs at the same *time* as spark A. In that frame a long stick records a separation in space between A and D identical with the spacelike interval, AD. This is called the **proper distance** between the two spacelike events.

In the Earth–laboratory frame in Figure 6-2, Meredith waves her wand (event D) at the same time as Sun explodes (event A). In the right-moving rocket frame Sun explodes *after* Meredith waves her wand. In the left-moving rocket frame Sun explodes *before* the wand wave. For a pair of events separated by a spacelike interval, labels "before" and "after" have no invariant meaning: they are frame-dependent. To allow the wand to control Sun would be to scramble cause and effect!

No particle—not even a flash of light—can move between two events connected by a spacelike interval. To do so would require it to cover a distance greater than the time available to cover this distance (space separation greater than time separation). In brief, it would have to travel faster than light. This is alternative evidence that two events separated by a spacelike interval cannot be causally connected: one of them cannot "get at" the other one by any possible signal.

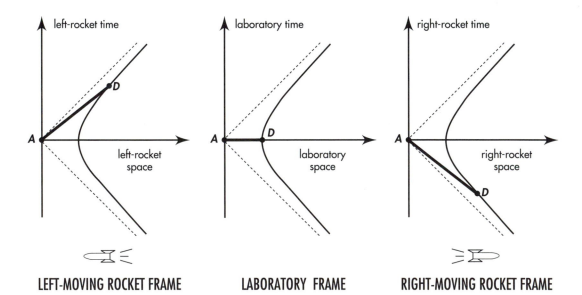

LEFT-MOVING ROCKET FRAME   **LABORATORY FRAME**   **RIGHT-MOVING ROCKET FRAME**

FIGURE 6-2. *The spacelike pair of events* A *and* D *(with event* A *arbitrarily chosen as reference event) as recorded in the spacetime maps of three free-float frames.* Point D *lies on a hyperbola opening along the space axis in every rocket and laboratory frame. The shortest distance between these events is recorded in the laboratory frame, the frame in which the two events occur at the same time. A heavy line represents the spacetime separation* AD. *No particle can travel along this line; the speed would be greater than light speed—and would be infinitely great as measured in the laboratory frame, since the particle would have to cover the distance from* A *to* D *in zero time!*

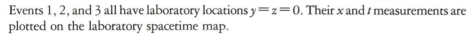

SAMPLE PROBLEM 6-1
RELATIONS BETWEEN EVENTS

Events 1, 2, and 3 all have laboratory locations $y = z = 0$. Their x and t measurements are plotted on the laboratory spacetime map.

a. Classify the interval between events 1 and 2: timelike, spacelike, or lightlike.

b. Classify the interval between events 1 and 3.

c. Classify the interval between events 2 and 3.

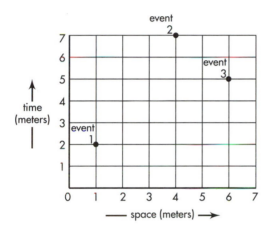

SOLUTION

a. For event 1, $t = 2$ meters and $x = 1$ meter. For event 2, $t = 7$ meters and $x = 4$ meters. The squared interval between them: $(\text{interval})^2 = (7 - 2)^2 - (4 - 1)^2 = 5^2 - 3^2 = 25 - 9 = 16$ (meters)^2. The time part is greater than the space part, so the interval between these two events is *timelike:* $\tau = 4$ meters.

b. For event 1, $t = 2$ meters and $x = 1$ meter. For event 3, $t = 5$ meters and $x = 6$ meters. The squared interval between them: $(\text{interval})^2 = (5 - 2)^2 - (1 - 6)^2 = 3^2 - 5^2 = 9 - 25 = -16$ (meters)^2. The space part is greater than the time part, so the interval is *spacelike:* $s = 4$ meters. (For spacelike intervals, we subtract the squared time part from the squared space part before taking the square root.)

c. For event 2, $t = 7$ meters and $x = 4$ meters. For event 3, $t = 5$ meters and $x = 6$ meters. The squared interval between them: $(\text{interval})^2 = (7 - 5)^2 - (4 - 6)^2 = 2^2 - 2^2 = 4 - 4 = 0$ (meters)^2. The time part equals the space part, so the interval is *lightlike*: it is a null interval.

Lightlike Interval (Null Interval): Two events stand in a lightlike relation when the interval between them is zero:

$$(\text{time separation})^2 - (\text{space separation})^2 = 0$$

or

magnitude of (separation in time) = (distance in space)   [for lightlike interval]   (6-3)

**Lightlike interval:
Time separation equals
space separation**

Lightlike interval:
Plotted along ±45 degree lines

An interval that is lightlike? A separation in time between two events, *A* and *G*, identical to the distance in space between them? What does this condition mean? This: A pulse of light can fly directly from event *A* and arrive with perfect timing at event *G*. How come? Distance in meters between the two locations measures the meters of time *required* for light to fly from one place to the other. Separation in time between the two events represents the time *available* for the trip. Time available equals time needed? Guarantee that the pulse from *A* arrives in coincidence with event *G*! More generally, whenever the influence of one event, spreading out at the speed of light, can directly affect a second event, then the interval between those two events rates as lightlike, zero, null.

Only light ("photons"), neutrinos, and gravitons can move directly between two events connected by a lightlike interval. Only by means of one of these light-speed particles can the one event in a lightlike pair cause the other.

The spherical out-going pulse of light from an event, *A*, may trigger two widely separated events, *E* and *G* (Figure 6-3). Does this common genesis imply that *E* and *G* occur at the same time? Yes and no! Yes, there's always a free-float reference frame in which the two daughter events appear as simultaneous. That frame — for no good reason — we call the laboratory frame in Figure 6-3. In other frames of reference — for example, the left-moving rocket frame in Figure 6-3 — the clocks show that *E* occurs before *G*. There are still other frames — the right-moving rocket frame is one — in which the clocks register *E* and *G* in the opposite order of time. But no frame shows either *E* or *G* in the past of *A*.

Hold it! Aren't spacelike separations impossible? I understand timelike and lightlike separations between two events, because a particle — or at least a light flash — can travel between them. Not even a light flash, however, can travel from one event to a second event separated from the first by an interval that is spacelike. The first event cannot possibly cause the second event in the spacelike case. Therefore a spacelike interval cannot arise in nature. So why talk about it?

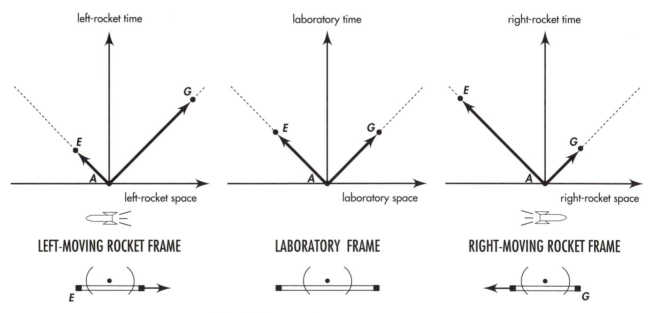

LEFT-MOVING ROCKET FRAME   **LABORATORY FRAME**   **RIGHT-MOVING ROCKET FRAME**

FIGURE 6-3. *Two lightlike pairs of events* AE *and* AG (*with event* A *arbitrarily chosen as reference event*) *as recorded in spacetime maps of three free-float frames. A flash originates at* A *and spreads outward from the center of a rod at rest in the laboratory frame. Events* E *and* G *are receptions of this flash at the two ends of the rod as recorded by different observers. In the laboratory frame, reception events* E *and* G *occur at the same time. In the right-moving rocket frame, the rod moves to the left, so event* G *occurs sooner than event* E. *In the left-moving rocket frame, the rod moves to the right, so event* E *occurs sooner than event* G.

Oops! A spacelike interval between two events certainly can and does arise in nature. Signals from the supernova labeled 1987A reported that event to us in 1987, which was 150,000 years after the explosion occurred. Yet occur it did! No astronomer of Babylonian, Egyptian, or Greek days reported it, nor could they even know of it. Yet it had already happened for them. That event separated itself from each of them by a spacelike interval. Only the advance of time to the year 1987 brought down the interval between that explosion and Earthbound observers from spacelike to lightlike. In that year a light pulse carried the earliest possible report of that explosion to our eyes. And look today? See no explosion at that location in the sky. The light from it has passed us by. Our present relation to that event? Timelike!

6.3 LIGHT CONE: PARTITION IN SPACETIME

invariance of the interval preserves cause and effect

Thus far in dealing with the interval between two events, *A* and *B*, we have considered primarily the situation in which these events lie along a single direction in space — on the reference line where the laboratory and rocket reference clocks are located. In contrast, the surveyors in our imaginary kingdom made use of two space dimensions — northward and eastward. We know, however, that Euclidean space is truly three-dimensional. A surveyor measuring hilly terrain soon appreciates the need for a third dimension: the direction vertically upward! The measure of distance in three dimensions requires a simple extension of the expression for distance in two dimensions: The square of the distance becomes the sum of the squares of *three* mutually perpendicular separations:

Interval generalized to three space dimensions

$$(\text{distance})^2 = (\text{north separation})^2 + (\text{east separation})^2 + (\text{up separation})^2$$

Euclidean space requires three dimensions. In contrast, spacetime, which includes the time dimension, demands four. The expression for the square of a timelike interval now has four terms: a positive term (the square of the time separation) and three negative terms (the squares of the separations in three space dimensions).

$$(\text{interval})^2 = (\text{time separation})^2 - (\text{north separation})^2$$
$$- (\text{east separation})^2 - (\text{up separation})^2$$

The three space terms can be represented by the single distance term in the equation above, yielding

$$(\text{timelike interval})^2 = (\text{time separation})^2 - (\text{distance})^2$$
$$(\text{spacelike interval})^2 = (\text{distance})^2 - (\text{time separation})^2$$
$$(\text{lightlike interval})^2 = 0 = (\text{time separation})^2 - (\text{distance})^2$$

or, for the lightlike interval,

$$\text{magnitude of (separation in time)} = (\text{distance in space}) \quad \text{[lightlike interval]} \quad (6\text{-}3)$$

For pairs of events with lightlike separation, the interval equals zero. The zero interval is a unique feature of Lorentz geometry, new and quite different from

SAMPLE PROBLEM 6-2

EXPLETIVE DELETED

At 12:00 noon Greenwich Mean Time (GMT) an astronaut on Moon drops a wrench on his toe and shouts "Damn!" into his helmet microphone (event A), carried by a radio signal toward Earth. At one second after 12:00 noon GMT a short circuit (event D) temporarily disables the receiving amplifier at Mission Control on Earth. Take Earth and Moon to be 3.84×10^8 meters apart in the Earth frame and assume zero relative motion.

a. Does Mission Control on Earth hear the astronaut's expletive?

b. Could the astronaut's strong language have caused the short circuit on Earth?

c. Classify the spacetime separation between events A and D: timelike, spacelike, or lightlike.

d. Find the proper distance or proper time between events A and D.

e. For all possible rocket frames passing between Earth and Moon, find the shortest possible distance between events A and D. In the rocket frame for which this distance is shortest, determine the time between the two events.

SOLUTION

a. In one second, electromagnetic radiation (light and radio waves) travels 3.0×10^8 meters in a vacuum. Therefore the radio signal does not have time to travel the 3.84×10^8 meters between Moon and Earth in the one second available between the events A and D as measured in the Earth frame. So Mission Control does not hear the exclamation.

b. No signal travels faster than light. So the astronaut's strong language cannot have caused the short circuit.

c. The space part of the separation between events (3.84×10^8 meters) dominates the time part (one second $= 3.0 \times 10^8$ meters). Therefore the separation is spacelike.

d. The square of the proper distance s comes from the expression

$$\begin{aligned}
s^2 &= (\text{space separation})^2 - (\text{time separation})^2 \\
&= (3.84 \times 10^8 \text{ meters})^2 - (3.00 \times 10^8 \text{ meters})^2 \\
&= (14.75 - 9.00) \times 10^{16} \text{ (meters)}^2 \\
&= 5.75 \times 10^{16} \text{ (meters)}^2
\end{aligned}$$

The proper distance equals the square root of this value: $s = 2.40 \times 10^8$ meters

e. The proper distance equals the shortest distance between two spacelike events as measured in any rocket frame moving between them (Figure 6-2, laboratory map). Hence 2.40×10^8 meters equals the shortest possible distance between events A and D. In the particular rocket frame for which the distance is shortest, the time between the two events has the value zero—events A and D are simultaneous in this frame.

SAMPLE PROBLEM 6-3

SUNSPOT

Bradley grabs his sister's wand and waves it, shouting "Sunspot!" At that very instant his father, Lloyd, who is operating a home solar observatory, sees a spot appear on the face of Sun. Let event E be Bradley waving the wand and event A be eruption of the sunspot at the surface of Sun itself. The Earth–Sun distance equals approximately 1.5×10^{11} meters. Neglect relative motion between Earth and Sun.

a. Is it possible that Bradley's wand waving caused the sunspot to erupt on Sun?

b. Is it possible that the sunspot erupting on Sun caused Bradley to wave his wand?

c. Classify the spacetime separation between events A and E: timelike, spacelike, or lightlike.

d. Find the value of proper distance or proper time between events A and E.

e. For all possible rocket frames passing between Earth and Sun, find the shortest possible distance or the shortest possible time between events A and E.

SOLUTION

a. Light travels 1 meter of distance in 1 meter of time — or 1.5×10^{11} meters of distance in 1.5×10^{11} meters of time. Hence in the Earth-Sun frame, eruption of the sunspot (event A) occurred 1.5×10^{11} meters of time *before* Bradley waved the wand (event E). So Bradley's wand waving could not have caused the eruption on Sun.

b. On the other hand, it is possible that eruption of the sunspot caused Bradley to wave his wand: He raises the wand in the air, looks over his father's shoulder, and waves the wand as the spot appears on the projection screen. (We neglect his reaction time.)

c. Events A and E are connected by one light pulse; their space and time separations both have the value 1.5×10^{11} meters in the Earth frame. Therefore the spacetime separation between them is lightlike.

d. Space and time separations between events A and E are equal. Therefore the interval between them has value zero. Hence proper time between them — equal to proper distance between them — also has value zero.

e. The interval is invariant. Therefore all possible free-float rocket frames passing between Earth and Sun reckon zero interval between events A and E. This means each of them measures space separation between events A and E equal to the time separation between these events. The common value of the space and time separations are not the same for all rocket frames, but they are equal to one another in every individual rocket frame. We are asked to find the shortest possible value for this time.

Think of a rocket just passing Sun as the sunspot erupts, the rocket headed toward Earth at nearly light speed with respect to Earth. Rocket lattice clocks record the light flash from the sunspot moving away from the rocket at standard speed unity. However, these clocks record that Earth lies very close to Sun (Lorentz contraction of distance) and that Earth rushes toward the rocket at nearly light speed. Therefore light does not travel far to get to Earth in this rocket frame; neither does it take much time. For a rocket moving arbitrarily close to light speed, this distance between A and E approaches zero, and so does the time

SAMPLE PROBLEM 6-3

between A and E. Hence the shortest possible distance between A and E—equal to the shortest possible time between A and E—has the value zero. But this constitutes a limiting case, since rocket speed may approach but cannot equal the speed of light in any free-float frame.

Light flash traces out light cone in spacetime diagram

anything in Euclidean geometry. In Euclidean geometry it is never possible for distance AG between two points to be zero unless all three of the separations (northward, eastward, and upward) equal zero. In contrast, interval AG between two events can vanish even when separation in space and separation in time are individually quite large. Equation (6-3) describes the separation between lightlike events, but now separation in space may show up in two or three space dimensions as well as one time dimension. The distance in space is always positive.

It is interesting to plot on an appropriate map locations of all events, G, G_1, G_2, G_3, . . . , that can be connected with one given event A by a single spreading pulse of light. Every such future event has a distance in space from A identical to its delay in time after A. Only so can it satisfy the requirement (6-3) for a null interval. For it:

$$\text{(future time with respect to } A) = + \text{ (distance in space from } A) \quad \text{[lightlike interval]} \quad (6\text{-}4)$$

It is equally interesting to display—and on the same diagram—all the events H, H_1, H_2, H_3, . . . that can send a light pulse to A. Every such event fulfills the condition

$$\text{(past time relative to } A) = - \text{ (distance in space from } A) \quad \text{[for lightlike interval]} \quad (6\text{-}5)$$

Both of these equations satisfy the magnitude equation (6-3).

In Figure 6-4 we suppress display of a third space dimension in the interest of simplicity. We limit attention to future events G, G_1, G_2, . . . and past events H, H_1, H_2, . . . that lie on a north–south/east–west plane in space. A flash emitted from event A expands as a circle on this space plane. As it spreads out from event A, this circle of light traces out a cone opening upward in the spacetime map of Figure 6-4. This is called the **future light cone** of event A. The cone opening downward traces the history of an in-coming circular pulse of radiation so perfectly focused that it converges toward event A, collapsing exactly at event A at time zero. This downward-opening cone has the name **past light cone** of event A. All the events G, G_1, G_2, . . . lie on the future light cone of event A, all events H, H_1, H_2, . . . on its past light cone.

Numerous as the events may be that lie on the light cone, typically there are many more that don't! Look, for example, at all the events that occur 7 meters of time later than the zero time of event A. On the spacetime map, these events define a plane 7 meters above the $t = 0$ plane in which event A lies, and parallel to that plane. The light cone intersects this plane in a circle (circle in the present map; a sphere in a full spacetime map with three space dimensions). An event on the plane falls into one or another of three categories, relative to event A, according as it lies inside the circle (as does B in Figure 6-4), on it (as does G), or outside it (as does D).

The light cone is unique to Lorentz geometry. It gives nature a structure beyond any power of Euclidean geometry. The light cone does more than divide events on a single plane into categories. It classifies every event, everywhere in spacetime, into one or another of five distinct categories according to the causal relation that event bears to the chosen event, A:

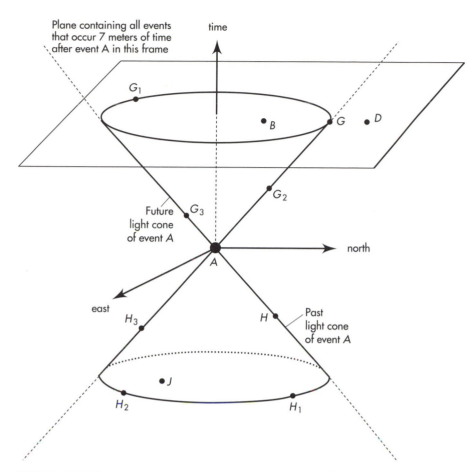

FIGURE 6-4. *Light cone as partition in spacetime; perspective three-dimensional spacetime map showing eastward, northward, and time locations of events occurring on a flat plane in space. Events* G, G₁, G₂, *and* G₃ *are on the future light cone of event* A; *events* H, H₁, H₂, *and* H₃ *are on its past light cone. See also Figure 6-5.*

1. Can a material **particle** emitted at *A* affect what **is going** to happen at *B*? If so, *B* lies *inside the future light cone* of *A* and forms a timelike pair with event *A*.

2. Can a **light ray** emitted at *A* affect — with no time to spare — what **is going** to happen at *G*?
   If so, *G* lies *on the future light cone* of *A* and forms a lightlike pair with event *A*.

3. Can **no effect whatever** produced at *A* affect what happens at *D*?
   If so, *D* lies outside the future and past light cones of *A* and forms a spacelike pair with event *A*. It lies in the *absolute elsewhere* of *A*.

4. Can a material **particle** emitted at *J* affect what **is happening** at *A*?
   If so, *J* lies *inside the past light cone* of *A* and forms a timelike pair with event *A*.

5. Can a **light ray** emitted at *H* affect — with no time to spare — what **is happening** at *A*?
   If so, *H* lies *on the past light cone* of *A* and forms a lightlike pair with event *A*.

Cause and effect preserved by light cone

Nature reveals a cause-and-effect structure beyond the vision of Euclidean geometry. The causal relation between an event *B* and another event *A* falls into one or the

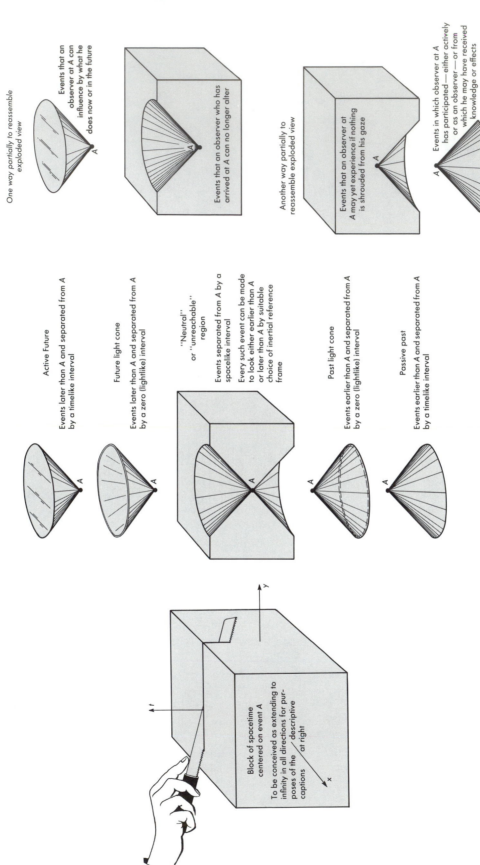

One way partially to reassemble exploded view

Events that an observer at A can influence by what he does now or in the future

Events that an observer who has arrived at A can no longer alter

Another way partially to reassemble exploded view

Events that an observer at A may yet experience if nothing is shrouded from his gaze

Events in which observer at A has participated—either actively or as an observer—or from which he may have received knowledge or effects

Active Future

Events later than A and separated from A by a timelike interval

Future light cone

Events later than A and separated from A by a zero (lightlike) interval

"Neutral" or "unreachable" region

Events separated from A by a spacelike interval

Every such event can be made to look either earlier than A or later than A by a suitable choice of inertial reference frame

Past light cone

Events earlier than A and separated from A by a zero (lightlike) interval

Passive past

Events earlier than A and separated from A by a timelike interval

Block of spacetime centered on event A

To be conceived as extending to infinity in all directions for purposes of the descriptive captions at right

FIGURE 6-5. *Exploded view of the regions into which the events of spacetime fall apart when classified with respect to a selected event* A.

other of five categories picked out by the light cone of A. That light cone and those categories have an existence in spacetime quite apart from any space and time measurements that may be used to describe them. Zero interval between events in one free-float frame means zero interval between the same events in every overlapping free-float frame. The light cone is the light cone is the light cone!

Event A appears at the origin of every spacetime map in this chapter. What's so special about event A?

Nothing whatever is special about event A! On the contrary, we have not captured the full story of the causal structure of spacetime until for *every* event A (A_1, A_2, A_3, . . .) we have classified every *other* event B (B_1, B_2, B_3, . . .) into the appropriate category — timelike! lightlike! spacelike! — with respect to that event.

Figure 6-5 summarizes the relations between a selected event A and all other events of spacetime. 🖎

CHAPTER 6 EXERCISES

PRACTICE

6-1   relations between events

This is a continuation of Sample Problem 6-1. Events 1, 2, and 3 all have the laboratory coordinates $y = z = 0$. Their x- and t-coordinates are plotted on the laboratory spacetime diagram.

a   Answer the following questions three times: once for the timelike pair of events 1 and 2, once for the spacelike pair of events 1 and 3, and once for the lightlike pair of events 2 and 3.

   (1) What is the proper time (or proper distance) between the two events?
   (2) Is it possible that one of the events caused the other event?
   (3) Is it possible to find a rocket frame in which the spatial order of the two events is reversed? That is, is it possible to find a rocket frame in which the event that occurs to the right of the other event in the laboratory frame will occur to the left of the other event in the rocket frame?

(4) Is it possible to find a rocket frame in which the temporal order of the two events is reversed? That is, is it possible to find a rocket frame in which the event that occurs before the other event in the laboratory frame occurs after the other event in the rocket frame?

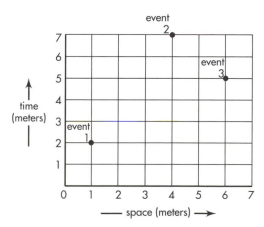

EXERCISE 6-1. *Laboratory spacetime map.*

b For the timelike pair of events, find the speed and direction of a rocket frame with respect to which the two events occurred at the same place. For the spacelike pair of events, find the speed and direction of a rocket frame with respect to which the two events occurred at the same time.

6-2 timelike, lightlike, or spacelike?

The first table lists the space and time coordinates of three events plus the reference event (event 0) as observed in the laboratory frame.

> ⬭ **EXERCISE 6-2** ⬭

LABORATORY COORDINATES OF THREE EVENTS

	t (years)	x (years)	y (years)
Event 0	0	0	0
Event 1	3	4	0
Event 2	6	5	0
Event 3	8	8	3

a Copy the second table. In the top half of each box in the second table, write the nature of the interval — timelike, lightlike, or spacelike — between the two corresponding events.

b In the bottom half of each box in the second table, write "yes" if it is possible that one of the events caused the other and "no" if it is not possible.

c Find the speed (with respect to the laboratory frame) of a rocket frame in which event 1 and event 2 in the first table occur at the same place.

d Find the speed (with respect to the laboratory frame) of a rocket frame moving along the x-axis in which event 2 and event 3 in the first table occur at the same time.

6-3 proper time and proper distance

Note: This exercise uses the Lorentz transformation equations.

a Two events P and Q have a spacelike separation. Show in general that a rocket frame can be found in which the two events occur at the same time. Also show that in this rocket frame the distance between the two events is equal to the proper distance between them. (One method: assume that such a rocket frame exists and then use the Lorentz transformation equations to show that the relative velocity of this rocket frame is less than the speed of light, thus justifying the assumption made.)

b Two events P and R have a timelike separation. Show in general that a rocket frame can be found in which the two events occur at the same place. Also show that in this rocket frame the time between the two events is equal to the proper time between them.

PROBLEMS

6-4 autobiography of a photon

A photon emitted by a star on one side of our galaxy is absorbed near a star on the other side of our galaxy,

> ⬭ **EXERCISE 6-2** ⬭

INTERVAL BETWEEN EVENTS: TIMELIKE, LIGHTLIKE, OR SPACELIKE?

	Event 1	Event 2	Event 3
Event 0	------------------	------------------	------------------
Event 1	⟶	------------------	------------------
Event 2		⟶	------------------

100,000 light-years away from its point of origin as measured in the frame of the galaxy. How does the photon experience its own birth and death? That is to say, what are the space and time separations between the birth and death of the photon in the frame of the photon?

Discussion: We cannot answer this question, because we cannot move along with the photon. No matter how fast the unpowered rocket in which we ride, we still measure light to move past us with the speed of light! Still, we can try to answer the question as a limiting case in the galaxy frame. Think of extremely energetic PROTONS traveling the same path. As protons of greater and greater energy are emitted by the first star and are absorbed near the second star at the other side of the galaxy, what happens to the distance between these two events in the frame of the proton? What happens to the time between these events in the frame of the proton? Come in this way to a limiting case in which the PROTON is moving arbitrarily close to the speed of light in the galaxy frame. In this limit, what would you expect the distance and time to be between birth and death in the frame of a PHOTON traveling the same path in space?

a You are the photon. Using the above argument, write the first few sentences of your autobiography.

At the end of the trip, near a star at the fringe of our galaxy, a galaxy-spanning photon travels 10 kilometers vertically through the atmosphere of a planet before it enters a telescope and is absorbed in the eye of an astronomer.

The average **index of refraction** of the atmosphere of this planet is $n = 1.00030$. The speed of the photon in such an atmosphere is $v = v_{conv}/c = 1/n$. (The speed of light *in a vacuum* is unity.)

b What is the proper time for this last leg of the trip—the time in the rest frame of the "slowed-down" photon? How far apart is the top of the atmosphere and the astronomer's eye in the frame of the photon? ;

c Complete your photon autobiography with an additional couple of sentences.

Discussion: Relativity is a classical theory—that is, a nonquantum theory—in which photons are postulated to move at light speed in a vacuum and at a speed $v = 1/n$ in air, where n is the index of refraction. **Quantum electrodynamics** (QED), the quantum theory of interactions between light and matter, tells us that it is incorrect to talk of a single photon moving through air. Rather, one thinks of an initial photon being absorbed by an atom in the air and a second photon emitted, the second photon then absorbed by another atom, which emits a third photon, and so forth. The classical relativistic analysis is not correct when viewed from the quantum perspective. For more on quantum electrodynamics, read Richard P. Feynman, *QED: The Strange Theory of Light and Matter* (Princeton, Princeton University Press, 1985).

6-5 the detonator paradox

A U-shaped structure made of the strongest steel contains a detonator switch connected by wire to one metric ton (1000 kilograms) of the explosive TNT, as shown in the figure. A T-shaped structure made of the same strong steel fits inside the U, with the long arm of the T not quite long enough to reach the detonator switch when both structures are at rest in the laboratory.

Now the T structure is removed far to the left and accelerated to high speed. It is Lorentz-contracted along its direction of motion. As a result, its long arm is not long enough to reach the detonator switch when the two collide. Therefore there will be no explosion.

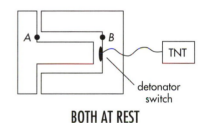

BOTH AT REST

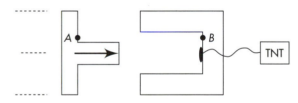

REST FRAME OF U STRUCTURE

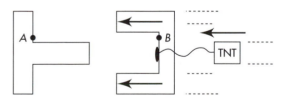

REST FRAME OF T STRUCTURE

EXERCISE 6-5. *Both at rest: The leg of the T almost reaches the detonator switch when both the T and the U are at rest. Points A and B are used in part **b** of the exercise.* **Rest frame of U structure:** *The leg of the moving T is Lorentz contracted in the rest frame of the U. Does this mean that the explosion will not take place?* **Rest frame of T structure:** *The legs of the moving U are Lorentz-contracted in the rest frame of the T. Does this mean explosion will take place?*

However, look at the same situation in the rest frame of the T structure. In this frame the arm of the T has its rest length, while the two arms of the U structure are Lorentz-contracted. Therefore the arm of the T will certainly strike the detonator switch and there will be a terrible explosion.

a   Make a decisive prediction: Will there be an explosion or not? Your life depends on it!

b   The wire from the detonator switch to the TNT is restrung through point B on the U structure when both structures are at rest, and a laser is installed at point A on the T structure. Later, when the two structures collide at A, the laser fires a pulse at B that cuts the detonator wire. Does this new apparatus change your prediction about detonation of the TNT?

Acknowledgment: A paper describing this paradox crossed the desk of one of the authors, but the paper and the name of its author have been lost. The laser inhibitor device was devised by Gordon Roesler.

6-6   how fast can you walk?

Webster's Eighth says that to "walk" means to "go on foot without lifting one foot clear of the ground before the other touches the ground." In other words, at least one foot must be on the ground at all times. Use this definition to discover the maximum speed of walking imposed by relativity.

We assume advanced technology here! A walking robot moves its free foot forward at nearly the speed of light. Then one might argue (ambiguously) as follows: While the free foot is moving forward, the planted foot is on the ground, ready to be picked up *when* [look out!] the free foot comes down in front. Half the time each foot is in motion at nearly light speed and half the time it is at rest. Therefore the average speed of each foot, equal to the maximum possible speed of the walking robot, is half the speed of light.

Why is this argument ambiguous? Because of the relativity of simultaneity. The word *when* applied to separated events should always unfurl a red flag. The event "front foot down" (label FrontDown) and the event "rear foot up" (label RearUp) occur at different places along the line of motion. Observers in relative motion will disagree about whether or not events FrontDown and RearUp occur at the same time. Therefore they will disagree about whether or not the robot has one foot on the ground at all times in order to satisfy the dictionary definition of walking.

How to remove the ambiguity in the definition of walking? One way is to make the conventional definition frame-independent: One foot must be on the ground at all times *as observed in every free-float frame of reference.* What limits does this place on the two events FrontDown and RearUp? The rear foot must leave the ground after, or at least simultaneous with, the front foot touching the ground, as observed by all free-float observers. Use the following outline to derive the consequences of this definition for the maximum speed of walking.

a   Consider the three possible relationships between events FrontDown and RearUp: timelike, lightlike, and spacelike. For each of these three relationships, write down answers to the following three questions:

(1)   Will the temporal order of the two events be the same for all observers?
(2)   Does this relationship adequately satisfy the frame-independent definition of walking?
(3)   If so, does this relationship give the maximum possible speed for walking?

Show that you answer "yes" to all three questions only for a lightlike relationship between the two events.

b   A lightlike relationship between events FrontDown and RearUp means that light can just travel from one event to the other with no time left over. Let the distance between these events — the length of one step in the Earth frame — be the unit of distance and time. Show that for the limiting speed in this frame, each foot spends two units of time moving forward, then waits one unit while the light signal propagates to the other foot, then waits three units while the other foot goes through the same process. Summary: Out of six units of time, each foot moves forward at (nearly) the speed of light for two units. What is the average speed of each foot, and therefore the speed of the walker, as measured in the Earth frame?

c   Draw a spacetime diagram for the Earth frame, showing worldlines for each of the robot's feet and worldlines for the connecting light flashes. Add a worldline showing the averaged motion of the torso, always located halfway between the two feet in the Earth frame. Demonstrate that this torso moves at the speed of the walker reckoned above.

d   Paul Horwitz says, "We determined the value of a maximum walking speed by finding a frame-independent definition of *walking.* Therefore this walking robot moves at the same speed as observed in every frame." Is Paul right?

Reference: George B. Rybicki, *American Journal of Physics,* Volume 59, pages 368–369 (April 1991).

6-7   the flickering bulb paradox: a project

Note: The following is too long for a regular exercise, but it has many insights worth pursuing as a longer activity. Therefore we call it a project.

Two long parallel conducting rails are open at one end but connected electrically at the other end

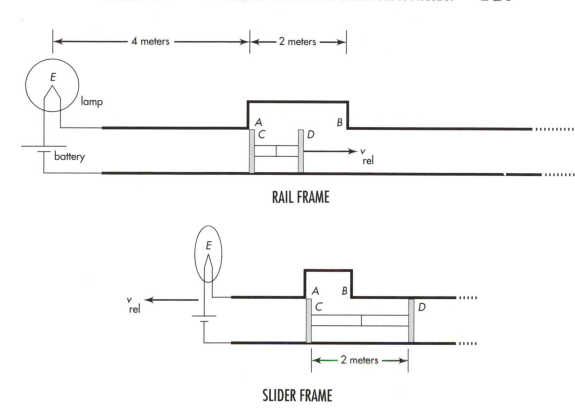

RAIL FRAME

SLIDER FRAME

EXERCISE 6-7. *Rail frame: Configuration at* t = 0 *in the rest frame of the rails. Slider* CD *moves to the right with speed* v_{rel} *such that the Lorentz-contraction factor equals 2. The vertical legs of the slider are conductors; the horizontal crosspiece is an insulator. Slider frame: Configuration at* t' = 0 *in the rest frame of the slider. The rails and lamp move to the left with speed* v_{rel} *such that the Lorentz-contraction factor is 2.*

through a lamp and battery, as shown in the figure (rail frame). One of the rails has a square vertical offset 2 meters long. Between the rails moves (without friction) an H-shaped slider, whose vertical legs are conductors but whose horizontal crosspiece is an insulator. (Assume that the vertical legs are not perfect conductors so that, with a sufficiently powerful battery, a voltage is maintained between the rails even when they are connected by the vertical legs of the slider.) If either vertical leg of the slider connects the two rails, the electrical circuit is completed, permitting the lamp to light.

The rest (proper) length of the slider is also 2 meters, but it moves at such a speed that its Lorentz-contracted length is 1 meter in the rail frame. Hence in the rail frame there is a lapse of time during which neither leg of the slider is in contact with the upper rail. Since the circuit is open during this period, the bulb should switch off for a time and then on again —it should flicker.

The figure (slider frame) shows the configuration at $t' = 0$ in the slider frame. In this frame the slider is at rest, its length is equal to its rest length, 2 meters, while the rails, the lamp, and the battery all move to the left with a speed such that their lengths along the direction of motion are reduced by a factor of 2. In

particular the offset in the upper rail is Lorentz-contracted to a length of one meter. Therefore, in the slider frame, one or the other of the slider conductors always spans the rails, so the circuit is never broken and the bulb should never switch off—it should NOT flicker!

Those trying to disprove relativity shout, "Paradox! In the rest frame of the rails the lamp switches off and then on again—it flickers. In contrast, in the rest frame of the slider the lamp stays on—it does not flicker. Yet all observers must agree: The lamp either flickers or it does not flicker. Relativity must be wrong!"

Analyze the system in sufficient detail either to demonstrate conclusively the correctness of this objection or to pinpoint its error.

Reference: G. P. Sastry, *American Journal of Physics*, Volume 55, pages 943–946 (October 1987).

6-8 the contracting train paradox: a project

Note: The following is too long for a regular exercise, but it has many insights worth pursuing as a longer activity. Therefore we call it a project.

Kerwin Warnick writes in with the following paradox. A spaceship of proper length L_o accelerates from rest. Its front end travels a distance x_F in time t_F to a final speed at which the ship is contracted to half its rest length. In the same time t_F the rear end moves the same distance x_F as the front end plus the distance $L_o/2$ by which the ship has contracted. Distance traveled by the rear end $x_F + (L_o/2)$ in time t_F means an average speed $[x_F + (L_o/2)]/t_F$. Since the proper length L_o can be arbitrarily large, this average speed can be arbitrarily great, even greater than the speed of light. "This disproves relativity!" he exclaims.

Analyze this thought experiment in sufficient detail either to demonstrate conclusively the correctness of Warnick's objection or to pinpoint its error.

Reference: Edwin F. Taylor and A. P. French, *American Journal of Physics*, Volume 51, pages 889–893 (October 1983).

CHAPTER 7

MOMENERGY

Every physical quantity is represented by a geometric object.

Theme of Herman Weyl

7.1 MOMENERGY: TOTAL CONSERVED IN A COLLISION

**momentum conserved.
energy conserved.
momenergy conserved!**

Paradoxically, few examples of motion are more complicated than a collision, and few are simpler. The complication shows nowhere more clearly than in the slow-motion videotape of the smashup of two automobiles. Millisecond by millisecond the fender of one colliding car deforms another fraction of a centimeter. Millisecond by millisecond the radiator grille of the other car bends inward a little more on the way to total collapse: steel against steel, force against force, crumpling surface against crumpling surface. What could be more complex?

For the drivers of the colliding cars the experience is shattering. They are hardly aware of noise and complicated damage. A single impression overpowers their senses: the inevitability of the crash. Call it what we will—inertia, momentum, the grip of spacetime on mass—something is at work that drives the two vehicles together as the frantic drivers jam their brake pedals down, locking the wheels as the cars slither over the glassy ice, crash into one another, then slide apart.

Does mass lose its inertia during the collision? No. Inertia does its best to keep each car going as it was, to keep its **momentum** constant in magnitude and direction. Momentum: we can think of it loosely as an object's will to hold its course, to resist deflection from its appointed way. The higher the object's momentum, the more violently it hits whatever stands in its way. But the momentum of a single object is not all-powerful. The two vehicles exchange momentum. But spacetime insists and demands that whatever momentum one car gains the other car must lose. Regardless of all complications of detail and regardless of how much the momentum of any one

Smashup complicated?

Smashup is simple!

Momentum conservation simplifies description

189

Energy too is conserved

object may change, the combined momentum of the two objects remains constant: the total is unchanged in the collision. A like statement applies to energy, despite a conversion of energy of motion into heat energy and fender crumpling.

A collision thus manifests a wonderful simplicity: the combination of the motion-descriptive quantities (momentum and energy) of the two colliding bodies does not change. That combination is identical before and after the collision. In a word, it is *conserved*. This conserved combination we call **momentum–energy** or, more briefly, **momenergy** (defined more carefully in Section 7.2). We will use the two terms interchangeably in this book.

Momenergy is conserved!

A collision cannot be elevated from mere talk to numbers without adopting, directly or indirectly, the principle of conservation of momentum and energy. In the enterprise of identifying the right numbers, using them, and understanding them, no concept is more powerful than what relativity smilingly holds forth: momenergy.

Wait a minute. Apparently you are going to find new expressions for momentum and energy, then combine them in some way to form a unity: momenergy. But I have three complaints. (1) What is wrong with what good old-fashioned secondary school physics textbooks give us, the Newtonian expressions for momentum — $p_{Newton} = mv_{conv}$ — and kinetic energy $K_{Newton} = \frac{1}{2}mv^2_{conv}$ — where v_{conv} is expressed in conventional units, say meters/second? (2) Momentum and energy do not even have the same units, as these formulas make clear. How can you combine quantities with different units? (3) Momentum and energy are different things entirely; why try to combine them at all?

Take your questions in order.

1. **Newtonian Expressions:** Only for slow-moving particles do we get correct results when we use Newtonian expressions for momentum and energy. For particle speeds approaching that of light, however, total energy and momentum of an isolated system, as Newton defined momentum and energy, are *not* conserved in a collision. In contrast, when momentum and energy are defined relativistically, then total momentum and total energy of particles in an isolated system *are* conserved, no matter what their observed speeds.

2. **Units:** It is easy to adopt identical units for momentum and energy. As a start we adopt identical units for space and time. Then the speed of a particle is expressed in unit-free form, v, in meters of distance per meter of light-travel time (Section 2.8). This choice of units, which we have already accepted earlier in this book, gives even Newtonian expressions for momentum — $p_{Newton} = mv$ — and kinetic energy — $K_{Newton} = \frac{1}{2}mv^2$ — the same unit: **mass.** These are not relativistic expressions, but they do agree in their units, and agree in units with the correct relativistic expressions.

3. **Momentum and Energy Different:** Yes, of course, momentum and energy are different. Space and time are different too, but their combination, spacetime, provides a powerful unification of physics. Space and time are put on an equal footing, but their separate identities are maintained. Same for momenergy: We will see that its "space part" is momentum, its "time part" energy. We will also discover that its magnitude is the mass of the particle, reckoned using the good ol', ever-lovin', familiar minus sign: $m^2 = E^2 - p^2$.

Thus relativity offers us a wonderful unity. Instead of three separate motion-descriptive quantities — momentum, energy, and mass — we have a single quantity: momenergy.

7.2 MOMENERGY ARROW

a spacetime arrow pointing along the worldline

What lies behind the name *momentum – energy (momenergy)?* What counts are its properties. We most easily uncover three central properties of momenergy by combining everyday observation with momenergy's essential feature: Total momenergy is conserved in any collision.

First, think of two pebbles of different sizes moving with the same velocity toward the windshield of a speeding car. One bounces off the windshield without anyone noticing; the other startles the occupants and leaves a scratch. Five times the mass? Five times the punch-delivering capacity! Five times the momenergy. Momenergy, in other words, is proportional to mass.

Second, momentum-energy of a particle depends on its direction of travel. A pebble coming from the front takes a bigger chip out of the windshield than a pebble of equal mass and identical speed glancing off the windshield from the side. Therefore momenergy is not measurable by a mere number. It is a directed quantity. Like an arrow of a certain length, it has magnitude and direction.

Our experience with the unity of spacetime leads us to expect that the momenergy arrow will have three parts, corresponding to three space dimensions, plus a fourth part corresponding to time. In what follows we find that momenergy is indeed a four-dimensional arrow in spacetime, the **momenergy 4-vector** (Box 7-1). Its three "space parts" represent the momentum of the object in the three chosen space directions. Its "time part" represents energy. The unity of momentum and energy springs from the unity of space and time.

In what direction does the momenergy 4-vector of a particle point? It points in the "same direction in spacetime" as the worldline of the particle itself (Figure 7-1). There is no other natural direction in which it *can* point! Spacetime itself has no structure that indicates or favors one direction rather than another. Only the motion of the particle itself gives a preferred direction in spacetime. The particle moves from one event to a nearby event along its worldline. In so doing, it undergoes a **spacetime displacement,** small changes in the three space positions along with an accompanying small advance in the time. The spacetime displacement has four parts: it is a 4-vector. The momenergy arrow points in the direction of another arrow, the arrow of the particle's spacetime 4-vector displacement. Momenergy runs parallel to worldline!

Compare the worldline of an individual particle in spacetime with a single straw in a great barn filled with hay. This particular straw has a direction, an existence, and a meaning independent of any measuring method imagined by humans who stack the hay or by mice that live in it. Similarly, in the rich trelliswork of worldlines that course through spacetime, the arrowlike momenergy of the particle has an existence and definiteness independent of the choice — or even use — of any free-float frame of reference (Section 5.9).

No frame of reference? Then no clock available to time motion from here to there! Or rather no clock except one that the particle itself carries, its own wristwatch that records proper time. Proper time for what? Proper time for spacetime displacement between two adjacent events on the worldline of the particle. Proper time provides the only natural way to clock the rate of motion of the particle; that is the third and final feature of momenergy.

In brief, the momenergy of a particle is a 4-vector: Its magnitude is proportional to its *mass,* it points in the direction of the particle's *spacetime displacement,* and it is reckoned using the *proper time* for that displacement. How are these properties combined to form momenergy? Simple! Use the recipe for Newtonian momentum: mass times displacement divided by time lapse for that displacement. Instead of

Momenergy of particle proportional to its mass

Momenergy a directed quantity

Momenergy a 4-vector

Particle momenergy points along its worldline

Momenergy independent of reference frame

Particle wristwatch logs time for momenergy

BOX 7-1

WHAT IS A 4-VECTOR?

A vector is a mathematical object that has both *magnitude* and *direction*. The meanings of the terms *magnitude* and *direction*, however, differ between one geometry and another. Mathematics offers many geometries. The two geometries important to us in this book are Euclidean geometry and Lorentz geometry.

Euclidean geometry defines **3-vectors** located in 3-dimensional space. Let a speeding particle emit two sparks. The particle's spatial displacement from first spark to second spark is a 3-vector. Each of the **three components** (northward, eastward, and upward) of this 3-vector displacement has a value larger or smaller, depending on the orientation of the coordinate system chosen. In contrast, the magnitude of the displacement — the distance traveled (computed as the square root of the *sum* of the squares of the three components of displacement) — has the same value in all coordinate systems.

Lorentz geometry defines **4-vectors** located in 4-dimensional spacetime. Construct the 4-vector spacetime displacement from the three space components supplemented by the time component, the time between sparks emitted by the speeding particle. Each of these **four components** (including time) has a value larger or smaller, depending on the choice of free-float frame of reference from which it is measured. The square of the separation in time between the two sparks as so measured, diminished by the square of the separation in space in the chosen frame, yields the square of the spacetime interval between the two events. This interval has the same value in all free-float frames. It is also the proper time, the time between the two sparks read directly on the particle's wristwatch.

Newtonian mechanics combines (in various ways) time and mass of the particle with Euclidean 3-vector displacement of the particle to yield additional 3-vectors that describe particle motion: velocity, momentum, acceleration. Each 3-vector has magnitude and direction. The values of the three components of each 3-vector depend on the orientation of the chosen coordinate system. But for each 3-vector quantity, the 3-vector itself is the same, both in magnitude and direction in space, no matter what Euclidean coordinate system we choose. Every 3-vector exists even in the absence of any coordinate system at all! That is why the analysis of Newtonian mechanics can proceed in all its everyday applications independent of choice of coordinate system.

Relativistic mechanics combines (in various ways) proper time and mass of the particle with Lorentz 4-vector displacement of the particle to yield additional 4-vectors that describe particle motion. Central among these is the particle's **momentum – energy 4-vector**, or **momenergy.** Values of the four components of the momenergy 4-vector differ as measured in different free-float frames in relative motion. But the momenergy 4-vector itself is the same, both in magnitude (mass!) and direction in spacetime, no matter what the frame. The momenergy 4-vector of a particle exists even in the absence of any reference frame at all! That is why the analysis of relativistic mechanics can proceed in all its power independent of choice of free-float frame of reference.

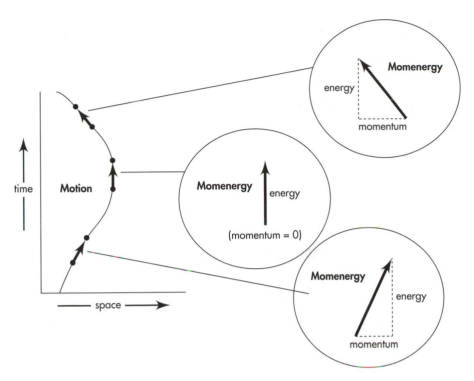

FIGURE 7-1. *Motion and momenergy seen as identically directed arrows.* The momenergy arrow is carried along the worldline with the particle. Under action of a force, the particle traces out a curved worldline. The momenergy arrow — its constant magnitude equal to the mass of the particle — continually alters its tilt to point in the same direction in spacetime as the worldline. (For the special case shown here, the particle moves in x and t, but not in y or z.)

Newtonian displacement in space, use Einstein's displacement in spacetime; instead of Newton's "universal time," use Einstein's proper time.

The result expresses the momenergy 4-vector in terms of the spacetime displacement 4-vector:

$$\text{(momenergy)} = \text{(mass)} \times \frac{\text{(spacetime displacement)}}{\text{(proper time for that displacement)}} \quad (7\text{-}1)$$

Particle momenergy
Magnitude: Mass
Direction: Along spacetime displacement

In any given free-float frame, the momentum of the particle is the three "space parts" of the momenergy and the particle's energy is the "time part." This expression for momenergy is simple, and it works — works as employed in the law of conservation of momenergy: Total momenergy before reaction equals total momenergy after reaction. Investigators have observed and analyzed more than a million collisions, creations, transformations, decays, and annihilations of particles and radiation. They have failed to discover a single violation of the relativistic law of conservation of momenergy.

To arrive at a formula as important as (7-1) so painlessly may at first sight create doubts. These doubts have to be dismissed. Fact is, there is no room for any alternative — as we see by going step by step through the factors in this equation.

Statement 1: *m* units of mass pursuing a given motion carry *m* times the momenergy of one unit of mass. Reasoning: *m* identical objects racing along side by side carry *m* times the momentum and *m* times the energy — and therefore *m* times the momenergy — of an object of unit mass.

Statement 2: Momenergy points in the same direction in spacetime as worldline. Reasoning: Where else *can* it point? Even the slightest difference in direction between momenergy and direction of motion along the worldline would bear witness to some crazy asymmetry in spacetime, for which no experiment in field-free space has ever given the slightest evidence.

Statement 3: The spacetime displacement between one event on the worldline and a nearby event on it specifies the direction of that worldline. Reasoning: The very concept of direction implies that there exists a segment, AB, of the worldline short enough to be considered straight. And to fix the direction of this spacetime displacement AB, it suffices to know the location of any two events, A and B, on this short segment.

Momenergy formula justified

Statement 4: Worldline direction — and therefore momenergy — is independent of the magnitude of the spacetime displacement. Reasoning: To pick an event B' on the worldline half as far from A as B along the short straight segment — thus to cut in half the spacetime displacement — makes no change in the direction of the worldline, therefore no change in the direction of the momenergy, therefore no change in the momenergy itself.

Statement 5: The unit 4-vector (spacetime displacement)/(proper time for that displacement) defines and measures the direction of the worldline displacement and therefore the direction of the momenergy 4-vector. Reasoning: What matters is not spacetime displacement individually, not proper time individually, but only their ratio. This ratio is the only directed quantity available to us to describe the *rate of motion* of the particle through spacetime.

The spacetime displacement, AB, has a magnitude equal to the interval (or proper time or wristwatch time) the particle requires to pass from A to B. That is why the ratio in question is a unit 4-vector.

Proper time provides the only natural way, the only frame-independent means, to clock the particle. If instead we should incorrectly put frame time into the denominator — frame time measured by the array of clocks in a particular free-float frame — the value of this time would differ from one frame to another. Divided into the spacetime displacement, it would typically not yield a unit vector. The vector's magnitude would differ from one frame to another. Therefore we must use in the denominator the proper time to go from A to B, a proper time identical to the magnitude of the spacetime displacement AB in the numerator.

Statement 6: The momenergy 4-vector of the particle is

$$(\text{momenergy}) = (\text{mass}) \times \frac{(\text{spacetime displacement})}{(\text{proper time for that displacement})} \quad (7\text{-}1)$$

Reasoning: There is no other frame-independent way to construct a 4-vector that lies along the worldline and has magnitude equal to the mass.

Units: In this book, as in more and more present-day writing, space and time appear in the same unit: meter. Numerator and denominator on the right side of equation (7-1) have the unit of meter. Therefore their quotient is unit-free. As a result, the right side of the equation has the same unit as the first factor: mass. So the left side, the momenergy arrow, must also have the unit of mass. As the oneness of spacetime is emphasized by measuring space and time in the same unit, so the oneness of momenergy is clarified by measuring momentum and energy in the same unit: mass. Table 7-1 at the end of the chapter compares expressions for momentum and energy in units of mass with expressions in conventional units.

Unit of momenergy: mass

You say that the equation for momenergy is

$$\text{(momenergy)} = \text{(mass)} \times \frac{\text{(spacetime displacement)}}{\text{(proper time for that displacement)}}$$

I thought that "spacetime displacement" was the interval, which is the proper time. I know, however, that I am wrong, because if spacetime displacement and proper time were the same, then the numerator and denominator of the fraction would cancel, and momenergy would simply equal mass. Surely you would have told us of such simplicity. What have I missed?

It is easy to confuse a vector — or a 4-vector — with its magnitude.

In the expression for momenergy, the spacetime displacement is a 4-vector (Box 7-1). In the laboratory frame this displacement 4-vector has four components, $\{dt, dx, dy, dz\}$. In a free-float rocket moving in an arbitrary direction, the displacement 4-vector has four components, $\{dt', dx', dy', dz'\}$, typically different, respectively, from those in the laboratory frame.

A vector in space (a 3-vector) has not only a magnitude but also a direction independent of any coordinate system. ("Which way did they go?" "That-a-way!" — pointing.) Similarly, the spacetime displacement has a magnitude and direction in spacetime independent of any reference frame. This **spacetime direction** distinguishes the 4-vector displacement (the numerator above) from its magnitude, which is the proper time for that displacement (the denominator). This proper time (interval) can be observed directly: it is the time lapse read off the wristwatch carried by the particle while it undergoes the spacetime displacement.

In summary the fraction

$$\frac{\text{(spacetime displacement)}}{\text{(proper time for that displacement)}}$$

has a numerator that is a 4-vector. This 4-vector numerator has the same *magnitude* as the denominator. The resulting fraction is therefore a *unit 4-vector* pointing along the worldline of the particle. This unit 4-vector determines the *direction* of the particle's momenergy in spacetime. And the magnitude of the momenergy? It is the mass of the particle, the first term on the right of the expression at the top of this page. In brief, the momenergy of a particle is 4-vector of magnitude m pointing along its worldline in spacetime. This description is independent of reference frame.

Unit 4-vector along worldline

7.3 MOMENERGY COMPONENTS AND MAGNITUDE

space part: momentum of the object
time part: energy of the object
magnitude: mass of the object

Accidents of history have given us not one word, momenergy, but two words, momentum and energy, to describe mass in motion. Before Einstein, mass and motion were described not in the unified context of spacetime but in terms of space and time separately, as that division shows itself in some chosen free-float frame. Often we still think in those separated terms. But the single concept *spacetime location* of an event unites the earlier two ideas of its position in space and the time of its happening. In the

Break down momenergy for examination

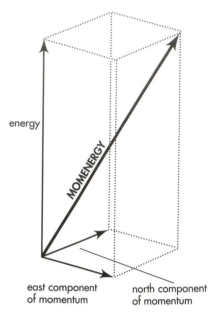

energy

MOMENERGY

east component
of momentum

north component
of momentum

FIGURE 7-2. *Momenergy arrow of a moving object translated into the language of momentum and energy, shown for the special case in which upward momentum (vertical momentum) equals zero. The momenergy arrow itself has an existence and direction (in that great haystack of worldlines and events that we call spacetime) independent of the choice, or even presence, of any free-float frame. In contrast, separate measures of momentum and energy do depend on choice of frame. They point parallel to, that is in the same direction as, the corresponding space and time directions of the chosen frame itself. See Figures 7-3 and 7-4 for a still more revealing representation of the proportion between momenergy and its components.*

same way we combine momentum and energy of a moving object into the single idea of momenergy arrow. Having assembled it, we now break momenergy down again, seeking new insight by examining its separate parts.

The unity of momenergy dissolves—in our thinking—into the separateness of momentum and energy when we choose a free-float frame, say the laboratory. In that laboratory frame the spacetime separation between two nearby events on the worldline of a particle resolves itself into four different separations: one in laboratory time and one in each of three perpendicular space directions, such as north, east, and upward. With each spacetime separation goes a separate part, a separate portion, a separate **component** of momenergy in the laboratory free-float frame (Figure 7-2).

The "space parts" of momenergy of a particle are its three components of momentum relative to a chosen frame. Their general form is not strange to us—mass times a velocity component. The "time part," however, is new to us, foreshadowing important insights into the nature of energy (Section 7.5). The four components are

$$\begin{pmatrix} \text{eastward} \\ \text{component} \\ \text{of} \\ \text{momenergy} \end{pmatrix} = \begin{pmatrix} \text{eastward} \\ \text{component} \\ \text{of} \\ \text{momentum} \end{pmatrix}$$

$$= (\text{mass}) \times \frac{(\text{eastward displacement})}{(\text{proper time for that displacement})}$$

$$\begin{pmatrix} \text{northward} \\ \text{component} \\ \text{of} \\ \text{momenergy} \end{pmatrix} = \begin{pmatrix} \text{northward} \\ \text{component} \\ \text{of} \\ \text{momentum} \end{pmatrix}$$

$$= (\text{mass}) \times \frac{(\text{northward displacement})}{(\text{proper time for that displacement})}$$

$$\begin{pmatrix} \text{upward} \\ \text{component} \\ \text{of} \\ \text{momenergy} \end{pmatrix} = \begin{pmatrix} \text{upward} \\ \text{component} \\ \text{of} \\ \text{momentum} \end{pmatrix}$$

$$= (\text{mass}) \times \frac{(\text{upward displacement})}{(\text{proper time for that displacement})}$$

$$\begin{pmatrix} \text{time} \\ \text{component} \\ \text{of} \\ \text{momenergy} \end{pmatrix} = (\text{energy}) = (\text{mass}) \times \frac{(\text{time displacement})}{(\text{proper time for that displacement})}$$

The calculus version of these equations is deliciously brief. Here, as in Section 6.2, tau (τ) stands for proper time:

Momenergy components of particle in a given frame

$$E = m \frac{dt}{d\tau}$$

$$p_x = m \frac{dx}{d\tau} \qquad \text{(7-2)}$$

$$p_y = m \frac{dy}{d\tau}$$

$$p_z = m \frac{dz}{d\tau}$$

The components of the momenergy 4-vector we now have before us in simple form, but how much is the absolutely-number-one measure of this physical quantity, its magnitude? This magnitude we reckon as we figure the magnitude of any Lorentz 4-vector: magnitude squared is the difference of squares of the time part and the space part:

(magnitude of momenergy arrow)^2
$$= \text{(energy)}^2 - \text{(east momentum)}^2 - \text{(north momentum)}^2 - \text{(up momentum)}^2$$
$$= E^2 - (p_x)^2 - (p_y)^2 - (p_z)^2$$
$$= m^2 \frac{(dt)^2 - (dx)^2 - (dy)^2 - (dz)^2}{(d\tau)^2} = m^2 \frac{(d\tau)^2}{(d\tau)^2} = m^2$$

In brief, the magnitude of the momenergy 4-vector, or its square,

$$\text{(magnitude of momenergy arrow)}^2 = E^2 - p^2 = m^2 \qquad (7\text{-}3)$$

Magnitude of momenergy 4-vector: mass!

is identical with the particle mass, or its square. Moreover, this mass is a quantity characteristic of the particle and totally independent of its state of motion.

It's worthwhile to translate this story into operational language. Begin with a particle that is at rest. Its 4-vector of energy and momentum points in the pure timelike direction, all energy, no momentum. Let an accelerator boost that particle. The particle acquires momentum. The space component of the 4-vector, originally zero, grows to a greater and greater value. In other words, the momenergy 4-vector tilts more and more from the "vertical," that is, from a purely timelike direction. However, its magnitude remains totally unchanged, at the fixed value m. In conse-

SAMPLE PROBLEM 7-1

MASS

The energy and momentum components of a particle, measured in the laboratory, are

$$E = 6.25 \text{ kilograms}$$
$$p_x = 1.25 \text{ kilograms}$$
$$p_y = p_z = 2.50 \text{ kilograms}$$

What is the value of its mass?

SOLUTION

We obtain a value for mass using equation (7-3):

$$m^2 = E^2 - (p_x)^2 - (p_y)^2 - (p_z)^2$$
$$= [(6.25)^2 - (1.25)^2 - (2.50)^2 - (2.50)^2] \text{ (kilograms)}^2$$
$$= [39.06 - 1.56 - 6.25 - 6.25] \text{ (kilograms)}^2$$
$$= [39.06 - 14.06] \text{ (kilograms)}^2$$
$$= 25.00 \text{ (kilograms)}^2$$

Hence

$$m = 5.0 \text{ kilograms}$$

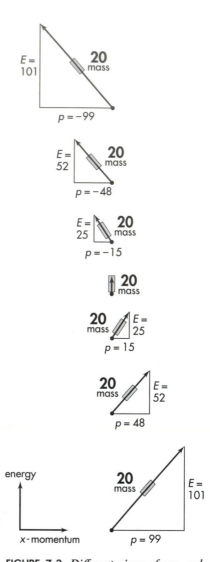

FIGURE 7-3. *Different views of one and the same momenergy 4-vector of a particle in seven different free-float frames. The y- and z-components of momentum are assumed to equal zero, and frames are chosen to give integer values for energy and x-momentum components. The mass of the particle equals 20 units as reckoned in every free-float frame:* $m^2 = E^2 - p^2$. *This invariant value of the mass is shown by the thick "handle" on each vector. For a frame in which the particle is at rest (center diagram), the energy is equal to the mass and the handle covers the vector.*

Does the momenergy 4-vector for this particle require for its existence any reference frame? No one would laugh more at such a misapprehension than the particle! The momenergy 4-vector has an existence in spacetime independent of any clocks and measuring rods. We, however, wish to assign to this 4-vector an energy and momentum. For that purpose we do require one or another free-float frame.

quence, the time component of that 4-vector, that is, the energy of the particle, undergoes a systematic alteration.

If the geometry of spacetime were Euclidean, this ever-growing tilt, this continuing rotation of the direction of the arrow of momenergy, would cause the vertical or time component to become ever shorter. However, spacetime is not Euclidean. It is Lorentzian, as appears in the minus sign in the equation for momenergy magnitude m: $m^2 = E^2 - p^2$. With momenergy magnitude, or particle mass m, being constant, and momentum p ever growing, Lorentz geometry itself tells us that the ever-growing tilt, the ever-larger momentum value, p, causes the time component of the momenergy — the energy E — not to shorten, as in a Euclidean spacetime, but to lengthen as the acceleration proceeds:

$$E = (m^2 + p^2)^{1/2} = \text{an increasing function of momentum, } p$$

This marvelously simple relation between energy and momentum, full of geometric as well as physical content, has by now been tried and verified in so many thousands of experiments of such varied kinds that it counts today as battle-tested.

Energy, momentum, and mass, expressed so far in the language of algebra, let themselves be displayed even more clearly in the language of pictures. Only one obstacle stands in the way. The paper is Euclidean and the vertical leg of a right triangle typically is shorter than the hypotenuse. In contrast, spacetime is Lorentzian, and the timelike dimension (the energy) is typically longer than the "hypotenuse" (the mass). We are indebted to our colleague William A. Shurcliff for a way to have our cake and eat it too, a device to employ Euclidean paper and yet display Lorentzian length. How? By laying over the hypotenuse of the Euclidean triangle a fat line or **handle** of length adjusted to the appropriate Lorentzian magnitude (Figure 7-3). The length of the handle represents the invariant value of the particle mass. This length remains the same, whatever the values of energy and momentum, values that differ as the particle is observed from one or another frame of reference in relative motion.

Figure 7-3 shows a few of the infinitely many different values of energy and momentum that one and the same particle can have as measured in different free-float frames. Each arrow, being depicted on a Euclidean sheet of paper, necessarily appears with an apparent length that increases with slope or particle speed. The handle on the arrow, by contrast, has the length appropriate to Lorentz geometry. This length represents particle mass, $m = 20$, a quantity independent of particle speed. The momenergy 4-vector of a material particle is always timelike. Why timelike? Because the momenergy 4-vector lies in the same spacetime direction as the worldline of the particle (Section 7-2). The events along the worldline have a timelike relationship: Time displacement between events is greater than the space displacement. One

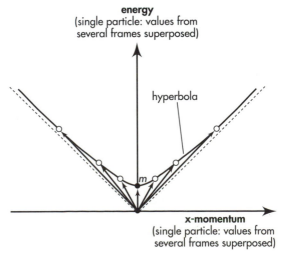

FIGURE 7-4. *Momenergy 4-vector for the single particle of Figure 7-3 as observed in seven free-float frames, these plots then superposed on a composite momenergy diagram. Frames are chosen so that y- and z-components of momentum equal zero. Locus of the tips of the arrows traces out a hyperbola. The central short vertical arrow pointing to the dot labeled m represents momenergy as measured in the particle rest frame. In this frame momentum has value zero and energy — "rest energy" — equals the mass of the particle. For clarity, the handles have been omitted from the 4-vectors, which all have identical invariant magnitude* m = 20.

consequence is that the particle moves at less than the speed of light in every possible free-float frame.

E, p, m of particle in different frames related by hyperbola

The equation $E^2 - p^2 = m^2 = $ (constant) is the formula for a hyperbola. Figure 7-4 generates this hyperbola by superposing on the same figure spacetime vectors that represent energy and momentum of the same particle in different free-float frames. For visual clarity the handles are omitted from these 4-vectors. However, each momenergy 4-vector has the same magnitude, equal to the particle mass, $m = 20$. 🖋

7.4 MOMENTUM: "SPACE PART" OF MOMENERGY

simply use proper time instead of Newton's so-called "universal" time

Newton called momentum "quantity of motion." The expressions for momentum that spacetime physics gives us, the last three equations in (7-2), seem at first sight to distinguish themselves by a trivial difference from the expressions for momentum given to us long ago by Newton's followers:

$$p_{x\,\text{Newton}} = m\frac{dx}{dt}, \quad p_{y\,\text{Newton}} = m\frac{dy}{dt}, \quad p_{z\,\text{Newton}} = m\frac{dz}{dt} \qquad \text{[valid for low velocity]}$$

Newtonian versus relativistic expressions for momentum

That difference? Today, proper time $d\tau$ between nearby events on the worldline of the particle. Laboratory time, in older days, when the concepts of proper time and interval were unknown. The percentage difference between the two, trivial or even negligible under everyday circumstances, becomes enormous when the speed of the object approaches the speed of light.

We explore most simply the difference between relativistic and Newtonian predictions of momentum by analyzing a particle that travels with speed v in the x-direction only. Then the relation between displacement of this particle and its speed is $x = vt$. For small displacements, for example between two nearby spark events on the worldline, this becomes, in the mathematical limit of interest in calculus notation, $dx = vdt$.

The proper time between the two nearby sparks is always less than the laboratory time:

$$d\tau = [(d\tau)^2]^{1/2} = [(dt)^2 - (dx)^2]^{1/2} = [(dt)^2 - (vdt)^2]^{1/2}$$

$$= (dt)(1 - v^2)^{1/2} = \frac{dt}{\gamma} \qquad (7\text{-}4)$$

where gamma, $\gamma = 1/(1 - v^2)^{1/2}$ is the time stretch factor (Section 5.8). This figure for the interval, or proper time, between the two nearby sparks we now substitute into equations (7-2) in order to learn how the relativistic expressions for energy and momentum depend on particle speed:

$$E = m\frac{dt}{d\tau} = \frac{m}{(1 - v^2)^{1/2}} = m\gamma$$

$$p_x = m\frac{dx}{d\tau} = \frac{m\,(dx/dt)}{(1 - v^2)^{1/2}} = \frac{mv_x}{(1 - v^2)^{1/2}} = mv_x\gamma$$

$$(7\text{-}5)$$

Low speed: Newton and Einstein agree on value of momentum

The momentum expression is the same as for Newtonian mechanics — mass m times velocity (dx/dt) — except for the factor $(1 - v^2)^{1/2}$ in the denominator. That factor we can call 1 when the speed is small. For example, a commercial airliner moves through the air at approximately one millionth of the speed of light. Then the factor $(1 - v^2)^{1/2}$ differs from unity by only five parts in 10^{12}. Even for an alpha particle (helium nucleus) ejected from a radioactive nucleus with approximately 5 percent of the speed of light, the correction to the Newtonian figure for momentum is only a little more than one part in a thousand. Thus for low speeds the momentum expressed in equation (7-5) reduces to the Newtonian version.

High speed: Relativity reveals much larger momentum

At a speed close to that of light, however, the particle acquires a momentum enormous compared with the Newtonian prediction. The unusually energetic cosmic-ray protons mentioned at the end of Section 5.8 crossed the Milky Way in 30 seconds of their own time, but a thousand centuries or 3×10^{12} seconds of Earth time. The ratio $dt/d\tau$ between Earth time and proper time is thus 10^{11}. That is also the ratio between the correct relativistic value of the protons' momentum and the Newtonian prediction.

Units: Both Newtonian and relativistic expressions for momentum contain speed, a ratio of distance to time. From the beginning we have measured distance and time in the same unit, for example meter. Therefore the ratio of distance to time is unit-free. In Section 2.8, we expressed speed as a dimensionless quantity, the fraction of light speed:

Unit of momentum: mass

$$v = \frac{\text{(meters of distance covered by particle)}}{\text{(meters of time required to cover that distance)}}$$

$$= \frac{\text{(particle speed in meters/second)}}{\text{(speed of light in meters/second)}} = \frac{v_{\text{conv}}}{c} \qquad \text{(7-6)}$$

In terms of speed v (called beta, β, by some authors), Newtonian and relativistic expressions for the magnitude of the momentum have the forms

$$p_{\text{Newton}} = mv \qquad \text{[valid for low speed]} \quad \text{(7-7)}$$
$$p = mv/(1 - v^2)^{1/2} \qquad \text{[good at any speed]} \quad \text{(7-8)}$$

More Units: In order to convert momentum in units of mass to momentum in conventional units, such as kilogram meters/second, multiply expressions (7-6), (7-7), and (7-8) by the speed of light c and use the subscript "conv" for "conventional":

$$p_{\text{conv Newton}} = p_{\text{Newton}}\, c = mvc = m(v_{\text{conv}}/c)\, c = mv_{\text{conv}} \qquad \text{[low speed]} \quad \text{(7-9)}$$

Conversion to conventional momentum units

$$p_{\text{conv}} = pc = \frac{mvc}{[1 - v^2]^{1/2}} = \frac{m(v_{\text{conv}}/c)c}{[1 - (v_{\text{conv}}/c)^2]^{1/2}}$$

$$= \frac{mv_{\text{conv}}}{[1 - (v_{\text{conv}}/c)^2]^{1/2}} \qquad \text{[any speed]} \quad \text{(7-10)}$$

Thus conversion from momentum in units of mass to momentum in conventional units is always accomplished by multiplying by the conversion factor c. This is true whether the expression for momentum being converted is Newtonian or relativistic. Table 7-1 at the end of the chapter summarizes these comparisons. ✒

7.5 ENERGY: "TIME PART" OF MOMENERGY

energy has two parts: rest energy (= mass) plus kinetic energy

What about the "time part" of the momentum–energy of a particle—the part we have called its energy? This is certainly a strange-looking beast! As measured in a particular free-float frame, say the laboratory, this time component as given in equation (7-5) is

$$E = m\frac{dt}{d\tau} = \frac{m}{(1 - v^2)^{1/2}} = m\gamma \qquad \text{(7-11)}$$

Relativistic expression for energy

Compare this with the Newtonian expression for kinetic energy, using K as the symbol for kinetic energy:

$$K_{\text{Newton}} = \frac{1}{2} mv^2 \qquad \text{[valid for low speed]} \quad \text{(7-12)}$$

How does the relativistic expression for energy, equation (7-11), compare with the Newtonian expression for kinetic energy (7-12)? To answer this question, first look at the behavior of these two expressions when particle speed equals zero. The Newtonian kinetic energy goes to zero. In contrast, at zero speed $1/(1 - v^2)^{1/2} = 1$ and the relativistic value for energy becomes equal to mass of the particle,

$$E_{\text{rest}} = m \qquad \text{(7-13)}$$

Rest energy of a particle equals its mass

where E_{rest} is called **rest energy of the particle.** Rest energy of a particle is simply its mass. So the relativistic expression for energy does not go to zero at zero speed, while the Newtonian expression for kinetic energy does go to zero.

Is this an irreconcilable difference? The Newtonian formula does not contain an expression for rest energy, equal to the mass of the particle. But here is the distinction: The relativistic expression gives the value for *total energy* of the particle, while the Newtonian expression describes *kinetic energy* only (valid for low speed). However, in Newtonian mechanics any constant energy whatever can be added to the energy of a particle without changing the laws that describe its motion. One may think of the zero-speed limit of the relativistic expression for energy as providing this previously undetermined constant.

When we refer to energy of a particle we ordinarily mean total energy of the particle. As measured in a frame in which the particle is at rest, this total energy equals rest energy, the mass of the particle. As measured from frames in which the particle moves, total energy includes not only rest energy but also kinetic energy.

This leads us to define kinetic energy of a particle as energy above and beyond its rest energy:

$$\text{(energy)} = \text{(rest energy)} + \text{(kinetic energy)}$$

Kinetic energy defined

or

$$E = m + K \qquad \text{(7-14)}$$

SAMPLE PROBLEM 7-2
MOTION IN THE X-DIRECTION

An object of mass 3 kilograms moves 8 meters along the x-direction in 10 meters of time as measured in the laboratory. What is its energy and momentum? Its rest energy? Its kinetic energy?

What value of kinetic energy would Newton predict for this object? Using relativistic expressions, verify that the velocity of this object equals its momentum divided by its energy.

SOLUTION

From the statement of the problem:

$$m = 3 \text{ kilograms}$$
$$t = 10 \text{ meters}$$
$$x = 8 \text{ meters}$$
$$y = 0 \text{ meters}$$
$$z = 0 \text{ meters}$$

From this we obtain a value for the speed:

$$v = \frac{x}{t} = \frac{8 \text{ meters of distance}}{10 \text{ meters of time}} = 0.8$$

Use v to calculate the factor $1/(1 - v^2)^{1/2}$ in equation (7-8):

$$\frac{1}{(1 - v^2)^{1/2}} = \frac{1}{(1 - (0.8)^2)^{1/2}} = \frac{1}{(1 - 0.64)^{1/2}} = \frac{1}{(0.36)^{1/2}} = \frac{1}{0.6} = \frac{5}{3}$$

From equation (7-11) the energy is

$$E = m/(1 - v^2)^{1/2} = (3 \text{ kilograms}) (5/3) = 5 \text{ kilograms}$$

From equation (7-8) momentum has the magnitude

$$p = mv/(1 - v^2)^{1/2} = (5/3) \times (3 \text{ kilograms}) \times 0.8 = 4 \text{ kilograms}$$

Rest energy of the particle just equals its mass:

$$E_{\text{rest}} = m = 3 \text{ kilograms}$$

From equation (7-15) kinetic energy K equals total energy minus rest energy:

$$K = E - m = 5 \text{ kilograms} - 3 \text{ kilograms} = 2 \text{ kilograms}$$

The Newtonian prediction for kinetic energy is

$$K_{\text{Newton}} = \frac{1}{2}mv^2 = \frac{1}{2} \times 3 \times (0.8)^2 = 0.96 \text{ kilogram}$$

which is a lot smaller than the correct relativistic result. Even at the speed of light, the Newtonian prediction would be $K_{\text{Newton}} = 1$ kilogram, whereas relativistic value would increase without limit.

SAMPLE PROBLEM 7-2

Equation (7-16) says that velocity equals the ratio (magnitude of momentum)/(energy):

$$v = \frac{p}{E} = \frac{4 \text{ kilograms}}{5 \text{ kilograms}} = 0.8$$

This is the same value as reckoned directly from the given quantities.

From this comes the relativistic expression for kinetic energy K:

$$K = E - E_{rest} = E - m = \frac{m}{(1 - v^2)^{1/2}} - m = m\left[\frac{1}{(1 - v^2)^{1/2}} - 1\right] \qquad (7\text{-}15)$$

Box 7-2 elaborates the relation between this expression and the Newtonian expression (7-12). Notice that if we divide the respective sides of the momentum equation (7-8) by corresponding sides of the energy equation (7-11), the result gives particle speed:

$$v = \frac{p}{E} \qquad (7\text{-}16)$$

We could have predicted this directly from the first figure in this chapter, Figure 7-1. Speed v is the tilt (slope) of the worldline from the vertical: (space displacement)/(time for this displacement). Momenergy points along the worldline, with space component p and time component E. Therefore momenergy slope p/E equals worldline slope v.

Still More Units: In order to convert energy in units of mass to energy in conventional units, such as joules, multiply the expressions above by the square of light speed, c^2, and use subscript "conv":

$$E_{conv} = Ec^2 = \frac{mc^2}{[1 - (v_{conv}/c)^2]^{1/2}} \qquad \text{[good at any speed]} \quad (7\text{-}17)$$

$$E_{conv \ rest} = mc^2 \qquad \text{[particle at rest]} \quad (7\text{-}18)$$

$$K_{conv} = (E - E_{rest})c^2 = mc^2\left[\frac{1}{[1 - (v_{conv}/c)^2]^{1/2}} - 1\right] \qquad \text{[good at any speed]} \quad (7\text{-}19)$$

$$K_{conv \ Newton} = \frac{1}{2}mv^2c^2 = \frac{1}{2}m\left(\frac{v_{conv}}{c}\right)^2 c^2 = \frac{1}{2}mv_{conv}^2 \qquad \text{[low speed only]} \quad (7\text{-}20)$$

Conversion to conventional energy units

Thus conversion from energy in units of mass to energy in conventional units is always accomplished by multiplying by conversion factor c^2. This is true whether the expression for energy being converted is Newtonian or relativistic. Table 7-1 at the end of the chapter summarizes these comparisons.

Equation (7-18) is the most famous equation in all physics. Historically, the factor c^2 captured the public imagination because it witnessed to the vast store of energy available in the conversion of even tiny amounts of mass to heat and radiation. The units of mc^2 are joules; the units of m are kilograms. However, we now recognize that joules and kilograms are units different only because of historical accident. The

SAMPLE PROBLEM 7-3

MOMENERGY COMPONENTS

For each of the following cases, write down the four components of the momentum–energy 4-vector in the given frame in the form $[E, p_x, p_y, p_z]$. Each particle has mass m.

a. A particle moves in the positive x-direction in the laboratory with kinetic energy equal to three times its rest energy.

b. The same particle is observed in a rocket in which its kinetic energy equals its mass.

c. Another particle moves in the y-direction in the laboratory frame with momentum equal to twice its mass.

d. Yet another particle moves in the negative x-direction in the laboratory with total energy equal to four times its mass.

e. Still another particle moves with equal x, y, and z momentum components in the laboratory and kinetic energy equal to four times its rest energy.

SOLUTION

a. Total energy of the particle equals rest energy m plus kinetic energy $3m$. Therefore its total energy E equals $E = m + 3m = 4m$. The particle moves along the x-direction, so $p_y = p_z = 0$ and $p_x = p$, the total momentum. Substitute the value of E into the equation $m^2 = E^2 - p^2$ to obtain

$$p^2 = E^2 - m^2 = (4m)^2 - m^2 = 16m^2 - m^2 = 15m^2$$

Hence $p_x = (15)^{1/2}m$.

In summary, the components of the momenergy 4-vector are

$$[E, p_x, p_y, p_z] = [4m, (15)^{1/2}m, 0, 0]$$

Of course the magnitude of this momenergy 4-vector equals the mass of the particle m — true whatever its speed, its energy, or its momentum.

b. In this rocket frame, total energy — rest energy plus kinetic energy — has the value $E = 2m$. As before, $p^2 = E^2 - m^2 = (2m)^2 - m^2 = 4m^2 - m^2 = 3m^2$. Hence $p_x = 3^{1/2}m$ and components of the 4-vector are $[E, p_x, p_y, p_z] = [2m, 3^{1/2}m, 0, 0]$.

c. In this case $p_x = p_z = 0$ and $p_y = p = 2m$. Moreover, $E^2 = m^2 + p^2 = m^2 + (2m)^2 = 5m^2$. So, finally, $[E, p_x, p_y, p_z] = [5^{1/2}m, 0, 2m, 0]$.

d. We are given directly that $E = 4m$, the same as in part **a,** except here the particle travels in the negative x-direction so has negative x-momentum. Hence:

$$[E, p_x, p_y, p_z] = [4m, -(15)^{1/2}m, 0, 0]$$

e. Total energy equals $E = 5m$. All momentum components have equal value, say

$$p_x = p_y = p_z = P$$

In this case we use the full equation that relates energy, momentum, and mass:

$$(p_x)^2 + (p_y)^2 + (p_z)^2 = 3P^2 = E^2 - m^2 = (5m)^2 - m^2 = 24m^2$$

or $P^2 = 8m^2$ and hence $[E, p_x, p_y, p_z] = [5m, 8^{1/2}m, 8^{1/2}m, 8^{1/2}m]$.

BOX 7-2

ENERGY IN THE LOW-VELOCITY LIMIT

Energy at relativistic speeds and energy at everyday speeds: How are expressions for these two cases related?

Energy in Terms of Momentum: In the limit of velocities low compared with the speed of light, the relativistically accurate expression for energy $E = (m^2 + p^2)^{1/2}$ reduces to $E = m + p^2/(2m) +$ corrections. To see why and how, and to estimate the corrections, it is convenient to work in dimensionless ratios. Thus we focus on the accurate expression in the form $E/m = [1 + (p/m)^2]^{1/2}$, or even simpler, $y = [1 + x]^{1/2}$, and on the approximation to this result, in the form

$$E/m = 1 + (1/2)(p/m)^2 + \text{corrections, or } y = 1 + (1/2)x + \text{corrections}$$

Example: $x = 0.21$. Then our approximation formula gives $y = (1.21)^{1/2} = 1 + 0.105 + $ a correction. The accurate result is $y = 1.100$, which is the square root of 1.21. In other words, the correction is negative and extremely small: correction $= -0.005$.

Energy in Terms of Velocity: In the limit of velocities low compared with the speed of light, the relativistically accurate expression for energy $E = m/(1 - v^2)^{1/2}$, reduces to $E = m + (1/2)mv^2 + $ corrections. It is convenient again to work in dimensionless ratios. Thus we focus on the accurate expression in the form $E/m = [1 - v^2]^{-1/2}$, or even simpler, $y = [1 - x]^{-1/2}$, and on the approximation to this result, in the form

$$E/m = 1 + (1/2)v^2 + \text{corrections, or } y = 1 + (1/2)x + \text{corrections}$$

Example: $x = 0.19$. Then our approximation formula gives $y = 1 + (1/2) 0.19 + $ a correction $= 1.095 + $ a correction. The accurate result is $y = [1 - 0.19]^{-1/2} = (0.81)^{-1/2} = (0.9)^{-1} = 1.1111$ . . . In other words, the correction is positive and small: correction $= +0.01611$.

Another example: A jet plane. Take its speed to be exactly $v = 10^{-6}$. That speed, according to our approximation, brings with it a fractional augmentation of energy, a kinetic energy per unit mass, equal to $(1/2)v^2 = 5 \times 10^{-13}$ or 0.000 000 000 000 5. In contrast, the accurate expression $E/m = [1 - v^2]^{-1/2}$ gives the result $E/m = 1.000 000 000 000 500 000 000 009 375 000 000 000$ . . . The 5 a little less than halfway down the length of this string of digits is no trifle, as anyone will testify who has seen the consequences of the crash of a jet plane into a skyscraper. However, the 9375 further down the line is approximately a million million times smaller and totally negligible in its practical consequences.

In brief, low speed gives rise to a kinetic energy which, relative to the mass, is given to good approximation by $(1/2)v^2$ or by $(1/2)(p/m)^2$. Moreover, the same one or other unit-free number (a "fraction" because it is small compared to unity) automatically reveals to us the order of magnitude of the fractional correction we would have had to make in this fraction itself if we were to have insisted on a perfectly accurate figure for the kinetic energy.

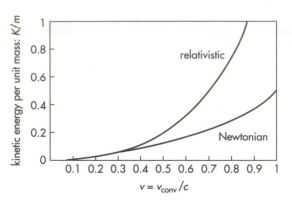

FIGURE 7-5. *Kinetic energy as a function of speed, as predicted by relativity [equation (7-19), valid for all speeds] and by Newtonian mechanics [equation (7-20), valid for low speeds only].*

conversion factor c^2, like the factor of conversion from seconds to meters or miles to feet, can today be counted as a detail of convention rather than as a deep new principle.

Central to an understanding of the equation $E_{rest} = m$ or its equivalent $E_{conv\ rest} = mc^2$ is the subscript "rest." Energy is not the same as mass! Energy is only the time part of the momenergy 4-vector. Mass is the magnitude of that 4-vector. The energy of an object, expressed in conventional units, has the value mc^2 *only* when that object is observed from a frame in which it is at rest. Observed from all other free-float frames, the energy of the object is greater than its rest energy, as shown by equation (7-17).

Energy: Time part of momenergy 4-vector
Mass: Magnitude of that 4-vector

Figure 7-5 compares relativistic and Newtonian predictions for kinetic energy per unit mass as a function of speed. At low speeds the values are indistinguishable (left side of the graph). When a particle moves with high speed, however, so that the factor $1/(1 - v^2)^{1/2}$ has a value much greater than one, relativistic and Newtonian expressions do not yield at all the same value for kinetic energy (right side of the graph). Then one must choose which expression to use in analyzing collisions and other high-speed phenomena. We choose the relativistic expression because it leads to the same value of the total energy of an isolated system before and after any interaction between particles in the system — it leads to conservation of total energy of the system.

All this talk of reconciliation at low speeds obscures an immensely powerful feature of the relativistic expression for total energy of an isolated system of particles. Total energy is conserved in *all* interactions among particles in the system: elastic and inelastic collisions as well as creations, transformations, decays, and annihilations of particles. In contrast, total kinetic energy of a system calculated using the Newtonian formula for low-speed interactions is conserved only for *elastic* collisions. Elastic collisions are *defined* as collisions in which kinetic energy is conserved. In collisions that are not elastic, kinetic energy transforms into heat energy, chemical energy, potential energy, or other forms of energy. For Newtonian mechanics of low-speed particles, each of these forms of energy must be treated separately: Conservation of energy must be invoked as a separate principle, as something beyond Newtonian analysis of mechanical energy.

Relativity: All forms of energy automatically conserved

In relativity, all these energies are included automatically in the single time component of total momenergy of a system — total energy — which is always conserved for an isolated system. Chapter 8 discusses more fully the momenergy of a system of particles and the effects of interactions between particles on the energy and mass of the system. 🐟

7.6 CONSERVATION OF MOMENERGY AND ITS CONSEQUENCES

total momenergy of an isolated system of particles is conserved

Momenergy puts us at the heart of mechanics. The relativity concept momenergy gives us the indispensable tool for mastering every interaction and transformation of particles.

What does it mean in practice to say in this language of momenergy components that the punch given to particle A by particle B in a collision is exactly equal in magnitude and opposite in spacetime direction to the punch given to B by A? That gain in momenergy of A is identical to loss of momenergy by B? That the sum of separate momenergies of A and B—this sum itself regarded as an arrow in spacetime, the arrow of *total momenergy* (Figure 7-6)—has the same magnitude and direction after the encounter that it had before? Or, in brief, how does the **principle of conservation of momenergy** translate itself into the language of components in a

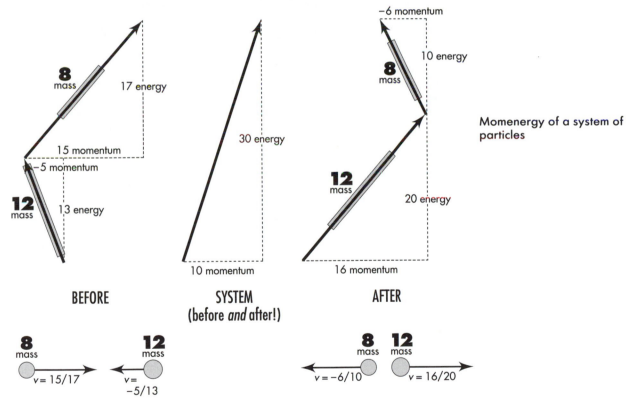

Momenergy of a system of particles

BEFORE   SYSTEM   AFTER
                    (before *and* after!)

FIGURE 7-6. *Conservation of total momenergy in a collision.* Before: *The lighter 8-unit mass, moving right with 15/17 light speed, collides with the slower and heavier 12-unit mass moving left (with 5 units of momentum to the left and 13 units of energy).* **System:** *Arrow of total momenergy of the system of two particles. Combined momentum of the colliding particles has value −5 + 15 = 10 units rightward. Combined energy of the two equals 13 + 17 = 30 units. The total system momenergy is conserved.* **After:** *One of many possible outcomes of this collision: The 8-unit mass bounces back leftward after collision, but the punch that it provided has reversed the direction of motion and increased the speed of the heavier 12-unit mass. The handle of the momenergy arrow of each particle gives the true magnitude of that momenergy, figured in the Lorentz geometry of the real physical world, as contrasted to the length of that 4-vector as it appears in the Euclidean—and therefore misleading—geometry of this sheet of paper. The scale of magnitudes in this figure is different from that of Figure 7-3.*

given free-float frame? Answer: Each *component* of the momenergy vector, when added together for particles A and B, has the same value after the collision as before the collision. In other words,

Energy of system conserved

$$\begin{pmatrix} \text{energy of } A \\ \text{before the} \\ \text{encounter} \end{pmatrix} + \begin{pmatrix} \text{energy of } B \\ \text{before the} \\ \text{encounter} \end{pmatrix} = \begin{pmatrix} \textbf{total } \text{energy} \\ \textbf{before } \text{the} \\ \text{encounter} \end{pmatrix}$$

$$= \begin{pmatrix} \textbf{total } \text{energy} \\ \textbf{after } \text{the} \\ \text{encounter} \end{pmatrix}$$

$$= \begin{pmatrix} \text{energy of } A \\ \text{after the} \\ \text{encounter} \end{pmatrix} + \begin{pmatrix} \text{energy of } B \\ \text{after the} \\ \text{encounter} \end{pmatrix}$$

called *conservation of the time part of momenergy.* Add to this three statements about the three space components of momenergy, of which the first one reads,

BOX 7-3

INVARIANT? CONSERVED? CONSTANT?

Is the speed of light a constant? An invariant? Is mass conserved in a collision? Is it an invariant? A constant? Many terms from everyday speech are taken over by science and applied to circumstances far beyond the everyday. The three useful adjectives *invariant, conserved,* and *constant* have distinct meanings in relativity .

Invariant

In relativity a quantity is invariant if it has the same value when measured by observers in different free-float frames — frames in relative motion. First among relativistic invariants is the speed of light: It has the same value when reckoned using data from the laboratory latticework of recording clocks as when figured using data from the rocket latticework. A second central invariant is the interval between two events: All inertial observers agree on the interval (proper time or proper distance). A third mighty invariant is the mass of a particle. There are many other invariants, every one with its special usefulness.

Some very important quantities do *not* qualify as invariants. The time between two events is not an invariant. It differs as measured by observers in relative motion. Neither is the distance between events an invariant. It too differs from one frame to another. Neither the energy nor the momentum of a particle is an invariant.

Conserved

A quantity is conserved if it has the same value before and after some encounter or does not change during some interaction. The total momenergy of an isolated system of particles is conserved in an interaction among the particles. In a given free-float frame this means that the total energy is conserved. So is each component of total momentum. The magnitude of total momenergy of a system — the mass of that system — is also conserved in an interaction. On the other hand, the sum of the individual masses of the

$$\begin{pmatrix}\text{eastward component}\\\text{of momentum of }A\\\text{before the encounter}\end{pmatrix} + \begin{pmatrix}\text{eastward component}\\\text{of momentum of }B\\\text{before the encounter}\end{pmatrix}$$

$$= \begin{pmatrix}\text{eastward component}\\\text{of \textbf{total} momentum}\\\textbf{before} \text{ the encounter}\end{pmatrix}$$

$$= \begin{pmatrix}\text{eastward component}\\\text{of \textbf{total} momentum}\\\textbf{after} \text{ the encounter}\end{pmatrix}$$

$$= \begin{pmatrix}\text{eastward component}\\\text{of momentum of }A\\\text{after the encounter}\end{pmatrix} + \begin{pmatrix}\text{eastward component}\\\text{of momentum of }B\\\text{after the encounter}\end{pmatrix}$$

Momentum of system conserved

called *conservation of the space part of momenergy*. Figure 7-6 illustrates the conservation of momenergy in a recoil collision between two particles. Momentum is laid out

constituent particles of a system ordinarily is *not* conserved in a relativistic interaction. (For examples, see Chapter 8.)

Constant

Something that is constant does not change with time. The speed of the Great Pyramid with respect to the rock plateau of Giza is constant — equal to zero, or at least less than one millimeter per millennium. This speed may be constant, but it is not an invariant: As observed from a passing rocket, the Great Pyramid moves with blinding speed! Is the speed of the Great Pyramid conserved? Conserved during what encounter? There is no *before* or *after* to which the term "conserved" can refer. The term "conserved" simply does not apply to the speed of the Great Pyramid.

It is true that the speed of light in a vacuum is constant — it does not change with time. It is also true, but an entirely different statement, that the speed of light is an invariant — has the same value measured by different observers in uniform relative motion. It is true that total momenergy of an isolated system is constant — does not change with time. It is also true, but an entirely different statement, that total momenergy of an isolated system is conserved in a collision or interaction among particles in that system.

> When anyone hears the word *invariant, conserved,* or *constant,* she is well-advised to listen for the added phrase *with respect to,* which should always be expressed or implied. Usually (but not always) *constant* means *with respect to* the passage of time. *Conserved* usually (but not always) means *with respect to* a collision or interaction. *Invariant* can have at least as many meanings as there are geometries to describe Nature: In Euclidean geometry, *distance* is invariant as measured *with respect to* relatively rotated coordinate axes. In Lorentz geometry, *interval* and *mass* are invariants as measured *with respect to* free-float frames in relative motion. The full meaning of the word *invariant* or *conserved* or *constant* depends on the condition under which this property is invoked.

right and left on the page; energy is marked off vertically. The left diagram shows two particles before collision and their momentum-energy vectors. The right diagram shows the corresponding display after the collision.

The center diagram shows total momenergy of the **system** of two particles. The momenergy vectors of the two particles *before* the collision add up to this total; the momenergy vectors of the two particles *after* the collision add up to the same total. Total momenergy of the system has the same value after as before: it is *conserved* in the collision.

Momenergy of system conserved!

Well, you've done it again: You've given us a powerful tool that seems impossible to visualize. How can one think about this momenergy 4-vector, anyway? Can you personally picture it in your mind's eye?

We can *almost* visualize the momenergy arrow, by looking at Figure 7-6 for example. There momentum and energy *components* of a given momenergy vector have their correct relative values. And the *direction* of the momenergy arrow in spacetime is correctly represented in the diagram.

However, the *magnitude* of this arrow — mass of the particle — does *not* correspond to its length in the momenergy diagram. This is because mass is reckoned from the *difference* of squares of energy and momentum, whereas length of a line on the Euclidean page of a book is computed from the *sum* of squares of horizontal and vertical dimensions. The handle or thickened region on the typical arrow and the big, boldface number for mass remind us of the failure — the lie — that results from trying to represent momenergy on such a page.

To observe a given momenergy 4-vector first from one free-float frame, then from another, and then from another (Figure 7-3) is to see the apparent direction of the arrow changing. The change in frame brings with it changes in the energy and momentum components. However, magnitude does not change. Mass does not change. To examine the momenergy 4-vector of a particle in different frames is to gain improved perspective on what momenergy is and does.

See if this analogy helps: The momentum–energy 4-vector is like a tree. The tree has a location for its base and for its tip whether or not we choose this, that, or the other way to measure it. The shadow the tree casts on the ground, however, depends upon the tilt of the tree and the location of Sun in the sky.

Likewise, momenergy of a particle as it passes through a given event on its worldline has a magnitude and direction, a fixity in spacetime, independent of any choice we make of free-float frame from which to observe and measure it. No means of reporting momenergy is more convenient for everyday purposes than separate specification of momentum and energy of the object in question in some chosen free-float frame. Those two quantities separately, however, are like the shadow of the tree on the ground. As Sun rises the shadow shortens. Similarly the momentum of a car or spaceship depends on the frame in which we see it. In one frame, terrifying. In another frame, tame. In a comoving frame, zero momentum, as the tree's shadow disappears when Sun lies in exactly that part of the sky to which the tilted tree points. In such a special frame of reference, the time component of an object's momenergy — that is, its energy — takes on its minimum possible value, which is equal to the mass itself of that object. However, in whatever free-float frame we observe it, the arrow of momenergy clings to the same course in spacetime, maintains the same length, manifests the same mass.

7.7 SUMMARY

momenergy of an object unifies energy, momentum, and mass

The **momenergy 4-vector** of a particle equals its mass multiplied by the ratio of its spacetime displacement to proper time—wristwatch time—for that displacement (Section 7.2):

$$\begin{pmatrix} \text{momenergy} \\ \text{4-vector} \end{pmatrix} = (\text{mass}) \begin{pmatrix} \dfrac{\begin{array}{c}\text{spacetime} \\ \text{displacement} \\ \text{4-vector}\end{array}}{\begin{array}{c}\text{proper time} \\ \text{for that} \\ \text{displacement}\end{array}} \end{pmatrix} \qquad (7\text{-}1)$$

Momenergy of a particle is a 4-vector. It possesses magnitude equal to the particle's mass. The momenergy at any given event in the motion of the particle points in the direction of the worldline at that event (Section 7.2).

The momenergy of a particle has an existence independent of any frame of reference.

The terms momenergy, momentum, and energy, as we deal with them in this book, all have a common unit: mass. In older times mass, momentum, and energy were all conceived of as different in nature and therefore were expressed in different units. The conventional units are compared with mass units in Table 7-1.

The magnitude of the momenergy 4-vector of a particle is reckoned from the difference of the squares of energy and momentum components in any given frame (Section 7.3):

$$m^2 = E^2 - (p_x)^2 - (p_y)^2 - (p_z)^2$$

or, more simply,

$$m^2 = E^2 - p^2 = (E')^2 - (p')^2 \qquad (7\text{-}3)$$

Mass m of the particle is an *invariant,* has the same numerical value when computed using energy and momentum components in the laboratory frame (unprimed components) as in any rocket frame (primed components).

In a given inertial frame, the momenergy 4-vector of a particle has four **components.** Three space components describe the momentum of the particle in that frame (Sections 7.3 and 7.4):

$$p_x = m\,\frac{dx}{d\tau}$$

$$p_y = m\,\frac{dy}{d\tau} \qquad (7\text{-}2)$$

$$p_z = m\,\frac{dz}{d\tau}$$

The magnitude of the momentum can be expressed as the factor $1/(1 - v^2)^{1/2}$ times the Newtonian expression for momentum mv. The result is

$$p = mv/(1 - v^2)^{1/2} \qquad (7\text{-}8)$$

The "time part" of the momenergy 4-vector in a given inertial frame equals energy of the particle in that frame (Sections 7.3 and 7.5):

$$E = m \frac{dt}{d\tau} = \frac{m}{(1 - v^2)^{1/2}} \qquad \text{(7-2), (7-11)}$$

For a particle at rest, the energy of the particle has a value equal to its mass:

$$E_{\text{rest}} = m \qquad \text{(7-13)}$$

For a moving particle, the energy combines two parts: rest energy—equal to mass of the particle—plus the additional kinetic energy K that the particle has by virtue of its motion:

$$E = E_{\text{rest}} + K = m + K \qquad \text{(7-14)}$$

From these equations comes an expression for **kinetic energy**:

$$K = E - m = m \left[\frac{1}{(1 - v^2)^{1/2}} - 1 \right] \qquad \text{(7-15)}$$

The momenergy 4-vector derives from **conservation** its power to analyze particle interactions. Conservation states that the total momenergy 4-vector of an isolated system of particles is conserved, no matter how particles in the system interact with one another or transform themselves. This conservation law holds independent of choice of the free-float frame in which we employ it (Section 7.6).

In any given inertial frame, conservation of total momenergy of an isolated system breaks apart into four conservation laws:

1. Total energy of the system before an interaction equals total energy of the system after the interaction.

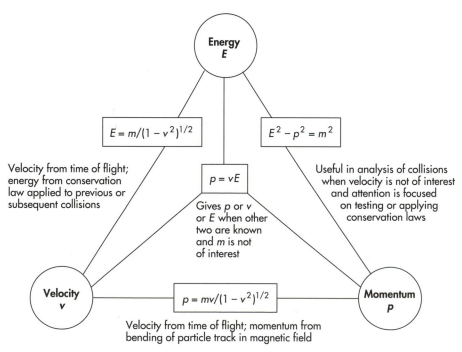

FIGURE 7-7. *Formulas that relate momentum, energy, mass, and velocity of an object, and notes about their uses in analyzing experiments. In this diagram, p is the magnitude of the momentum.*

TABLE 7-1
QUANTITIES RELATING TO MOMENERGY

In units of mass (for example, E and p both in kilograms; x, y, z, t, τ in meters)	Reference equations	In conventional units (for example, E_{conv} in joules, p_{conv} in kilogram meters/second; t_{conv} in seconds)
Energy $\quad E = m\dfrac{dt}{d\tau} = \dfrac{m}{(1-v^2)^{1/2}}$	(7-2, 5, 11, 17)	$E_{\text{conv}} = \dfrac{mc^2}{[1-(v_{\text{conv}}/c)^2]^{1/2}}$
Rest energy $\quad E_{\text{rest}} = m$	(7-13, 18)	$E_{\text{conv rest}} = mc^2$
Kinetic energy $\quad K = m\left(\dfrac{1}{(1-v^2)^{1/2}} - 1\right)$	(7-15, 19)	$K_{\text{conv}} = mc^2\left(\dfrac{1}{[1-(v_{\text{conv}}/c)^2]^{1/2}} - 1\right)$
Momentum $\quad p = \dfrac{mv}{(1-v^2)^{1/2}}$	(7-8, 10)	$p_{\text{conv}} = \dfrac{mv_{\text{conv}}}{[1-(v_{\text{conv}}/c)^2]^{1/2}}$
Momentum components $\quad p_x = m\dfrac{dx}{d\tau} = \dfrac{mv_x}{(1-v^2)^{1/2}}$	(7-2, 5)	$p_{x\text{ conv}} = \dfrac{mv_{x\text{ conv}}}{[1-(v_{\text{conv}}/c)^2]^{1/2}}$
$\quad p_y = m\dfrac{dy}{d\tau} = \dfrac{mv_y}{(1-v^2)^{1/2}}$	(7-2, 5)	$p_{y\text{ conv}} = \dfrac{mv_{y\text{ conv}}}{(1-(v_{\text{conv}}/c)^2)^{1/2}}$
$\quad p_z = m\dfrac{dz}{d\tau} = \dfrac{mv_z}{(1-v^2)^{1/2}}$	(7-2, 5)	$p_{z\text{ conv}} = \dfrac{mv_{z\text{ conv}}}{(1-(v_{\text{conv}}/c)^2)^{1/2}}$
Mass $\quad m^2 = E^2 - p^2$	(7-3)	$m^2c^4 = E_{\text{conv}}^2 - p_{\text{conv}}^2 c^2$
Particle speed $\quad v = \dfrac{p}{E}$	(7-16)	$v_{\text{conv}} = \dfrac{p_{\text{conv}} c^2}{E_{\text{conv}}}$
Newtonian low-speed limit **Kinetic energy** $\quad K_{\text{Newton}} = \dfrac{1}{2}mv^2$	(7-12, 20)	$K_{\text{conv Newton}} = \dfrac{1}{2}mv_{\text{conv}}^2$
Momentum $\quad p_{\text{Newton}} = mv$	(7-7, 9)	$p_{\text{conv Newton}} = mv_{\text{conv}}$
Momentum components $\quad p_{x\text{ Newton}} = mv_x$		$p_{x\text{ conv Newton}} = mv_{x\text{ conv}}$
$\quad p_{y\text{ Newton}} = mv_y$		$p_{y\text{ conv Newton}} = mv_{y\text{ conv}}$
$\quad p_{z\text{ Newton}} = mv_z$		$p_{z\text{ conv Newton}} = mv_{z\text{ conv}}$

2. Total x-momentum of the system is the same before and after the interaction.
3. Total y-momentum of the system is the same before and after the interaction.
4. Total z-momentum of the system is the same before and after the interaction.

In this chapter we have developed expressions that relate energy, momentum, mass, and velocity. Which of these expressions is useful depends upon circumstances and the system we are trying to analyze. Figure 7-7 summarizes these equations and circumstances under which they may be useful. Table 7-1 compares energy and momentum in units of mass and in conventional units. ✒

ACKNOWLEDGMENT

The authors are grateful to William A. Shurcliff for the idea of a "handle" that displays on a Euclidean page the invariant magnitude of a particle's momenergy 4-vector—a magnitude equal to the mass of the particle.

CHAPTER 7 EXERCISES

PRACTICE

7-1   momenergy 4-vector

For each of the following cases, write down the four components of the momentum–energy (momenergy) 4-vector in the given frame in the form $[E, p_x, p_y, p_z]$. Assume that each particle has mass m. You may use square roots in your answer.

a   A particle moves in the positive x-direction in the laboratory with total energy equal to five times its rest energy.

b   Same particle as observed in a frame in which it is at rest.

c   Another particle moves in the z-direction with momentum equal to three times its mass.

d   Yet another particle moves in the negative y-direction with kinetic energy equal to four times its mass.

e   Still another particle moves with total energy equal to ten times its mass and x-, y-, and z-components of momentum in the ratio 1 to 2 to 3.

7-2   system mass

Determine the mass of the system of particles shown in Figure 7-6. Is this system mass equal to the sum of the masses of the individual particles in the system? Does the mass of this system change as a result of the interaction? Does the momenergy 4-vector of the system change as a result of the interaction? (In Chapter

8 there is a lot more discussion about the mass of a system of particles.)

7-3   much ado about little

Two freight trains, each of mass 5×10^6 kilograms (5000 metric tons) travel in opposite directions on the same track with equal speeds of 42 meters/second (about 100 miles/hour). They collide head on and come to rest.

a   Calculate in milligrams the kinetic energy for each train $(1/2)mv^2$ before the collision. (Newtonian expression OK for 100 mph!) (1 milligram $= 10^{-3}$ gram $= 10^{-6}$ kilogram)

b   After the collision, the mass of the trains plus the mass of the track plus the mass of the roadbed has increased by what number of milligrams? Neglect energy lost in the forms of sound and light.

7-4   fast protons

Each of the protons described in the table emits a flash of light every meter of its own (proper) time $d\tau$. Between successive flash emissions, each proton travels a distance given in the left column. Complete the table. Take the rest energy of the proton to be equal to 1 GeV $= 10^9$ eV and express momentum in the same units. Hints: Avoid calculating or using the speed v in relativistic particle problems; it is too close to unity to distinguish between protons of radically different energies. An accuracy of two significant fig-

---

EXERCISE 7-4

FAST PROTONS

Lab distance Δx traveled between flashes (meters)	Momentum $m\,dx/d\tau$ (GeV)	Energy (GeV)	Time stretch factor γ	Lab time between flashes (meters)
0				
0.1				
1				
5				
10				
10^3				
10^6				

ures is fine; don't give more. Recall: $E^2 - p^2 = m^2$ and $E = m\,dt/d\tau = m\gamma$ [note tau!].

PROBLEMS

7-5   Lorentz transformation for momenergy components

The rocket observer measures energy and momentum components of a particle to have the values E' and p_x', p_y', and p_z'. What are the corresponding values of energy and momentum measured by the laboratory observer? The answer comes from the Lorentz transformation, equation (L-10) in the Special Topic following Chapter 3.

The moving particle emits a pair of sparks closely spaced in time as measured on its wristwatch. The rocket latticework of clocks records these emission events; so does the laboratory latticework of clocks. The rocket observer constructs components of particle momentum and energy, equation (7-2), from knowledge of particle mass m, the spacetime displacements dt', dx', dy', and dz' derived from the event recordings, and the proper time $d\tau$ computed from these spacetime components. Laboratory momenergy components come from transforming the spacetime displacements. The Lorentz transformation, equation (L-10), for incremental displacements gives

$$dt = v\gamma dx' + \gamma dt'$$
$$dx = \gamma dx' + v\gamma dt'$$
$$dy = dy'$$
$$dz = dz'$$

a   Multiply both sides of each equation by the invariant mass m and divide through by the invariant proper time $d\tau$. Recognizing the components of the momenergy 4-vector in equation (7-2), show that the transformation equations for momenergy are

$$E = v\gamma p'_x + \gamma E'$$
$$p_x = \gamma p'_x + v\gamma E'$$
$$p_y = p'_y$$
$$p_z = p'z$$

b   Repeat the process for particle displacements dt, dx, dy, and dz recorded in the laboratory frame to derive the inverse transformations from laboratory to rocket.

$$E' = -v\gamma p_x + \gamma E$$
$$p'_x = \gamma p_x - v\gamma E$$
$$p'_y = p_y$$
$$p'_z = p_z$$

7-6   fast electrons

The Two-Mile Stanford Linear Accelerator accelerates electrons to a final kinetic energy of 47 GeV (47×10^9 electron-volts; one electron-volt $= 1.6 \times 10^{-19}$ joule). The resulting high-energy electrons are used for experiments with elementary particles. Electromagnetic waves produced in large vacuum tubes ("klystron tubes") accelerate the electrons along a straight pipelike structure 10,000 feet long (approximately 3000 meters long). Take the rest energy of an electron to be $m \approx 0.5$ MeV $= 0.5 \times 10^6$ electron-volts.

a   Electrons increase their kinetic energy by approximately equal amounts for every meter traveled along the accelerator pipe as observed in the laboratory frame. What is this energy gain in MeV/meter? Suppose the Newtonian expression for kinetic energy were correct. In this case how far would the electron travel along the accelerator before its speed were equal to the speed of light?

b   In reality, of course, even the 47-GeV electrons that emerge from the end of the accelerator have a speed v that is less than the speed of light. What is the value of the difference $(1 - v)$ between the speed of light and the speed of these electrons as measured in the laboratory frame? [Hint: For v very near the value unity, $1 - v^2 = (1 + v)(1 - v) \approx 2(1 - v)$.] Let a 47-GeV electron from this accelerator race a flash of light along an evacuated tube straight through Earth from one side to the other (Earth diameter 12,740 kilometers). How far ahead of the electron is the light flash at the end of this race? Express your answer in millimeters.

c   How long is the "3000-meter" accelerator tube as recorded on the latticework of rocket clocks moving along with a 47-GeV electron emerging from the accelerator?

7-7   super cosmic rays

The Haverah Park extensive air shower array near Leeds, England, detects the energy of individual cosmic ray particles indirectly by the resulting shower of particles this cosmic ray creates in the atmosphere. Between 1968 and 1987 the Haverah Park array detected more than 25,000 cosmic rays with energies greater than 4×10^{17} electron-volts, including 5 with an energy of approximately 10^{20} electron-volts. (rest energy of the proton $\approx 10^9$ electron-volts $= 1.6 \times 10^{-10}$ joule)

a   Suppose a cosmic ray is a proton of energy 10^{20} electron-volts. How long would it take this proton to cross our galaxy as measured on the proton's wristwatch? The diameter of our galaxy is approximately

10^5 light-years. How many centuries would this trip take as observed in our Earth-linked frame?

b The research workers at Haverah Park find no evidence of an upper limit to cosmic ray energies. A proton must have an energy of how many times its rest energy for the diameter of our galaxy to appear to it Lorentz-contracted to the diameter of the proton (about 1 fermi, which is equal to 10^{-15} meters)? How many metric tons of mass would have to be converted to energy with 100-percent efficiency in order to give a proton this energy? One metric ton equals 1000 kilograms.

Reference: M. A. Lawrence, R. J. O. Reid, and A. A. Watson, *Journal of Physics G: Nuclear and Particle Physics,* Volume 17, pages 733-757 (1991).

7-8 rocket nucleus

A radioactive decay or "inverse collision" is observed in the laboratory frame, as shown in the figure.

Suppose that $m_A = 20$ units, $m_C = 2$ units, and $E_C = 5$ units.

a What is the total energy E_A of particle A?

b From the conservation of energy, find the total energy E_D (rest plus kinetic) of particle D.

c Using the expression $E^2 - p^2 = m^2$ find the momentum p_C of particle C.

d From the conservation of momentum, find the momentum p_D of particle D.

e What is the mass m_D of particle D?

f Does $m_C + m_D$ after the collision equal m_A before the collision? Explain your answer.

g Draw three momenergy diagrams for this reaction similar to those of Figure 7-6: BEFORE, SYSTEM, and AFTER. Plot positive and negative momentum along the positive and negative horizontal direction, respectively, and energy along the vertical direction. On the AFTER diagram draw the momenergy vectors for particles C and D head to tail so that they they add up to the momenergy vector for the system. Place labeled mass handles on the arrows in all three diagrams, including the arrow for the system.

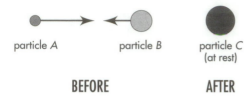

EXERCISE 7-9. *Two particles collide to form a third at rest in the laboratory frame.*

7-9 sticky collision

An inelastic collision is observed in the laboratory frame, as shown in the figure. Suppose that $m_A = 2$ units, $E_A = 6$ units, $m_C = 15$ units.

a From the conservation of energy, what is the energy E_B of particle B?

b What is the momentum p_A of particle A? Therefore what is the momentum p_B of particle B?

c From $m^2 = E^2 - p^2$ find the mass m_B of particle B.

d Quick guess: Is the mass of particle C after the collision less than or greater than the sum of the masses of particles A and B before the collision? Validate your guess from the answer to part **c**.

7-10 colliding putty balls

A ball of putty of mass m and kinetic energy K streaks across the frozen ice of a pond and hits a second identical ball of putty initially at rest on the ice. The two stick together and skitter onward as one unit. Referring to the figure, find the mass of the combined particle using parts **a–e** or some other method.

a What is the total energy of the system before the collision? Keep the kinetic energy K explicitly, and don't forget the rest energies of both particles A and B. Therefore what is the total energy E_C of particle C after the collision?

b Using the equation $m^2 = E^2 - p^2 = (m + K)^2 - p^2$ find the momentum p_A of particle A before the collision. What is the total momentum of the system before the collision? Therefore what is the momentum p_C of particle C after the collision?

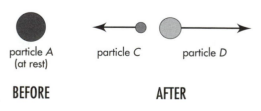

EXERCISE 7-8. *Radioactive decay of a particle.*

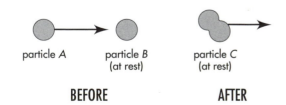

EXERCISE 7-10. *Two putty balls stick together.*

c Again use the equation $m^2 = E^2 - p^2$ to find the mass m_C of particle C. Show that the result satisfies the equation

$$m_C^2 = (2m)^2 + 2mK = (2m)^2\left(1 + \frac{K}{2m}\right)$$

d Examine the result of part **c** in two limiting cases. (1) The value of m_C in the Newtonian low-velocity limit in which kinetic energy is very much less than mass: $K/m \ll 1$. Is this what one expects from everyday living? (2) What is the value of m_C in the highly relativistic limit in which $K/m \gg 1$? What is the upper limit on the value of m_C? Discussion: Submicroscopic particles moving at extreme relativistic speeds rarely stick together when they collide. Rather, their collision often leads to creation of additional particles. See Chapter 8 for examples.

e Discussion question: Are the results of part **c** changed if the resulting blob of putty rotates, whirling like a dumbbell about its center as it skitters along?

7-11 limits of Newtonian mechanics

a One electron-volt (eV) is equal to the increase of kinetic energy that a singly charged particle experiences when accelerated through a potential difference of one volt. One electron-volt is equal to 1.60×10^{-19} joules. Verify the rest energies of the electron and the proton (masses listed inside the back cover) in units of million electron-volts (MeV).

b The kinetic energy of a particle of a given velocity v is not correctly given by the expression $1/2\ mv^2$. The error

$$\frac{\left(\begin{array}{c}\text{relativistic expression}\\\text{for kinetic energy}\end{array}\right) - \left(\begin{array}{c}\text{Newtonian expression}\\\text{for kinetic energy}\end{array}\right)}{\left(\begin{array}{c}\text{Newtonian expression}\\\text{for kinetic energy}\end{array}\right)}$$

is one percent when the Newtonian kinetic energy has risen to a certain fraction of the rest energy. What fraction? Hint: Apply the first three terms of the binomial expansion

$$(1 + z)^n = 1 + nz + \frac{1}{2}n(n - 1)z^2 + \dots$$

to the relativistic expression for kinetic energy, an accurate enough approximation if $|z| \ll 1$. Let this point — where the error is one percent — be arbitrarily called the "limit of Newtonian mechanics." What

is the speed of the particle at this limit? At what kinetic energy does a proton reach this limit (energy in MeV)? An electron?

c An electron in a modern color television tube is accelerated through a voltage as great as 25,000 volts and then directed by a magnetic field to a particular pixel of luminescent material on the inner face of the tube. Must the designer of color television tubes use special relativity in predicting the trajectories of these electrons?

7-12 derivation of the relativistic expression for momentum — a worked example

A very fast particle interacts with a very slow particle. If the collision is a glancing one, the slow particle may move as slowly after the collision as before. Reckon the momentum of the slow-moving particle using the Newtonian expression. Now demand that momentum be conserved in the collision. From this derive the relativistic expression for momentum of the fast-moving particle.

The top figure shows such a glancing collision. After the collision each particle has the same speed as before the collision, but each particle has changed its direction of motion.

Behind this figure is a story. Ten million years ago, and in another galaxy nearly ten million light-years distant, a supernova explosion launched a proton toward Earth. The energy of this proton far exceeded anything we can give to protons in our earthbound particle accelerators. Indeed, the speed of the proton so nearly approached that of light that the proton's wristwatch read a time lapse of only one second between launch and arrival at Earth.

We on Earth pay no attention to the proton's wristwatch. For our latticework of Earth-linked observers, ages have passed since the proton was launched. Today our remote outposts warn us that the streaking proton approaches Earth. Exactly one second on our clocks before the proton is due to arrive, we launch our own proton at the slow speed one meter/second almost perpendicular to the direction of the incoming proton (BEFORE part of the top figure). Our proton saunters the one meter to the impact point. The two protons meet. So perfect is our aim and timing that after the encounter our proton simply reverses direction and returns with the same speed we gave it originally (AFTER part of the top figure). The incoming proton also does not change speed, but it is deflected upward at the same angle at which it was originally slanting downward.

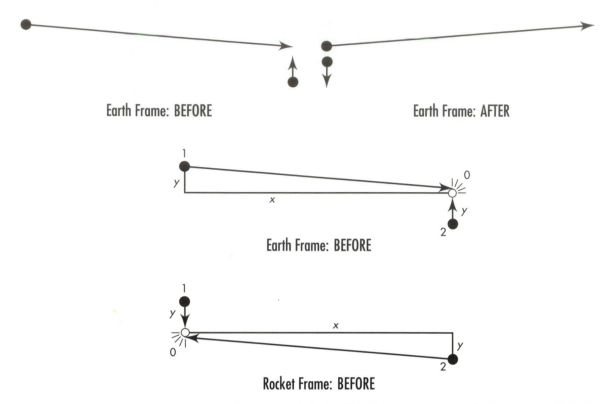

Earth Frame: BEFORE   Earth Frame: AFTER

Earth Frame: BEFORE

Rocket Frame: BEFORE

EXERCISE 7-12. *Top: A symmetric elastic collision between a fast proton and a slow proton in which each proton changes direction but not speed as a result of the encounter.* **Center:** *Events and separations as observed in Earth frame before the collision. Here* x $=$ *10 million light-years and* y $=$ *1 meter, so these figures are not to scale!* **Bottom:** *Events and separations as observed in the rocket frame before the collision.*

How much does y-momentum of our slow-moving proton change during this encounter? Newton can tell us. At a particle speed of one meter/second, his expression for momentum, mv, is accurate. Our proton simply reverses its direction. Therefore the change in its momentum is just 2 mv, twice its original momentum in the y-direction.

What is the change in the y-momentum of the incoming proton, moving at extreme relativistic speed? We demand that the change in y-momentum of the fast proton be equal in magnitude and opposite in direction to the change in y-momentum of our slow proton. In brief, y-momentum is conserved. This demand, plus a symmetry argument, leads to the relativistic expression for momentum.

Key events in our story are numbered in the center figure. Event 1 is the launching of the proton from the supernova ten million years (in our frame) before the impact. Event 2 is the quiet launch of our local proton one second (in our frame) before the impact. Event 0 is the impact itself. The x-direction is chosen so that y-displacements of both protons have equal magnitude between launch and impact, namely one meter.

Now view the same events from a rocket moving along the x-axis at such a speed that events 1 and 0 are

vertically above one another (bottom figure). For the rocket observer the transverse y-separations are the same as for the Earth observer (Section 3.6), so $y = 1$ meter in both frames. The order of events 1 and 2, however, is exactly reversed in time: For the rocket observer, we released our proton at high speed ten million years before impact and she releases hers one second before the collision. Otherwise the diagrams are symmetrical: To make the bottom figure look like the center one, exchange event numbers 1 and 2, then stand on your head!

Rocket observer and Earth observer do not agree on the time between events 1 and 0, but they agree on the proper time τ_{10} between them, namely one second. They also agree on the proper time τ_{20} between events 2 and 0. Moreover, because of the symmetry between the center and bottom figures, these two proper times have the same value: For the case we have chosen, the wristwatch (proper) time for each proton is one second between launch and impact.

$$\tau_{10} = \tau_{20}$$

We can use these quantities to construct expressions for the y-momenta of the two protons. Both are protons, so their masses m are the same and have the same invariant value for both observers. Because of the equality in magnitude of the y-displacements and the equality of τ_{20} and τ_{10}, we can write

$$m \frac{y}{\tau_{10}} = -m \frac{y}{\tau_{20}} \qquad \text{[both frames]}$$

The final key idea in the derivation of the relativistic expression for momentum is that the slow-moving proton travels between events 2 and 0 in an Earth-measured time that is very close in value to the proper time between these events. The vertical separation y between events 2 and 0 is quite small: one meter. In the same units, the time between them has a large value in the Earth frame: one second, or 300 million meters of light-travel time. Therefore, for such a slow-moving proton, the proper time τ_{20} between events 2 and 0 is very close to the Earth time t_{20} between these events:

$$\tau_{20} \approx t_{20} \qquad \text{[Earth frame only]}$$

Hence rewrite the both-frames equation for the Earth frame:

$$m \frac{y}{\tau_{10}} = -m \frac{y}{t_{20}} \qquad \text{[Earth frame only]}$$

The right side of this equation gives the y-momentum of the slow proton before the collision, correctly calculated using the Newtonian formula. The change in momentum of the slow proton during the collision is twice this magnitude. Now look at the left side. We claim that the expression on the left side is the y-momentum of the very fast proton. The y-momentum of the fast proton also reverses in the collision, so the change is just twice the value of the left side. In brief, this equation embodies the conservation of the y-component of total momentum in the collision. Con-

clusion: The left side of this equation yields the relativistic expression for y-momentum: mass times displacement divided by *proper* time for this displacement.

What would be wrong with using the Newtonian expression for momentum on the left side as well as on the right? That would mean using earth time t_{10} instead of proper time τ_{10} in the denominator of the left side. But t_{10} is the time it took the fast proton to reach Earth from the distant galaxy as recorded in the Earth frame—ten million years or 320 million million seconds! With this substitution, the equation would no longer be an equality; the left side would be 320 million million times smaller in value than the right side (smaller because t_{10} would appear in the denominator). Nothing shows more dramatically than this the radical difference between Newtonian and relativistic expressions for momentum—and the correctness of the relativistic expression that has proper time in the denominator.

This derivation of the relativistic expression for momentum deals only with its y-component. But the choice of y-direction is arbitrary. We could have interchanged y and x axes. Also the expression has been derived for particles moving with constant velocity before and after the collision. When velocity varies with time, the momentum is better expressed in terms of incremental changes in space and time. For a particle displacement dr between two events a proper time $d\tau$ apart, the expression for the magnitude of the momentum is

$$p = m \frac{dr}{d\tau}$$

One-sentence summary: In order to preserve conservation of momentum for relativistic collisions, simply replace Newton's "universal time" t in the expression for momentum with Einstein's invariant proper time τ.

CHAPTER 8

COLLIDE. CREATE. ANNIHILATE.

8.1 THE SYSTEM

an isolated island of violence

Particle physics is one of the great adventures of our time. No one can venture into the heart of it without momenergy as guide and lamp. Particles clash, yes. But however cataclysmic the encounter, it always displays one great simplicity. It takes place on a local stage, an island of violence, apart from all happenings in the outside world. In other isolated arenas of action football players form a team, actors a troupe, soldiers a platoon; but in a battle of matter and energy, the participants receive the name **system**.

What the action starts with, what particles there are, what speeds they have, what directions they take: that's the story of the system at the start of the action. We may or may not pursue in all detail every stage of every encounter, as we view the scenes of a play or watch the episodes of a game. However, nothing that claims to be an account of the clash, brief though it may be, is worthy of the name unless it reports every participant that leaves the scene with its speed and its direction. Departing, they still belong to the system. Moreover, at every step of the way from entry to departure we continue to use for the collection of participants the name *system*.

The child keeps count of who wins and who loses in the shoot-out before he or she learns to ask questions of right and wrong, of why and wherefore. We likewise keep tabs on what goes into an encounter and what comes out only to the extent of broadcasting the participants' momenergies before and after the act of violence. We do not open up in this book the more complex story of the forces, old and new, that govern the chances for this, that, and the other outcome of a given encounter. We limit ourselves to the ground rules of momenergy conservation in an isolated system. 🖎

Keep score of momenergy for the system

221

8.2 THREE MODEST EXPERIMENTS

elastic glass balls; inelastic wads of gum; weighing heat

A collision does not have to be violent to qualify for attention nor be exotic to make momenergy scorekeeping interesting. It is fun to begin with momenergy scorekeeping for three encounters of everyday kinds before strolling out onto the laboratory floor of high-energy particle physics.

Elastic collision: Momenergy automatically conserved

First Experiment: Elastic Collision. Suspend two identical glass marbles from the ceiling by two threads of the same length so that the marbles hang, at rest, just barely touching. Draw one back with the finger and release it (Figure 8-1). The released marble gathers speed. The speed peaks just as the first marble collides with the second. The collision is elastic: Total kinetic energy before the collision equals total kinetic energy after the collision. The elastic collision brings the first marble to a complete stop. The impact imparts to the second all the momentum the first one had. Conservation of momentum could not be clearer:

$$
\begin{pmatrix} \text{total momentum} \\ \text{to the right just} \\ \textbf{before} \text{ the collision,} \\ \text{all of it resident on} \\ \text{the } \textbf{first} \text{ marble} \end{pmatrix}
=
\begin{pmatrix} \text{total momentum} \\ \text{to the right just} \\ \textbf{after} \text{ the collision,} \\ \text{all of it resident on} \\ \text{the } \textbf{second} \text{ marble} \end{pmatrix}
$$

And energy? In the collision the two particles exchange roles. The first particle comes to a halt. The second particle moves exactly as the first one did before the collision. Hence energy too is clearly conserved.

Just before the collision and just after: How do conditions compare? Same total momentum. Same total energy. Therefore same total momenergy.

Inelastic collision: Momenergy also conserved

Second Experiment: Inelastic Collision. Replace the two glass marbles by two identical balls of putty, wax, or chewing gum (Figure 8-2). Pull them aside by equal amounts and release.

Both released balls of chewing gum gather speed, moving toward one another. The equal and opposite velocities peak just before they collide with each other. By symmetry, the momentum of the right-moving particle has the same magnitude as the momentum of the left-moving particle. However, these momenta point in opposite directions. Regarded as vectors, they sum to zero. The momentum of the *system* therefore equals zero just before the collision.

Just after the collision? The two balls have stuck together. They are both at rest; each has zero momentum. Their combined momentum is also zero. In other words,

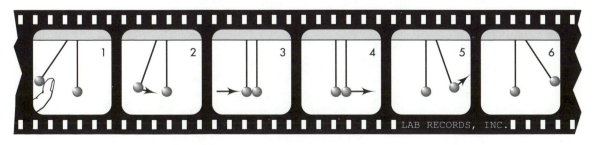

FIGURE 8-1. *One marble collides elastically with another.*

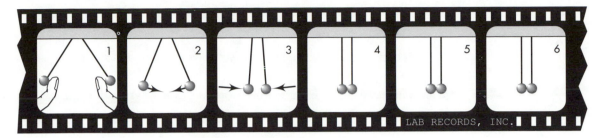

FIGURE 8-2. *One ball of chewing gum locks onto the other.*

the momentum of the *system* is zero after the collision. Zero it was also before the collision. Thus the momentum of the system is conserved.

For system energy the outcome is more perplexing. Just before the collision, each ball has an energy consisting of its mass m and its kinetic energy K. These energies add to make the total energy of the system: $E_{system} = 2m + 2K$.

After the collision? Both balls of chewing gum are at rest, stuck together as a single blob, which now constitutes the entire system. The energy of that stationary blob must be its rest energy, equal to the mass of the system: $E_{system} = E_{rest} = M_{system}$. What is the value of that system energy? It must be the same as the energy of the system before the collision, equal to $2m + 2K$, where m is the mass of each ball before the collision. Hence, if energy is conserved, $M_{system} = 2m + 2K$. This is greater than the sum of masses of the incoming particles.

Where does this extra mass come from? The energy of relative motion of the incoming particles gets converted, during the collision, into energy of plastic deformation and heat. Each of these forms of locked-in energy yields an increment of mass. In consequence the mass of the pair of balls, stuck together as one, exceeds the sum of masses of the two balls before impact.

Third Experiment: Weighing Heat. If warmed and distorted balls of gum have more mass than cool and undistorted balls, then maybe we can measure directly the increased mass simply by heating an object and weighing it. In this case the system consists of a single large object, such as a tub of water, stationary and therefore with zero total momentum. System energy consists of the summed individual masses of all water molecules plus the summed kinetic energies of their random motions. This summed kinetic energy increases as we add heat to the water; hence its mass should increase. Can we detect the corresponding increase in weight as we heat the water in the tub?

Alas, never yet has anyone succeeded in weighing heat. In 1787 Benjamin Thompson, Count Rumford (1753 – 1814), tried to detect an increase in weight of barrels of water, mercury, and alcohol as their temperature rose from $29°$ F to $61°$ F (in which range ice melts). He found no effect. He concluded "that ALL ATTEMPTS TO DISCOVER ANY EFFECTS OF HEAT UPON THE APPARENT WEIGHTS OF BODIES WILL BE FRUITLESS" (capital letters his). Professor Vladimir Braginsky of the University of Moscow once described to us a new idea for weighing heat. Let a tiny quartz pellet hang on the end of a long thin near-horizontal quartz fiber, like a reeled-in fish at the end of a long supple fishing rod. A fly that settles on the fish increases its weight; the fishing rod bends a little more. Likewise heat added to the pellet will increase its mass and will bend the quartz-fiber "fishing rod" a little more. That is the idea. The sensitivity required to detect a bending so slight unfortunately surpasses the present limit of technology. Braginsky himself already has invented, published, and made available to workers all over the world a now widely applied scheme to measure very small effects. There is real hope that he — or someone else — will weigh heat and confirm what we already confidently expect. ✒

Kinetic energy converted to mass

Can we weigh heat? Not yet!

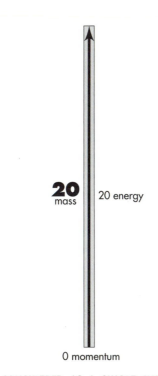

CONSIDERED AS TWO PARTS

20
mass

20 energy

0 momentum

CONSIDERED AS A SINGLE SYSTEM

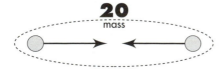

FIGURE 8-3. *Two noninteracting particles, each of mass 8, are in relative motion. Taken together, they constitute a system of mass 20. Where does the mass 20 reside? In the system!*

8.3 MASS OF A SYSTEM OF PARTICLES

energies add. momenta add. masses do not add.

No one with any detective instincts will rest content with the vague thought that heat has mass. Where within our stuck-together wads of chewing gum or Rumford's barrel of water or Braginsky's quartz pellet is that mass located? In random motions of the atoms? Nonsense. Each atom has mass, yes. But does an atom acquire additional mass by virtue of any motion? Does motion have mass? No. Absolutely not. Then where, and in what form, does the extra mass reside? Answer: Not in any part, but in the *system*.

Heat resides not in the particles individually but in the system of particles. Heat arises not from motion of one particle but from relative motions of two or more particles. Heat is a **system property.**

The mass of a system is greater when system parts move relative to each other. Of this central point, no simpler example offers itself than a system composed of a single pair of masses. Our example? Two identical objects (Figure 8-3). Each has mass 8. Relative to the laboratory frame of reference each object has momentum 6, but the two momenta are opposite in direction. The energy of each object is $E = (m^2 + p^2)^{1/2} = (8^2 + 6^2)^{1/2} = 10$.

The total momentum of the two-object system is $p_{\text{system}} = 6 - 6 = 0$. The energy of the system is $E_{\text{system}} = 10 + 10 = 20$. Therefore the mass of the system is $M_{\text{system}} = (E_{\text{system}}^2 - p_{\text{system}}^2)^{1/2} = [(20)^2 - 0^2]^{1/2} = 20$. Thus the mass of the system exceeds the sum of the masses of the two parts of the system. The mass of the system does not agree with the sum of the masses of its parts.

Energy is additive. Momentum is additive. But mass is *not* additive.

Ask where the *extra* $20 - 16 = 4$ units of mass are located? Silly question, any answer to which is also silly!

Ask where the 20 units of mass are located? Good question, with a good answer. The 20 units of mass belong to the system as a whole, not to any part individually.

Where is the life of a puppy located? Good question, with a good answer. Life is a property of the *system* of atoms we call a puppy, not a property of any part of the puppy.

Where is the *extra* ingredient added to atoms to yield a live puppy? Unacceptable question, any answer to which is also unacceptable. Life is not a property of any of the individual atoms of which the puppy is constituted. Nor is it a property of the space between the atoms. Nor is it an ingredient that has to be added to atoms. Life is a property of the puppy *system*.

Life is remarkable, but in one respect the two-object system that we are talking about is even more remarkable. Life requires organization, but the two-object system of Figure 8-3 lacks organization. Neither mass interacts with the other. Yet the total energy of the two-object system, and its total momentum, regarded from first one frame of reference, then another, then another, take on values identical in every respect to the values they would have were we dealing throughout with a single object of mass 20 units. Totally unlinked, the two objects, viewed as a system, possess the dynamic attributes — energy, momentum, and mass — of a single object.

This wider idea of mass — the mass of an isolated *system* composed of disconnected objects: what right have we to give it the name "mass"? Nature, for whatever reason, demands conservation of total momenergy in every collision. Each collision, no matter how much it changes the momenergy of each participant, leaves unchanged the sum of their momenergies, regarded as a directed arrow in spacetime — a 4-vector. Encounter or no encounter, and however complex any encounter, system momenergy does not alter. Neither in spacetime direction nor in magnitude does it ever change. But the magnitude — the length of the arrow of total momenergy, figured as we figure any spacetime interval — is system mass. Whether the system consists of a single object or of many objects, and whether these objects do or do not collide or otherwise interact with each other, this system mass never changes. That's why the concept of system mass makes sense!

An example? Again, two objects of mass 8, again each moving toward a point midway between them at $v = (\text{momentum})/(\text{energy}) = (p=6)/(E=10) = 3/5$ the speed of light. Now, however, we analyze the two motions in a frame moving with the right-hand object (Figure 8-4). In this new frame the right-hand object is at rest: mass, $m = 8$; momentum, $p = 0$; energy, $E = [m^2 + p^2]^{1/2} = 8$. The left-hand object is approaching with a speed (addition of velocities: Section L.7 of the Special Topic following Chapter 3; also Exercise 3-11)

Different free-float frames. Same system mass.

$$v = \frac{3/5 + 3/5}{1 + (3/5)(3/5)} = \frac{6/5}{34/25} = \frac{15}{17}$$

It has energy $E = m/[1 - v^2]^{1/2} = 8/[1 - (15/17)^2]^{1/2} = 17$ and momentum $p = vE = 15$. So much for the parts of the system! Now for the system itself. For the system the energy is $E_{\text{system}} = 8 + 17 = 25$ and the momentum is $p_{\text{system}} = 0 + 15 = 15$.

Now for the test! Does the concept of system mass make sense? In other words, does system mass turn out to have the same value in the new frame as in the original frame? It does:

$$M_{\text{system}} = (E_{\text{system}}^2 - p_{\text{system}}^2)^{1/2} = [(25)^2 - (15)^2]^{1/2} = [625 - 225]^{1/2}$$

$$= [400]^{1/2} = 20$$

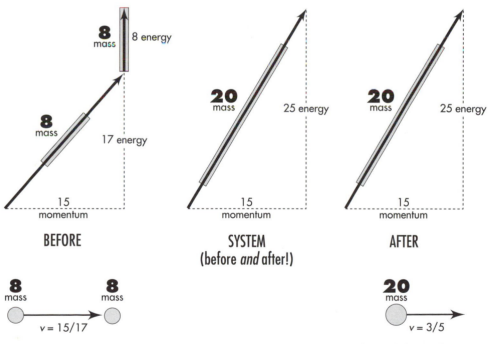

BEFORE   SYSTEM   AFTER
                    (before *and* after!)

FIGURE 8-4. *System of Figure 8-3 observed from a frame moving with the right-hand object.* The *right-hand object is therefore initially at rest.* **Before:** *Arrows of momenergy for two objects before collision. Each object has a mass of eight units (shaded handles). The upper, vertical, arrow belongs to the particle originally at rest, the slanted arrow to the incoming particle.* **System:** *Addition of the two momenta (one of them zero!) gives the total momentum before collision. Similarly, addition of the two energies gives the total energy. Mass of the* system — *even before the two particles interact!* — *comes from the expression for the "hypotenuse" of a spacetimelike triangle. Result: 20 units of mass (shaded handle on center 4-vector):*

$$(mass)^2 = (energy)^2 - (momentum)^2 = (25)^2 - (15)^2 = 625 - 225 = 400 = (20)^2$$

After: *The two particles now collide and amalgamate to form one particle. Arrow of total momenergy after the amalgamation is identical to arrow of total momenergy before the collision. Mass of this two-object system exceeds the mass of one object plus the mass of the other, not only after the collision but also before. Mass is not an additive quantity.*

SAMPLE PROBLEM 8-1

MASS OF A SYSTEM OF MATERIAL PARTICLES

Compute M_{system} for each of the following systems. The particles that make up these systems do not interact with one another. Express the system mass in terms of the unit mass m; do not use momenta or velocities in your answers. [Note: In the following diagrams, arrows represent (3-vector) momenta.]

System a

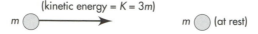

(kinetic energy = K = 3m)

m ○   m ○ (at rest)

System b

(kinetic energy = K = 5m)   (kinetic energy = K = 5m)

m ○   m ○

System c

(energy = E = 7m)

3m ○   m ○ (at rest)

System d

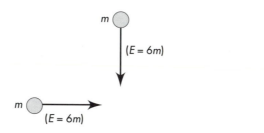

m ○

(E = 6m)

m ○

(E = 6m)

SOLUTION

System a: System energy equals the rest energy of the two particles (the sum of their masses) plus the kinetic energy of the moving particle: $E_{system} = (m + m) + 3m = 5m$. Squared momentum of the system equals that of the moving particle: $p_{system}^2 = p^2 = E^2 - m^2 = (4m)^2 - m^2 = 15m^2$. Mass of the system is reckoned from the difference between the squares of energy and momentum:

$$M_{system} = [E_{system}^2 - p_{system}^2]^{1/2} = [25m^2 - 15m^2]^{1/2} = [10]^{1/2}m = 3.162\ m$$

Moreover, if the two objects collide and amalgamate, the system energy remains at the value 25, the system momentum remains at the value 15, and the system mass remains 20, as illustrated in Figure 8-4.

In summary, the mass of an isolated system has a value independent of the choice of frame of reference in which it is figured. System mass remains unchanged by encounters between the constituents of the system. And why? Because the system mass is the length (in the sense of spacetime interval) of the arrow of total momentum-energy. This momenergy total is unaffected by collisions among the parts or by any transformations, decays, or annihilations they may undergo. System mass *does* make sense!

System! System! You keep talking about "system," even when the particles do not interact, as in the system of Figure 8-3. It seems to me that you are totally arbitrary in the way you define a system. Who chooses which particles are in the system?

System b: System energy equals rest energy of the two particles plus kinetic energy of the two particles: $E_{system} = 2m + 10m = 12m$. Squared momentum of each particle is $p^2 = E^2 - m^2 = (6m)^2 - m^2 = 35m^2$ yielding $p = (35)^{1/2}m$. System momentum is twice this: $p_{system} = 2(35)^{1/2} m$. The mass of the system is

$$M_{system} = [E_{system}^2 - p_{system}^2]^{1/2} = [(12m)^2 - \{2(35)^{1/2}m\}^2]^{1/2}$$
$$= [144 - 140]^{1/2}m = [4]^{1/2}m = 2m$$

In this one special case the mass of the system equals the sum of masses of the objects that make up the system. We could have seen this result immediately by observing the system from a reference frame that moves along with the particles. In this frame the particles are at rest and have zero total momentum; the total energy is identical to the sum of the individual rest energies (the individual masses). So in this case the mass of the system is equal to its energy, which is equal to the sum of masses. Moreover, system mass is an invariant. Thus $2m$ is the mass of the system as reckoned in all reference frames, including the one in which System b is pictured.

System c: Total energy = system energy = $E_{system} = 7m + m = 8m$. System momentum equals the momentum of the moving particle: $p_{system}^2 = E^2 - m^2 = (7m)^2 - (3m)^2 = 49m^2 - 9m^2 = 40m^2$. Hence the system mass is

$$M_{system} = [64m^2 - 40m^2]^{1/2} = [24]^{1/2}m = 4.899m$$

System d: This part of the problem serves as a reminder that momentum is a Euclidean 3-vector. The squared momentum of each particle is $p^2 = E^2 - m^2 = 36m^2 - m^2 = 35m^2$. Their total momentum is *not* the algebraic sum of the momenta, because they are vectors pointing in perpendicular directions. This perpendicular orientation allows us to equate the squared system momentum to the sum of the squares of the individual momenta: $p_{system}^2 = 35m^2 + 35m^2 = 70m^2$. System energy is the sum of the energies (energy is a scalar and adds like a scalar!): $E_{system} = 6m + 6m = 12m$. Hence system mass is

$$M_{system} = [144m^2 - 70m^2]^{1/2} = [74]^{1/2}m = 8.602m$$

Compare this result with that of System b, which also contained two particles, each of total energy $6m$.

We do! We can draw the dashed line around any collection of objects whatever, subject to this one restriction: no object in our system may interact with any external object or experience a force from outside the system. Our system must be *isolated*. With that single limitation, the system we choose is arbitrary, has a conserved total energy, a conserved total momentum, and a system mass that is invariant — a mass that has the same value no matter in which free-float frame it is reckoned.

I can't believe the story you tell. Those two mass-8 objects, you say, may fly past each other. Then your talk about the system mass is just talk, terminology. Or they may whang into each other and amalgamate. Then your talk is all wrong, and for an obvious reason. As the objects collide they slow and come to rest relative to each other. At that instant and in that "rest frame" (the frame of Figure 8-3), each has zero momentum, and energy equal to its mass. So the total momentum of the system is zero, and its total energy is $8 + 8 = 16$. That means a mass of 16. Yet you claim 20.

TABLE 8-1

CLEOPATRA'S VASE, HER BATH, AND INTERSTELLAR VACUUM: ILLUSTRATIVE FRACTIONAL CHANGES IN MASS OF SYSTEMS

System before	System after	Fractional increase in system mass (to nearest power of 10)
One-kilogram vase	Vase smashed into so many fragments that 100 centimeters² of glass-to-glass bonds are broken	10^{-18}
Bath water at 15° C	Bath water at 40° C	10^{-12}
Water (H_2O)	Atomic hydrogen (H) and oxygen (O)	10^{-9}
Earth	All molecules of Earth lifted against the pull of their mutual gravity to infinite separation from one another	10^{-9}
Hydrogen atom in lowest energy state	Electron withdrawn to infinite separation from nucleus	10^{-8}
Deuteron	Deuteron separated into proton and neutron	10^{-3}
Neutron star	Widely separated iron atoms at rest with respect to each other	10^{-1}
A vacuum, before it is zapped by converging photons	Electron–positron pair bound as a positronium atom	Infinite fractional increase

Slow and come to rest? Yes. But that means force: "elastic," gravitational, electromagnetic, or nuclear force. That's the new and valuable point you make here. And those particles, pushing against that force, store up energy. This energy, too, has to be put into the bookkeeping. When amalgamating particles come to rest relative to one another, the energy of interaction "balances the books" — it so happens — and leads to a final mass of 20, greater than the sum of masses of the original objects. For the figuring of system mass, however, we really don't have to get into this detail. It is enough for us to know that total momentum is conserved, $p_{system} = 0$ in Figure 8-3, and total energy — in whatever way it is apportioned between the objects and the fields of force that act between them — is also conserved, $E_{system} = 20$. The length, in the sense of interval, of the 4-vector of momenergy for the system remains unchanged: $M_{system} = 20$.

System energy increase? System mass can increase.

What about a system that is *not* isolated? A system that has — and keeps — zero momentum, but receives an increment of energy? Then its mass rises by an amount exactly equal to that input of energy. The increase in mass is the same whether that energy goes into altering the relative motion of the parts of the system or increasing the energy of interaction between them or some combination of motion and interaction. Supply energy to a system by heating it or setting it into internal vibration or fracturing the bonds between its parts? Each is a guaranteed way to increase the mass of the system (Table 8-1)!

8.4 ENERGY WITHOUT MASS: PHOTON

**light moves with zero aging.
photons move with zero mass.**

A striking example of the primacy of momenergy over mass is furnished by a **quantum of light** colliding with an electron.

Quantum? A quantum of luminous energy of a given color or, in more technical terms, light of a given wavelength or frequency of vibration. Max Planck discovered in 1900 that light of a given color comes only in quanta — "hunks" — of energy of a standard amount, an amount completely determined by the color (Table 8-2). We can have one quantum, one hunk, one **photon**, of green light, or two, or fifteen, but never two and a half.

Nothing did more to raise the light quantum, the hunk of luminous energy, the photon, to the status of a particle than experiments carried out by 28-year-old Arthur Holly Compton at Washington University, St. Louis, in 1920. Shining X-rays of known wavelength (and hence of known frequency and known quantum energy) on a variety of different substances, he measured the wavelength (and hence the quantum energy) of the emergent "scattered" X-rays. He got identical changes in wavelength at identical angles of observation from many kinds of materials. There was no way he could explain this result except to say that the scattering object was in every case the same, an electron, whatever the atom in which the electron happened to reside.

But why did the change of wavelength have a unique value, the same for all materials at a given angle of scattering? Every idea of classical physics failed to fit, Compton found. "Compton arrived at his revolutionary quantum theory for the scattering process rather suddenly in late 1922," a biographer tells us. "He now treated the interaction as a simple collision between [an X-ray quantum] and a free electron . . . [He] found that [this hypothesis gave results] which agreed perfectly with his data . . . When Compton reported his discovery at meetings of the American Physical Society, it aroused great interest and strong opposition . . ." By 1927, however, his finding was generally accepted and in that year won him the Nobel Prize.

What does it mean to treat a photon on the same footing as a particle? It means this: attribution to the photon of an energy and a momentum, in other words momenergy.

Compton demonstrates quantum of radiation — photon!

TABLE 8-2

MOMENTUM AND ENERGY CARRIED BY ONE PHOTON, ONE QUANTUM, ONE HUNK OF LUMINOUS ENERGY OF VARIOUS "COLORS"

(Unit of energy used in this table: electron-volt or eV, the amount of energy given to an electron by accelerating it through an electrical potential difference of one volt)

Source of electromagnetic radiation	Momentum (and energy) of a single quantum	Frequency in vibrations per second	Wavelength in meters
KDKA, Pittsburgh: world's first radio broadcast station	4.22×10^{-9} eV	1.02×10^6	294
A sample infrared beam	1.24×10^{-2} eV	3×10^{12}	10^{-4}
Yellow radiation from a sodium arc lamp	2.11 eV	5.09×10^{14}	5.90×10^{-7}
Ultraviolet light from a mercury arc lamp	4.89 eV	1.18×10^{15}	2.54×10^{-7}
Ultraviolet star radiation of just barely sufficient quantum energy to strip a hydrogen atom of its electron	13.6 eV	3.29×10^{15}	0.91×10^{-7}
Each of two gamma rays given off in the mutual annihilation of a slow positron and a slow electron	5.11×10^5 eV	1.23×10^{20}	2.43×10^{-12}
Each of two gamma rays given out when a neutral pi meson, at rest, decays	6.75×10^7 eV	1.63×10^{22}	1.84×10^{-14}
Each of two gamma rays given off in the mutual annihilation of a slow proton and a slow antiproton	0.938×10^9 eV	2.27×10^{23}	1.32×10^{-15}

Photon momenergy points in lightlike direction

In what direction in spacetime does the photon's arrow of momenergy point? In a lightlike direction, because the photon—a quantum of light—travels with light speed!

When we turn from spacetime to a particular free-float frame of reference and observe a pulse of light at one event along its worldline and then observe it at a second event (Figure 8-5), we know in advance something important about the interval between the two events: It equals zero.

$$(\text{interval})^2 = (\text{distance between two events})^2 - (\text{time between two events})^2$$
$$= (\text{difference between two quantities of identical magnitude})$$
$$= 0$$

Photon momenergy: magnitude zero (photon mass = 0)

A photon in a pulse of light has a momenergy arrow with a tip and a tail, like the momenergy vector for any other particle. Between the tip and tail there is a magnitude. The magnitude for the photon, however, has the value zero—zero because this arrow points in the same direction in spacetime as the worldline of the light pulse (Figure 8-5). For that reason its space component (momentum) and its time component (energy) are equal. And, of course, we express the square of this magnitude as we express the square of any interval, as a *difference* between the squared timelike and spacelike separations between the two ends of the arrow:

$$(\text{magnitude of momenergy arrow of photon})^2$$
$$= (\text{photon energy})^2 - (\text{photon momentum})^2$$
$$= (\text{photon mass})^2 = 0$$

In brief, the lightlike character of the arrow of photon momenergy tells us that (1) photon mass equals zero and (2) the magnitude of momentum, or punch-delivering power, of the photon is identical in value with the energy of the photon:

$$(\text{photon energy}) = (\text{magnitude of photon momentum})$$

and

$$(\text{photon mass}) = 0$$

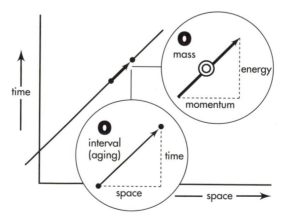

FIGURE 8-5. *Worldline of a photon.* Note its "unit slope in spacetime." **Insets:** *Unit slope of worldline means equal space and time separations between events on this worldline, hence zero interval between them—and zero aging for the photon. Momenergy of the same photon, also with unit slope, symbolizing three properties of the photon: it has zero mass (hence the big zero as an invariant "handle"), it travels with light speed, and it has a momentum identical in magnitude with its energy.*

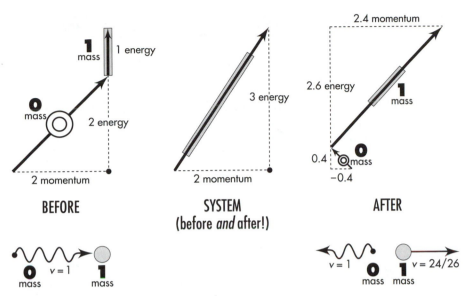

FIGURE 8-6. *Backscattering of a photon by a free electron. The wiggly arrow symbol represents a photon. Energy, momentum, and mass of all particles are expressed in units of electron mass.* **Before:** *The electron at rest has an energy equal to its mass (vertical arrow); the photon has an energy (and a momentum) of 2 electron masses (angled arrow).* **System:** *Arrow of total momenergy. (What is the mass of the* system?*)* **After:** *Arrows of momenergy of knocked-on electron (labeled 1) and backscattered photon (labeled 0) after the encounter. Arrow of total momenergy of the system remains the same (is conserved!) during this process.*

Figure 8-5 summarizes these features of the elementary quanta of visible light and other electromagnetic radiations. For a "handle" on the momenergy 4-vector of a photon—representative of its magnitude—we choose a stylized zero, 0.

Nothing shows these revolutionary features of light to better advantage than the very collision process studied by Arthur Compton: the encounter between a single photon and a single electron. We take the electron, loosely bound though it may be in one or another outer orbit of an atom, as essentially free and essentially at rest—at rest compared to the swift motion in which it finds itself after the high-energy photon hits it (Figure 8-6).

To simplify all numbers, we pick for the photon energy a value typical of gamma rays, considerably greater than that of the X-rays with which Compton worked but easily available today from various sources of radioactivity: 1.022 MeV (million electron-volts). We pick this number because we want to express all energies in units of electron mass, 9.11×10^{-31} kilograms or 0.511 MeV. Our choice of photon energy equals exactly two electron masses. Convenient!

Incoming photons of this energy, encountering an electron, are scattered by the electron sometimes in one direction, sometimes in another, and sometimes straight backward. In that most extreme of encounters—backward scattering—an interchange of momentum takes place that nevertheless preserves total momentum and also total energy, as illustrated in Figure 8-6. The electron is kicked forward with a momentum of $2.4 = 12/5$ times the electron mass, and the photon bounces backward with a momentum (and energy) of $0.4 = 2/5$ times the electron mass, much less than the two-electron masses of momentum (and energy) with which it approached. ✍

Compton collision analyzed

SAMPLE PROBLEM 8-2

MASS OF A SYSTEM THAT INCLUDES PHOTONS

A photon has no rest energy — that is, no mass of its own. However, a photon can contribute energy and momentum to a system of objects. Hence the presence of one or more photons in a system can increase the mass of that system. More: A system consisting entirely of zero-mass photons can itself have nonzero mass!

Find system mass M_{system} for each of the following systems. The particles that make up these systems do not interact with one another. Express the system mass in terms of the unit mass m (or the unit energy E in the photons-only systems). Use only energy and mass in your answers: no momenta or velocities.

System a

(photon) $(E = 3m)$ m (at rest)

System b

(energy = $3E$)   (energy = E)

System c

(energy = $3E$)   (energy = E)

System d

(energy = E)

(energy = $3E$)

SOLUTION

System a: System energy equals the rest energy m of the material particle plus the energy $E = 3m$ of the photon: $E_{system} = m + 3m = 4m$. The momentum of the system is equal to the momentum of the photon, which is equal to its energy: $p_{system} = 3m$. The mass of the system is reckoned from the difference of the square of energy and momentum:

$$M_{system} = [E_{system}{}^2 - p_{system}{}^2]^{1/2} = [(4m)^2 - (3m)^2]^{1/2} = [16m^2 - 9m^2]^{1/2}$$
$$= [7]^{1/2}m = 2.646m$$

System b: System energy equals the sum of the energies of the two photons: $E_{system} = 3E + E = 4E$. System momentum equals sum of momenta of the two photons — which in this case also equals the sum of the energies of the two photons: $p_{system} = 3E + E = 4E$. Therefore system mass equals zero:

$$M_{system} = [E_{system}{}^2 - p_{system}{}^2]^{1/2} = [(4E)^2 - (4E)^2]^{1/2} = 0$$

We could have predicted this result immediately. Two photons moving along in step are, as regards momentum and energy, completely equivalent to a single photon of energy equal to the sum of energies of the separate photons. And a single photon has, of course, zero mass.

System c: Total energy = system energy = E_{system} = $3E + E = 4E$. System momentum equals the difference between the rightward momentum of the first particle and the leftward momentum of the second particle: $p_{system} = 3E - E = 2E$. Hence the system mass is

$$M_{system} = [16E^2 - 4E^2]^{1/2} = [12]^{1/2}E = 3.464m$$

Why can't we simply make a single photon by adding the energies of the two photons, as in system b? Because energies add as scalars, and momenta add as 3-vectors. In this case the total energy is $4E$ and the total momentum is $2E$. No way to make a single photon out of this; for a photon, energy and momentum must have equal magnitudes!

System d: This part serves as an additional reminder that momentum is a 3-vector. The system energy equals $E_{system} = E + 3E = 4E$. The squared momentum of the system equals the sum of squares of the momenta of the separate particles, since they move in perpendicular directions in this frame: $p_{system}^2 = E^2 + (3E)^2 = 10E^2$.

Hence system mass is:

$$M_{system} = [16E^2 - 10E^2]^{1/2} = [6]^{1/2}E = 2.449E$$

8.5 PHOTON USED TO CREATE MASS

photon hits electron, creates electron–positron pair

It should not be surprising that a photon can deliver energy without having any mass of its own. After all, an electron does have mass of its own; yet an electron traveling sufficiently close to light speed can impart to its target an amount of energy ten, a hundred, or a thousand times as great as its own mass. Not mass but momentum governs the size of punch that either photon or electron can deliver.

Incredibly, however, a photon in the presence of an electron can create matter out of empty space. To bring about this process, double the energy of the quantum of radiant energy shown in Figure 8-6. When a photon with energy equal to four electron masses hits an electron at rest, the photon most often recoils; in other words, it suffers backward scattering, an instance of the Compton process. Occasionally, however, the impacting photon produces out of empty space, near the struck electron, a new pair of electrons, one with a negative electric charge like all everyday electrons, the other with an identical amount of positive charge. The electron with positive charge has the name **positron** (Box 8-1).

This process goes on all the time high in Earth's atmosphere, where cosmic rays pour in from outer space. There, however, energies of cosmic-ray photons often far exceed four electron masses. In consequence, the struck electron and the two newly created electrons go off in slightly different directions and at different speeds. However, when the energy of the incoming photon is sufficiently finely tuned, in the immediate vicinity of an energy of four electron masses, the three particles can stick together as a super-light molecule, a **polyelectron**, a system analogous to what chemists call the hydrogen molecule ion (Figure 8-7).

Matter is born

FIGURE 8-7. *Comparison and contrast. **Left:** Two protons and an electron forming the hydrogen molecule ion of chemistry. (A proton is much more massive than an electron but can be envisioned as occupying less volume.) **Right:** Two electrons and a positron, forming a **polyelectron** created by impact of a properly tuned photon (about 4 MeV of energy) on an electron at rest.*

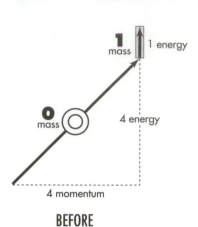
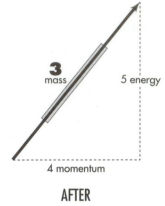

BEFORE   AFTER

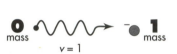
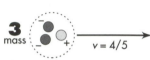

FIGURE 8-8. *Conservation of energy and momentum in the process of creating a pair (a positive and a negative electron) in the field of an electron.* **Before:** *A photon that has energy (and momentum) equal to four electron masses (sloping arrow) strikes an electron essentially at rest (vertical arrow).* **After:** *The photon has ceased to exist, and the two newly created particles have gone off in company with the original electron at 80 percent of light speed—a combined "particle" of three electron masses.*

Why does it take a light quantum with an energy of *four* electron masses to create (Figure 8-8) a polyelectron, a super-light hydrogen molecule ion, an object with a mass of three electron masses (in truth, a tiny bit less than three electron masses because of the negative binding energy among the three particles)? The question becomes all the more insistent when we recall that the electron that got hit already brought to the consummation of the deal a rest energy equal to one electron mass.

In brief, why do we have to put in five electron masses of energy to get out a three-electron-mass product? Simply asking this question points out where the explanation lies. The incident photon brings in a great momentum, and the electron with which it reacts has no momentum. So all that momentum has to go into the output product, the polyelectron. Since the polyelectron must have momentum, it must also have kinetic energy—energy not available for creating additional mass. In consequence, that object has so much energy of motion that only a much diminished part of the energy of the incident photon is available for the creation process itself.

System momentum means not all system energy available to create particles

8.6 MATERIAL PARTICLE USED TO CREATE MASS

proton hits proton, creates proton—antiproton pair

Any energetic particle can create other particles

Particles other than the photon can also create particles. A particle of any type can carry enough energy to create particles similar to or different from itself. Each such creation must not only follow momenergy conservation laws of special relativity, but it is also subject to the law of conservation of total electric charge and other conservation laws, as described in elementary particle physics.

BOX 8-1

BACKYARD ZOO OF PARTICLES

This is not a textbook of particle physics, but our examples include interactions between common particles. Here are brief descriptions of some of them.

Electron

Electrons form the outer structure of every atom and rattle around in approximately 99.99999999999 percent of its volume. The mass of the electrons of an atom, however, accounts for only about one two-thousandth of its mass or less. The electron carries a negative "elementary" electrical charge. Every accepted theory of particle physics treats the electron itself as an elementary particle — it is not made up of anything more fundamental. The **positron** is the antiparticle of the electron, with the same mass but a positive elementary charge. When positron and electron meet, sooner or later they mutually annihilate, yielding two or more high-energy photons (gamma rays). This will be the fate of the positron and one of the electrons in the polyelectron discussed in Section 8.5 soon after they begin to orbit one another.

Proton

The proton (Greek for "the first one") is, with the neutron, the most massive constituent of atomic nuclei. The simplest atom, hydrogen, in its most abundant form has a single proton as nucleus. The proton has a positive charge equal in magnitude to that of the electron, but a mass almost two thousand times as great as that of the electron. As far as we know the proton is stable; experiments have shown its lifetime to be greater than 10^{31} years — very much longer than the current age of the universe (about 10^{10} years). Particle physicists postulate that protons (and neutrons) are composed of still-more-elementary particles called quarks. The **antiproton**, antiparticle of the proton, has mass equal to that of the proton but negative unit charge. When it encounters a proton, the two particles annihilate, sometimes creating gamma rays but more often other particles not listed in this box.

Neutron

The neutron (from Latin *neuter* — "neither"; neither positively nor negatively charged) is similar to the proton but has no charge and has slightly greater mass. It is a constituent of all nuclei except for the most abundant form of elementary hydrogen. When not in a nucleus, the neutron decays into a proton, electron, and neutrino with half-life of about 10 minutes.

Photon

The photon, the quantum of light, has zero mass. Its properties are described in Section 8.4.

Neutrino

There are several kinds of neutrinos, all of which appear to have zero mass and to move at light speed. The neutrino (Italian for "little neutral one") has no charge and interacts only weakly with ordinary matter: Neutrinos of certain energies can pass through a block of lead one light-year thick with only a 50–50 chance of being absorbed! An immense flux of neutrinos passes continually through our bodies without injuring us. "Ten million trillion [10^{19}] neutrinos will speed harmlessly through your brain and body in the time it takes to read this sentence. By the time you have read this sentence, they will be farther away than the moon."

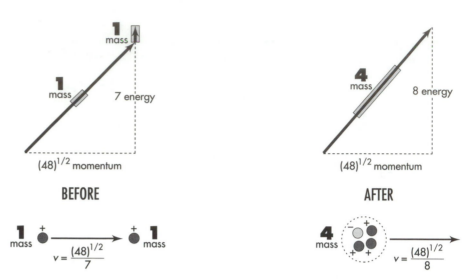

FIGURE 8-9. *Conservation of energy and momentum in the process of creation of a proton–antiproton pair by the impact of a proton on another proton.* **Before:** *The incoming proton (sloping arrow) moves with a speed* $(48)^{1/2}/7 = 99$ *percent that of light. The target proton initially stands at rest (vertical arrow).* **After:** *The resulting three protons and one antiproton are kicked to the right at* $(48)^{1/2}/8 = 87$ *percent of light speed.*

Threshold energy defined

Figure 8-9 shows "the creation of a proton–antiproton pair by a proton in the presence of another proton." The antiproton has mass equal to that of the proton but carries a negative unit charge (Box 8-1). The interaction shown leaves all four resulting particles moving along together. The resulting particles stay together when the incoming particle has the lowest energy that can create the additional pair. This minimum energy is called the **threshold energy.** We don't want the three particles to move apart after their creation. If we did, we would have to supply the incoming particle with additional kinetic energy. It would have to carry an energy greater than the threshold energy. We discuss here the threshold energy of the incoming proton.

Magnitudes of the momenergy vectors displayed in Figure 8-9 are expressed in "natural units" for the proton, namely the mass of the proton itself, 1.67×10^{-27} kilograms or 938.27 MeV. This time the numbers are not all integers: the momentum of the system has a value equal to the square root of 48, or 6.928 proton masses.

"Efficiency" of particle production

The creation of a proton–antiproton pair by a PROTON requires a total of eight proton units of energy to create two proton units of mass. In contrast the creation of an electron–antielectron pair by a PHOTON requires a total of only four electron units of energy to create two electron units of mass. Why is the photon process so much more efficient (in units of mass of the struck particle) than the proton process? Answer: The photon is annihilated in the creation process. In contrast, the incoming proton is not annihilated; the bookkeeper must keep the incoming proton on the payroll, providing momenergy after the collision to keep the proton in step with the other three particles. This after-collision momenergy of the proton is not available to be applied to other products of the collision. Therefore a proton of given total energy can create less mass than a photon of the same energy when each strikes a stationary target. ✒

8.7 CONVERTING MASS TO USABLE ENERGY: FISSION, FUSION, ANNIHILATION

fission and fusion both slide down the energy hill toward the minimum, iron. electron and positron annihilate to yield two energetic photons.

For a final perspective on the evanescence of mass and the preservation of momenergy, turn from processes where mass is created to three processes in which mass is destroyed: fission, fusion, and annihilation.

Anyone who first hears about the splitting of a nucleus (fission) as a source of energy, and the joining of two nuclei (fusion) also as a source of energy might gain the mistaken impression that a perpetual motion machine has been invented. Could we split and join the same nucleus over and over again, each time releasing energy? No. Here's why. Fission occurs in the splitting of uranium, for instance when a neutron strikes a uranium nucleus:

$$ {}_{0}^{1}n + {}_{92}^{235}U \longrightarrow {}_{92}^{236}U \longrightarrow {}_{37}^{95}Rb + {}_{55}^{141}Cs $$

In this equation the lower-left subscript tells the number of protons in the given nucleus and the upper-left superscript shows number of protons plus neutrons in the nucleus. The process described by this equation rearranges the 236 nucleons, that is, 92 protons plus 144 neutrons, into a configuration that comes a bit closer to that most stable of all available nuclear configurations, the iron nucleus:

$$ {}_{26}^{56}Fe $$

But fusion too, for example the process of uniting two rather light nuclei such as heavy hydrogen or deuterons to form a helium nucleus,

$$ {}_{1}^{2}D + {}_{1}^{2}D \longrightarrow {}_{2}^{4}He $$

can also be regarded as one step along the way toward rearranging nucleons (protons and neutrons) to achieve the iron configuration or something like it.

In brief, we can get energy out of nucleon rearrangement processes that move from looser binding of both heavier and lighter nuclei toward tighter binding of the (intermediate-mass) iron nucleus (Figure 8-10). In neither fission nor fusion, however, is the fraction of mass converted into energy as great as one percent. (For an example of fusion reaction in Sun, see Sample Problem 8-5, especially **c**.)

Annihilation is interesting because it can convert 100 percent of matter into radiation. Annihilation is interesting, too, because it has been demonstrated on the microscopic scale. A slow positive electron, a positron, joining up by chance to orbit with an everyday negative electron, eventually unites with it to annihilate them both and produce sometimes two, sometimes three light quanta (photons—called **gamma rays** in the case of these high energies):

$$ e^+ + e^- \longrightarrow 2 \text{ or } 3 \text{ photons} $$

Figure 8-11 displays the balance of energy and momentum in the two-quantum annihilation process.

Fission and fusion: Both go from looser to tighter binding

Annihilation converts 100% of matter into radiation

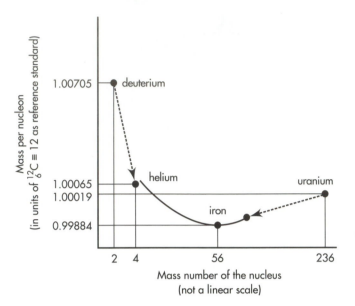

FIGURE 8-10. *Both the conversion of deuterium to the more massive helium in fusion and the conversion of uranium to lighter nuclei in fission decrease the mass per nucleon, both toward the most stable of nuclei, iron.*

Why 2 or 3 photons? Why can't just a single photon be emitted in this process?

Brief answer: Conservation of momentum. Fuller answer: Look at Figure 8-11. Before annihilation, the system has zero total momentum. A single photon remaining after the annihilation could not have zero momentum, no matter in which direction it moved! The presence of a single photon after the collision could not satisfy conservation of momentum. So annihilation never does and never can end up giving only a single photon. ✎

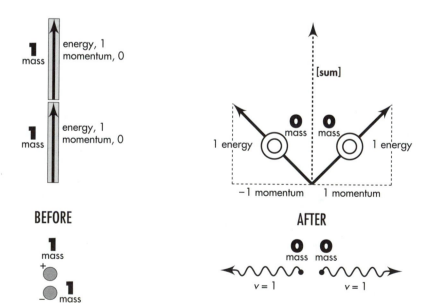

FIGURE 8-11. *Momenergy conservation in the two-photon electron–positron annihilation process.* **Before:** *Before annihilation each oppositely charged particle has rest energy and no momentum.* **After:** *The two particles have annihilated, creating two high-energy photons (gamma rays). The two photons fly apart in opposite directions; total momentum remains zero.*

BOX 8-2

ANALYZING A PARTICLE ENCOUNTER

Conservation of total momenergy! In any given free-float frame that means conservation of total energy and conservation of each of three components of total momentum. In no way does the power and scope of this principle make itself felt more memorably than the analysis of simple encounters of this, that, and the other kind in an isolated system of particles. "Analyzing an encounter" means using conservation laws and other relations to find unknown masses, energies, and momenta of particles in terms of known quantities. Sometimes a complete analysis is not possible; the information provided may be insufficient. Here are suggested steps in analyzing an encounter. Sample Problems 8-3 and 8-4 illustrate these methods.

1. **Draw a diagram** of particles *before* and particles *after* the interaction. Label particles entering with numbers or letters and particles leaving with different numbers or letters (even if they are the same particles). Use arrows to show particle directions of motion and label with symbols their masses, energies, and momenta, whether initially known or unknown.

2. Write down algebraically the **conservation of total energy.** Do not forget to include the rest energy — the mass m — of any particle not moving in the chosen free-float frame.

3. Write down algebraically the **conservation of total momentum.** Do not forget that momentum is a vector. In general this means demanding conservation of each of three components of total momentum.

4. Try to **solve** for unknowns in terms of knowns, still using symbols.

   a. Make liberal use of the relation $m^2 = E^2 - p^2$, where $p^2 = p_x^2 + p_y^2 + p_z^2$. For a photon or neutrino, mass equals zero and $E = p$ (in magnitude: Pay attention to the direction of the momentum vector p — or its sign if motion is in one space dimension).

   b. Do NOT use speed v of a particle unless forced to by requirements of the problem. Relativistic particles typically move with speeds very close to light speed, so speed proves to be a poor measure of significance. Increase by one percent the speed of a particle moving at $v = 0.99$ and you increase its energy by a factor of almost 10.

   c. Substitute numerical values into resulting equations as late as possible. Before substituting numerical values, check that all values are expressed in concordant units.

5. **Check** your result. Check units of the solution. Is the order of magnitude of numerical results reasonable? Substitute limiting values, for example letting energy of an incoming particle become very large (and very small). Is the limiting-case result reasonable?

Is there any **general conclusion** you can draw from your specific solution? Does this exercise illustrate a deep principle or lead to an even more interesting application of conservation laws?

SAMPLE PROBLEM 8-3

SYMMETRIC ELASTIC COLLISION

A proton of mass m and kinetic energy K in the laboratory frame strikes a proton initially at rest in that frame. The two protons undergo a *symmetric elastic* collision: the outgoing protons move in directions that make equal and opposite angles $\theta/2$ with the line of motion of the original incoming particle. Find energy and momentum of each outgoing particle and angle θ between their outgoing directions of motion for this symmetric case.

Historical note: When impact speed is small compared to the speed of light, this separation of directions, θ, is 90 degrees, according to Newtonian mechanics. Early cloud-chamber tracks sometimes showed symmetric collisions with angles of separation substantially less than 90 degrees, thereby giving evidence for relativistic mechanics and providing the first reliable measurements of impact energy.

SOLUTION, following steps in Box 8-2

1. **Draw a diagram** and label all four particles with letters:

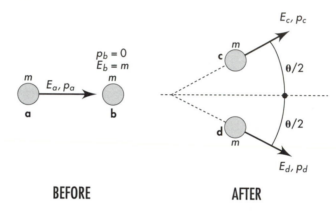

BEFORE   AFTER

Symmetry of this diagram implies that the two outgoing particles have equal energy and equal magnitude of momentum; that is, $E_c = E_d$ and (in magnitude) $p_c = p_d$.

2. **Conservation of energy:** Energy of each particle equals mass plus kinetic energy. And the masses don't change in this reaction. Therefore total kinetic energy after the encounter (divided equally between the two particles) equals the (known) total kinetic energy before the encounter, all localized on one particle. In brief: $K_c = K_d = K_a/2 = K/2$. Simple answer to one of the three questions we were asked!

3. **Conservation of momentum:** By symmetry, the vertical components of momenta of the outgoing particles cancel. Horizontal components add, leading to the relation

$$p_{tot} = p_a = p_c \cos(\theta/2) + p_d \cos(\theta/2) = 2p_d \cos(\theta/2)$$

or, in brief,

$$p_a = 2p_d \cos(\theta/2) \qquad \text{[conservation of momentum]}$$

4. **Solve for the unknown angle θ:** Along the way find the other requested quantity, the magnitude $p_c = p_d$ of the momenta after the collision. To that end, first find the momentum p_a before the collision, using the general formula for the momentum of an individual particle:

$$p = [E^2 - m^2]^{1/2} = [(K + m)^2 - m^2]^{1/2} = (K^2 + 2mK + m^2 - m^2)^{1/2}$$
$$= (K^2 + 2mK)^{1/2}$$

Therefore

$$p_a = (K^2 + 2mK)^{1/2}$$

From conservation of energy, $K_c = K_d = K/2$. Therefore

$$p_d = [(K/2)^2 + 2m(K/2)]^{1/2}$$

Substitute these expressions for p_a and p_d into the equation for conservation of momentum:

$$(K^2 + 2mK)^{1/2} = 2[(K/2)^2 + 2m(K/2)]^{1/2}\cos(\theta/2)$$

Square both sides and solve for $\cos^2(\theta/2)$ to obtain

$$\cos^2(\theta/2) = \frac{K + 2m}{K + 4m}$$

Now apply to this result the trigonometric identity

$$\cos^2(\theta/2) \equiv \frac{(\cos\theta + 1)}{2}$$

After some manipulation, obtain the desired result:

$$\cos\theta = \frac{(K/m)}{(K/m) + 4}$$

Here K is the kinetic energy of the incoming particle, m the mass of either particle, and θ the angle between outgoing particles. This result assumes (1) an elastic collision (kinetic energy conserved), (2) one particle initially at rest, (3) equal masses of the two particles, and (4) the symmetry of outgoing paths shown in the diagram.

5a. Limiting case: Low energy. In the case of low energy (Newtonian limit), the incoming particle has a kinetic energy K very much less than its rest energy m, so the ratio K/m approaches zero. In the limit, $\cos\theta$ becomes zero and $\theta = 90$ degrees. This is the accepted Newtonian result for low velocities (except for an exactly head-on collision, in which case the incoming particle stops dead and the struck particle moves forward with the same speed and direction as the original incoming particle).

5b. Limiting case: High energy. For extremely high-energy elastic collisions, the incident particle has a kinetic energy very much greater than its rest energy, so the ratio K/m increases without limit. In this case the quantity 4 in the denominator becomes negligible compared with K/m, so numerator and denominator both approach the value K/m, with the result $\cos\theta \rightarrow 1$ and $\theta \rightarrow 0$. This means that in the special symmetric case discussed here both resulting particles go forward in the same direction as the incoming particle, sharing equally the kinetic energy of the incoming particle.

For an incoming particle of very high energy, the elastic collision described here is only one of several possible outcomes. Alternative processes include creation of new particles.

ANNIHILATION

A positron of mass m and kinetic energy equal to its mass strikes an electron at rest. They annihilate, creating two high-energy photons. One photon enters a detector placed at an angle of 90 degrees with respect to the direction of the incident positron. What are the energies of both photons (in units of mass of the electron) and direction of motion of the second photon?

SOLUTION, following steps in Box 8-2

1. **Draw a diagram** and label the particles with letters.

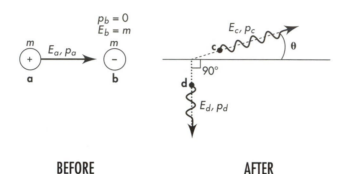

BEFORE   **AFTER**

2. **Conservation of energy** expressed in the symbols of the diagram, and including the rest energy of the initial stationary particle:

$$E_{tot} = E_a + m = E_c + E_d$$

3. **Conservation of each component of total momentum:**

$$p_{x\ tot} = p_a = p_c \cos \theta \qquad \text{[horizontal momentum]}$$
$$p_{y\ tot} = 0 = p_c \sin \theta - p_d \qquad \text{[vertical momentum]}$$

4. **Solve:** First of all, the problem states that the kinetic energy K of the incoming positron equals its rest energy m. Therefore its total energy $E_a = m + K = m + m = 2m$. Second, the outgoing particles are photons, for which $p_c = E_c$ and $p_d = E_d$ in magnitude, respectively. With these substitutions, the three conservation equations become

$$E_a + m = 2m + m = 3m = E_c + E_d \qquad \text{[conservation of energy]}$$
$$p_a = E_c \cos \theta \qquad \text{[conservation of horizontal momentum]}$$
$$E_d = E_c \sin \theta \qquad \text{[conservation of vertical momentum]}$$

CONVERSION OF MASS TO ENERGY IN SUN

Luminous energy from Sun pours down on the outer atmosphere of Earth at a rate of 1372 watts per square meter of area that lies perpendicular to the direction of this radiation. The figure 1372 watts per square meter has the name **solar con-** stant. The radius of Earth equals approximately 6.4×10^6 meters and the Earth-Sun distance equals 1.5×10^{11} meters. The mass of Sun is approximately 2.0×10^{30} kilograms.

These are three equations in three unknowns E_c and E_d and θ. Square both sides of the second and third equations, add them, and use a trigonometric identity to get rid of the angle θ:

$$p_d^2 + E_d^2 = E_c^2(\cos^2\theta + \sin^2\theta) = E_c^2$$

Substitute $p_d^2 = E_d^2 - m^2$ on the left side of this equation and again use $E_a = 2m$ to obtain a first expression for E_c^2:

$$E_c^2 = E_a^2 - m^2 + E_d^2 = 4m^2 - m^2 + E_d^2 = 3m^2 + E_d^2$$

Now solve the equation of conservation of energy for E_c and square it to obtain a second expression for E_c^2:

$$E_c^2 = (3m - E_d)^2 = 9m^2 - 6mE_d + E_d^2$$

Equate these two expressions for E_c^2 and subtract E_d^2 from both sides to obtain

$$3m^2 = 9m^2 - 6mE_d$$

Solve for unknown E_d:

$$E_d = \frac{9m^2 - 3m^2}{6m} = \frac{6m^2}{6m} = m$$

This yields our first unknown. Use this result and conservation of energy to find an expression for E_c:

$$E_c = 3m - E_d = 3m - m = 2m$$

Finally, angle θ comes from conservation of vertical momentum. For a photon $p = E$, so

$$\sin\theta = \frac{p_d}{p_c} = \frac{E_d}{E_c} = \frac{m}{2m} = \frac{1}{2}$$

from which $\theta = 30$ degrees. We have now solved for all unknowns: $E_c = 2m$, $E_d = m$, and $\theta = 30$ degrees.

5. **Limiting cases:** There is no limiting case here, since the energy of the incoming positron is specified fully in terms of the mass m common to electron and positron.

---

a. How much mass is converted to energy every second in Sun to supply the luminous energy that falls on Earth?

b. What *total* mass is converted to energy every second in Sun to supply luminous energy?

c. Most of Sun's energy comes from burning hydrogen nuclei (mostly protons) into helium nuclei (mostly a two-proton–two-neutron combination). Mass of the proton equals 1.67262×10^{-27} kilogram, while the mass of a helium nucleus of this kind equals 6.64648×10^{-27} kilogram. How many metric tons of hydrogen must Sun convert to helium every second to supply its luminous output? (One metric ton is equal to 1000 kilograms, or 2200 pounds.)

d. Estimate how long Sun will continue to warm Earth, neglecting all other processes in Sun and emissions from Sun.

SOLUTION

a. One watt equals one joule per second = one kilogram meter^2/second^2. We want to measure energy in units of mass—in kilograms. Do this by dividing the number of joules by the square of the speed of light (Section 7.5 and Table 7-1):

$$\frac{1372 \text{ joules}}{c^2} = \frac{1.372 \times 10^3 \text{ kilogram meters}^2/\text{second}^2}{9.00 \times 10^{16} \text{ meters}^2/\text{second}^2}$$

$$= 1.524 \times 10^{-14} \text{ kilograms}$$

Thus every second 1.524×10^{-14} kilogram of luminous energy falls on each square meter perpendicular to Sun's rays. The following calculations are based on a simplified model of Sun (see last paragraph of this solution). Therefore we use the approximate value 1.5×10^{-14} kilogram per second and two-digit accuracy.

What total luminous energy falls on Earth per second? It equals the solar constant (in kilograms per square meter per second) times some area (in square meters). But what area? Think of a huge movie screen lying behind Earth and perpendicular to Sun's rays (see the figure). The shadow of Earth on this screen forms a circle of radius equal to the radius of Earth. This shadow represents the zone of radiation removed from that flowing outward from Sun. Call the area of this circle the cross-sectional area A of Earth. Earth's radius $r = 6.4 \times 10^6$ meters, so the cross-sectional area A seen by incoming Sunlight equals $A = \pi r^2 = 1.3 \times 10^{14}$ meters^2. Hence a total luminous energy equal to $(1.5 \times 10^{-14}$ kilograms/meter$^2) \times (1.3 \times 10^{14}$ meters$^2) = 2.0$ kilograms fall on Earth every second. This equals the mass converted every second in Sun to supply the light incident on Earth.

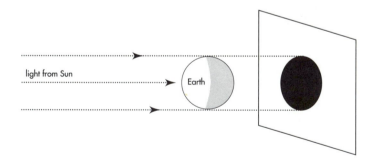

light from Sun

Earth

8.8 SUMMARY

mass: the magnitude of the 4-vector called momenergy

"Mass can be converted into energy and energy can be converted into mass"—this is a loose and sometimes misleading way to summarize some consequences of the two

b. Assume that Sun delivers sunlight at the same "solar-constant rate" to every part of a sphere surrounding Sun of radius equal to the Earth–Sun distance. The area of this large sphere has the value $4\pi R^2$ where $R = 1.5 \times 10^{11}$ meters, the average distance of Earth from Sun. This area equals 2.8×10^{23} meters^2. Therefore Sun converts a total of 2.8×10^{23} meters$^2 \times 1.5 \times 10^{-14}$ kilograms/meter^2 (from **a**) $= 4.2 \times 10^9$ kilograms of mass into luminous energy every second, or about 4 million metric tons per second.

c. Through a series of nuclear processes not described here, four protons transform into a helium nucleus consisting of two protons and two neutrons. The four original protons have a mass $4 \times 1.67262 \times 10^{-27} = 6.69048 \times 10^{-27}$ kilogram. The helium nucleus has a mass 6.64648×10^{-27} kilogram. The difference, 0.04400×10^{-27} kilogram, comes out mostly as light. (We cannot use two-digit accuracy here, because the important result is a difference between nearly equal numbers.)

The ratio of hydrogen burned to mass converted equals $6.69048/0.04400 = 150$ (back to two-digit accuracy!). So for each kilogram of mass converted to electromagnetic radiation, 150 kilograms of hydrogen burn to helium. In other words, about 0.7 percent of the rest energy (mass) of the original hydrogen is converted into radiation. Hence in order to convert 4.2×10^9 kilograms per second into radiation, Sun burns $150 \times 4.2 \times 10^9$ kilograms per second $= 6.3 \times 10^{11}$ kilograms of hydrogen into helium per second—about 630 million metric tons each second.

d. We can reckon Sun's mass by figuring how much Sun gravity it takes to guide our planet around in an orbit of 8 light-minute radius and one year time of circuit. Result: about 2.0×10^{30} kilograms. If Sun were all hydrogen, then the process of burning to helium at the present rate of 6.3×10^{11} kilograms every second would take $(2.0 \times 10^{30}$ kilograms$)/(6.3 \times 10^{11}$ kilograms/second$) = 3.2 \times 10^{18}$ seconds. At 32 million seconds per year, this would last about 10^{11} years, or 100 billion years.

Of course the evolution of a star is more complicated than the simple conversion of hydrogen into helium-plus-radiation. Other nuclear reactions fuse helium into more massive nuclei on the way to the most stable nucleus, iron-56 (Section 8.7). These other reactions occur at higher temperatures and typically proceed at faster rates than the hydrogen-to-helium process. Sun emits a flood of neutrinos (invisible; detected with elaborate apparatus; amount presently uncertain by a factor of 2, carry away less than 1 percent of Sun's output). Sun also loses mass as particles blown away from the surface, called the **solar wind**. And stars do not convert all their hydrogen to helium and other nuclei—or live for 100 billion years. According to current theory, the lifetime of a star like Sun equals approximately 10 billion years (10^{10} years). We believe Sun to be 4 to 5 billion years old. The remaining 6 billion years (6×10^9 years) or so should be sufficient time for our descendants to place themselves in the warmth of nearby stars.

principles that are basic and really accurate: (1) The total momenergy of an isolated system of particles remains unchanged in a reaction; (2) The invariant magnitude of the momenergy of any given particle equals the mass of that particle.

How much sound information about physics can be extracted from these basic principles? What troubles sometimes arise from accepting a too loose formulation of the "principle of equivalence of mass and energy"? Some answers to these questions appear in the dialog that follows, which serves also as a summary of this chapter.

DIALOG: USE AND ABUSE OF THE CONCEPT OF MASS

Does an isolated system have the same mass as observed in every inertial (free-float) reference frame?

Yes. Given in terms of energy E and momentum by $m^2 = E^2 - p^2$ in one frame, by $m^2 = (E')^2 - (p')^2$ in another frame. Mass of an isolated system is thus an *invariant*.

Does its *energy* have the same value in every inertial frame?

No. Energy is given by $E = (m^2 + p^2)^{1/2}$ or

$$E = m/(1 - v^2)^{1/2}$$

or

$$E = (\text{mass}) + (\text{kinetic energy}) = m + K$$

Value depends on the frame of reference from which the particle (or isolated system of particles) is observed. Value is lowest in the frame of reference in which the particle (or system) has zero momentum (zero *total* momentum in the case of an isolated system of particles). In that frame, and in that frame only, energy equals mass.

Does energy equal zero for an object of zero mass, such as a photon or neutrino or graviton?

No. Energy has value $E = (0^2 + p^2)^{1/2} = p$ (or in conventional units $E_{conv} = cp_{conv}$). Alternatively one can say—formally—that the entire energy resides in the form of *kinetic* energy ($K = p$ in this special case of *zero* mass), none at all in the form of rest energy. Thus,

$$E = (\text{mass}) + (\text{kinetic energy}) = 0 + K = K = p$$

(case of zero mass only!).

Can a photon—that has no mass—give mass to an absorber?

Yes. Light with energy E transfers mass $m = E$ ($= E_{conv}/c^2$) to a heavy absorber (Exercise 8.5).

Invariance of mass: Is that feature of nature the same as the principle that all electrons in the universe have the same mass?

No. It is true that all elementary particles of the same kind have the same mass. However, that is a fact totally distinct from the principle that the mass of an isolated system has identical value in whatever free-float frame it is figured (invariance of system mass).

Invariance of mass: Is that the same idea as the conservation of the momenergy of an isolated system?

No. Conservation of momenergy—the principle valid for an isolated system—says that the momenergy 4-vector figured *before* the constituents of a system have interacted is identical to the momenergy 4-vector figured *after* the constituents have interacted. In contrast, invariance of mass—the magnitude of the momenergy 4-vector—says that that mass is the same in *whatever* free-float frame it is figured.

Momenergy: Is that a richer concept than mass?

Yes. Momenergy 4-vector reveals mass and more: the motion of object or system with the mass

Conservation of the momenergy of an isolated system: Does this imply that collisions and interactions within an isolated system cannot change the system's mass?

Yes. Mass of an isolated system, being the magnitude of its momenergy 4-vector, can never change (as long as the system remains isolated).

Conservation of the momenergy of an isolated system: Does this say that the constituents that enter a collision are necessarily the same in individual mass and in number as the constituents that leave that collision?

No! The constituents often change in a high-speed encounter.

Example 1: Collision of two balls of putty that stick together—after collision hotter and therefore very slightly more massive than before.

Example 2: Collision of two electrons (e^-) with sufficient violence to create additional mass, a pair consisting of one ordinary electron and one positive electron (positron: e^+):

$$e^- \text{ (fast)} + e^- \text{ (at rest)} \rightarrow e^+ + 3e^-.$$

Example 3: Collision that radiates one or more photons:

$$e^- \text{ (fast)} + e^- \text{ (at rest)} \rightarrow$$

$$2 \begin{pmatrix} \text{electrons of} \\ \text{intermediate} \\ \text{speed} \end{pmatrix} + \begin{pmatrix} \text{electromagnetic} \\ \text{energy (photons)} \\ \text{emitted in the} \\ \text{collision process} \end{pmatrix}$$

In all three examples the *system* momenergy and *system* mass are each the same before as after.

Can I figure the mass of an isolated *system* composed of a number, n, of freely-moving objects by simply adding the masses of the individual objects? **Example:** Collection of fast-moving molecules.

Ordinarily NO, but yes in one very special case: Two noninteracting objects move freely and in step, side by side. Then the mass of the system *does* equal the sum of the two individual masses. In the general case, where the system parts are moving relative to each other, the relation between system mass and mass of parts is not additive. The length, in the sense of interval, of the 4-vector of total momenergy is not equal to the sum of the lengths of the individual momenergy 4-vectors, and for a simple reason: In the general case those vectors do not point in the same spacetime direction. *Energy* however, does add and *momentum* does add:

$$E_{\text{system}} = \sum_{i=1}^{n} E_i \quad \text{and} \quad p_{x,\text{system}} = \sum_{i=1}^{n} p_{x,i}$$

From these sums the mass of the system can be evaluated:

$$M_{\text{system}}^2 = E_{\text{system}}^2 - p_{x,\text{system}}^2 - p_{y,\text{system}}^2 - p_{z,\text{system}}^2$$

Can we simplify this expression for the mass of an isolated system composed of freely moving objects when we observe it from a free-float frame so chosen as to make the total momentum be zero?

Yes. In this case the mass of the system has a value given by the sum of energies of individual particles:

$$M_{\text{system}} = E_{\text{system}} = \sum_{i=1}^{n} E_i \qquad \text{[in zero-total momentum frame]}$$

Moreover, the energy of each particle can always be expressed as sum of rest energy m plus kinetic energy K:

$$E_i = m_i + K_i \qquad (i = 1, 2, 3, \ldots, n)$$

So the mass of the system exceeds the sum of the masses of its individual particles by an amount equal to the total kinetic energy of all particles (but only as observed in the frame in which *total* momentum equals zero):

$$M_{\text{system}} = \sum_{i=1}^{n} m_i + \sum_{i=1}^{n} K_i \qquad \text{[in zero-total momentum frame]}$$

For slow particles (Newtonian low-velocity limit) the kinetic energy term is negligible compared to the mass term. So it is natural that for years many thought that the mass of a system is the sum of the masses of its parts. However, such a belief leads to incorrect results at high velocities and is wrong as a matter of principle at all velocities.

What's the meaning of mass for a system in which the particles interact as well as move?

The energies of interaction have to be taken into account. They therefore contribute to the total energy, E_{system}, that gives the mass

$$M_{\text{system}} = (E_{\text{system}}^2 - p_{\text{system}}^2)^{1/2}$$

How do we find out the mass of a system of particles (Table 8-1) that are held — or stick — together?

Weigh it! Weigh it by conventional means if we are here on Earth and the system is small enough, otherwise by determining its gravitational pull on a satellite in free-float orbit about it.

Does mass measure "amount of matter"?

Nature does not offer us any such concept as "amount of matter." History has struck down every proposal to define such a term. Even if we could count number of atoms or by any other counting method try to evaluate amount of matter, that number would not equal mass. First, mass of the specimen changes with its temperature. Second, atoms tightly bonded in a solid weigh less — are less massive — than the same atoms free. Third, many of nature's atoms undergo radioactive decay, with still greater changes of mass. Moreover, around us occasionally, and continually in stars, the number of atoms and number of particles themselves undergo change. How then speak honestly? Mass, yes; "amount of matter," no.

Does the explosion in space of a 20-megaton hydrogen bomb convert 0.93 kilogram of mass into energy (fusion, Section 8.7)? [$\Delta m = \Delta E_{\text{rest, conv}}/c^2 = (20 \times 10^6 \text{ tons TNT}) \times (10^6 \text{ grams/ton}) \times (10^3 \text{ calories/gram of "TNT equivalent"}) \times (4.18 \text{ joules/calorie})/c^2 = (8.36 \times 10^{16} \text{ joules})/(9 \times 10^{16} \text{ meters}^2/\text{second}^2) = 0.93 \text{ kilogram}$]

Yes *and* no! The question needs to be stated more carefully. Mass of the system of expanding gases, fragments, and radiation has the *same* value immediately after explosion as before; mass M of the system has not changed. However, hydrogen has been transmuted to helium and other nuclear transformations have taken place. In consequence the *makeup* of mass of the system

$$M_{\text{system}} = \sum_{i=1}^{n} m_i + \sum_{i=1}^{n} K_i \qquad \text{[in zero-total momentum frame]}$$

has changed. The first term on the right—sum of masses of individual constituents—has *decreased* by 0.93 kilogram:

$$\left(\sum_{i=1}^{n} m_i\right)_{\text{after}} = \left(\sum_{i=1}^{n} m_i\right)_{\text{before}} - 0.93 \text{ kilogram}$$

The second term—sum of kinetic energies, including "kinetic energy" of photons and neutrinos produced—has *increased* by the same amount:

$$\left(\sum_{i=1}^{n} K_i\right)_{\text{after}} = \left(\sum_{i=1}^{n} K_i\right)_{\text{before}} + 0.93 \text{ kilogram}$$

The first term on the right side of this equation—the original heat content of the bomb—is practically zero by comparison with 0.93 kilogram. Thus part of the mass of *constituents* has been converted into energy; but the mass of the *system* has not changed.

The mass of the products of a nuclear *fission* explosion (Section 8.7: fragments of split nuclei of uranium, for example)—contained in an underground cavity, allowed to cool, collected, and weighed—is this mass less than the mass of the original nuclear device?

Yes! The key point is the waiting period, which allows heat and radiation to flow away until transmuted materials have practically the same heat content as that of original bomb. In the expression for the mass of the system

$$M_{\text{system}} = \sum_{i=1}^{n} m_i + \sum_{i=1}^{n} K_i \qquad \text{[in zero-total momentum frame]}$$

the second term on the right, the kinetic energy of thermal agitation—whose value rose suddenly at the time of explosion but dropped during the cooling period—has undergone no net alteration as a consequence of the explosion followed by cooling.

In contrast, the sum of masses

$$\sum m_i$$

has undergone a permanent decrease, and with it the mass M of *what one weighs* (after the cooling period) has dropped (see the figure).

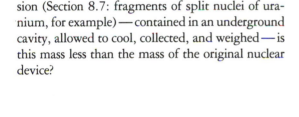

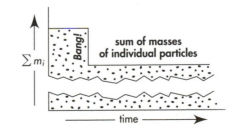

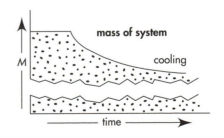

Does Einstein's statement that mass and energy are equivalent mean that energy is the *same* as mass?

No. Value of energy depends on the free-float frame of reference from which the particle (or isolated system of particles) is regarded. In contrast, value of mass is independent of inertial frame. Energy is only the *time* component of a momenergy 4-vector, whereas mass measures *entire magnitude* of that 4-vector. The time component gives the magnitude of the momenergy 4-vector only in the special case in which that 4-vector has no space component; that is, in a frame in which the momentum of the particle (or the total momentum of an isolated system of particles) equals zero. Only as measured in this special zero-momentum frame does energy have the same value as mass.

Then what *is* the meaning of Einstein's statement that mass and energy are equivalent?

Einstein's statement refers to the reference frame in which the particle is at rest, so that it has zero momentum p and zero kinetic energy K. Then $E = m + K \rightarrow m + 0$. In that case the energy is called the **rest energy** of the particle:

$$E_{\text{rest}} = m$$

In this expression, recall, the energy is measured in units of mass, for example kilograms. Multiply by the conversion factor c^2 to express energy in conventional units, for example joules (Table 7-1). The result is Einstein's famous equation:

$$E_{\text{rest, conv}} = mc^2$$

Many treatments of relativity fail to use the subscript "rest" — needed to remind us that this equivalence of mass and energy refers only to the *rest* energy of the particle (for a system, the total energy in the zero-total-momentum frame).

Without delving into all fine points of legalistic phraseology, how significant is the conversion factor c^2 in the equation $E_{\text{rest,conv}} = mc^2$?

The conversion factor c^2, like the factor of conversion from seconds to meters or miles to feet (Box 3-2), today counts as a detail of convention, rather than as a deep new principle.

If the factor c^2 is not the central feature of the relationship between mass and energy, what *is* central?

The distinction between mass and energy is this: Mass is the magnitude of the momenergy 4-vector and energy is the time component of the *same* 4-vector. Any feature of any discussion that emphasizes this contrast is an aid to understanding. Any slurring of terminology that obscures this distinction is a potential source of error or confusion.

Is the mass of a moving object greater than the mass of the same object at rest?

No. It is the same whether the object is at rest or in motion; the same in all frames.

Really? Isn't the mass, M, of a system of freely moving particles given, not by the sum of the masses m_i of the individual constituents, but by the sum of

Ouch! The concept of "relativistic mass" is subject to misunderstanding. That's why we don't use it. First, it applies the name mass — belonging to the

energies E_i (*but only in a frame in which total momentum of the system equals zero*)? Then why not give E_i a new name and call it "relativistic mass" of the individual particle? Why not adopt the notation

$$m_{i,\,rel} = E_i = m_i + K_i \quad ?$$

With this notation, can't one then write

$$M = \sum_{i=1}^{n} m_{i,\,rel} \quad ? \qquad \text{[in zero-total momentum frame]}$$

In order to make this point clear, should we call invariant mass of a particle its "rest mass"?

Can any simple diagram illustrate this contrast between mass and energy?

magnitude of a 4-vector—to a very different concept, the time component of a 4-vector. Second, it makes increase of energy of an object with velocity or momentum appear to be connected with some change in internal structure of the object. In reality, the increase of energy with velocity originates not in the object but in the geometric properties of space-time itself.

That is what we called it in the first edition of this book. But a thoughtful student pointed out that the phrase "rest mass" is also subject to misunderstanding: What happens to the "rest mass" of a particle when the particle moves? In reality mass is mass is mass. Mass has the same value in all frames, is invariant, no matter how the particle moves. [Galileo: "In questions of science the authority of a thousand is not worth the humble reasoning of a single individual."]

Yes. The figure shows the momentum-energy 4-vector of the same particle as measured in three different frames. Energy differs from frame to frame. Momentum differs from frame to frame. Mass (magnitude of 4-vector, represented by the length of handles on the arrows) has the same value, $m = 8$, in all frames.

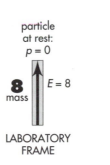

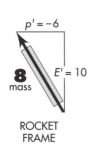

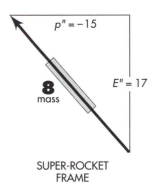

particle at rest:
$p = 0$

$E = 8$

LABORATORY FRAME

$p' = -6$

$E' = 10$

ROCKET FRAME

$p'' = -15$

$E'' = 17$

SUPER-ROCKET FRAME

REFERENCES

Quotation from Count Rumford in Section 8.2: Sanborn C. Brown, *Benjamin Thompson, Count Rumford* (MIT Press, Cambridge, Mass., 1979), page 220.

Reference to measurement of very small effects in Section 8.2: Vladimir Braginsky and A. B. Manukin, "Quantum non-demolition," in *Measurement of Weak Forces in Physics Experiments,* edited by David H. Douglas (University of Chicago Press, 1977).

Quotation from biography of A. H. Compton in Section 8.4: Robert S. Shankland, *Dictionary of Scientific Biography,* edited by Charles Coulston Gillespie, Volume III (Charles Scribner's Sons, New York, 1971).

Compton scattering reported in A. H. Compton, *Physical Review,* Volume 22, pages 409–413 (1923).

The polyelectron mentioned in Section 8.5 has been independently generated, through interaction of a slow positron with the electrons of a metal surface, by Alan Mills, Jr., at Bell Telephone Laboratories, as reported in *Physical Review Letters,* Volume 46, pages 717–720 (1981).

Final quotation in Box 8-1: Timothy Ferris, *Coming of Age in the Milky Way* (Anchor Books, Doubleday, New York, 1988), page 344.

Sample Problem 8-5 was suggested by Chet Raymo's science column in the *Boston Globe,* May 2, 1988, page 35.

Galileo quote in final dialog: Galileo Galilei, *Dialogo dei due massimi sistemi del mundo,* Landini, Florence. Translation by S. Drake, *Galileo Galilei — Dialogue Concerning the Two Chief World Systems — Ptolemaic and Copernican,* University of California Press, Berkeley and Los Angeles, 1953.

ACKNOWLEDGMENTS

We thank colleagues old and young for the comments that helped us clarify, formulate, and describe the concept of mass in this chapter and in the final dialog, and very specially Academician Lev B. Okun, Institute of Theoretical and Experimental Physics, Moscow, for correspondence and personal discussions. We believe that our approach agrees with that in two of his articles, both entitled "The Concept of Mass," which appeared in *Physics Today,* June 1989, pages 31–36, and *Soviet Physics-Uspekhi,* Volume 32, pages 629–638 (July 1989).

CHAPTER 8 EXERCISES

You now have at your disposal the power of special relativity to provide physical insight and accurate predictions about an immense range of phenomena, from nucleus to galaxy. The following exercises give only a hint of this range. Even so, there are too many to carry out as a single assignment or even several assignments. For this reason—and to anchor your understanding of relativity—we recommend that you continue to enjoy these exercises as your study moves on to other subjects. The following table of contents is intended to help organize this ongoing attention.

Reminder: In these exercises the symbol v (in other texts sometimes called β) stands for speed as a fraction of the speed of light c. Let v_{conv} be the speed in conventional units; then $v \equiv v_{conv}/c$.

CONTENTS

MASS AND ENERGY

8-1   examples of conversion

a   How much mass does a 100-watt bulb dissipate (in heat and light) in one year?

b   The total electrical energy generated on Earth during the year 1990 was probably between 1 and 2×10^{13} kilowatt-hours. To how much mass is this energy equivalent? In the actual production of this electrical energy is this much mass converted to energy? Less mass? More mass? Explain your answer.

c   Eric Berman, pedaling a bicycle at full throttle, produces one-half horsepower of *useful* power (1 horsepower = 746 watts). The human body is about 25 percent efficient; that is, 75 percent of the food burned is converted to heat and only 25 percent is converted to useful work. How long a time will Eric have to ride to lose one kilogram by the conversion of mass to energy? How can reducing gymnasiums stay in business?

8-2   relativistic chemistry

One kilogram of hydrogen combines chemically with 8 kilograms of oxygen to form water; about 10^8 joules of energy is released.

a   Ten metric tons (10^4 kilograms) of hydrogen combines with oxygen to produce water. Does the resulting water have a greater or less mass than the original hydrogen and oxygen? What is the magnitude of this difference in mass?

b   A smaller amount of hydrogen and oxygen is weighed, then combined to form water, which is weighed again. A very good chemical balance is able to detect a fractional change in mass of 1 part in 10^8. By what factor is this sensitivity more than enough — or insufficient — to detect the fractional change in mass in this reaction?

PHOTONS

8-3   pressure of light

a   Shine a one-watt flashlight beam on the palm of your hand. Can you feel it? Calculate the total force this beam exerts on your palm. *Should* you be able to feel it? A particle of what mass exerts the same force when you hold it at Earth's surface?

b   From the solar constant (1.372 kilowatts/square meter, Sample Problem 8-5) calculate the pressure of sunlight on an Earth satellite. Consider both reflecting and absorbing surfaces, and also "real" surfaces (partially absorbing). Why does the color of the light make no difference?

c   A spherical Earth satellite has radius $r = 1$ meter and mass $m = 1000$ kilograms. Assume that the satellite absorbs all the sunlight that falls on it. What is the acceleration of the satellite due to the force of sunlight, in units of g, the gravitational acceleration at Earth's surface? For a way to reduce this "disturbing" acceleration, see Figure 9-2.

d   It may be that particles smaller than a certain size are swept out of the solar system by the pressure of sunlight. This certain size is determined by the equality of the outward force of sunlight and the inward gravitational attraction of Sun. Estimate this critical particle size, making any assumptions necessary for your estimate. List the assumptions with your answer. Does your estimated size depend on the particle's distance from Sun?

Reference: For pressure of light measurement in an elementary laboratory, see Robert Pollock, *American Journal of Physics*, Volume 31, pages 901–904 (1963). Pollock's method of determining the pressure of light makes use of resonance to amplify a small effect to an easily measured magnitude. Dr. Pollock developed this experiment in collaboration with the same group of first-year students at Princeton University with whom the authors had the privilege to work out the presentation of relativity in the first edition of this book.

8-4   measurement of photon energy

A given radioactive source emits energetic photons (X-rays) or very energetic photons (gamma rays) with energies characteristic of the particular radioactive nucleus in question. Thus a precise energy measurement can often be used to determine the composition of even a tiny specimen. In the apparatus diagrammed in the figure, only those events are detected in which a count on detector A (knocked-on electron) is accompanied by a count on detector B (scattered photon). What is the energy of the incoming photons that are detected in this way, in units of the rest energy of the electron?

8-5   Einstein's derivation: equivalence of energy and mass — a worked example

Problem

From the fact that light exerts pressure and carries energy, show that this energy is equivalent to mass and hence — by extension — show the equivalence of all energy to mass.

Commentary: The equivalence of energy and mass is such an important consequence that Einstein very early, after his relativistic derivation of this result, sought and found an alternative elementary physical line of reasoning that leads to the same conclusion. He envisaged a closed box of mass M initially at rest, as shown in the first figure. A directed burst of electro-

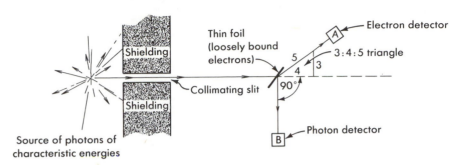

EXERCISE 8-4. *Measurement of photon energy.*

magnetic energy is emitted from the left wall. It travels down the length L of the box and is absorbed at the other end. The radiation carries an energy E. But it also carries momentum. This one sees from the following reasoning. The radiation exerts a pressure on the left wall during the emission. In consequence of this pressure the box receives a push to the left, and a momentum, p. But the momentum of the system as a whole was zero initially. Therefore the radiation carries a momentum p opposite to the momentum of the box. How can one use knowledge of the transport of energy and momentum by the radiation to deduce the mass equivalent of the radiation? Einstein got his answer from the argument that the center of mass of the system was not moving before the transport process and therefore cannot be in motion during the transport process. But the box obviously carries mass to the left. Therefore the radiation must carry mass to the right. So much for Einstein's reasoning in broad outline. Now for the details.

From relativity Einstein knew that the momentum p of a directed beam of radiation is equal to the energy E of that beam (Section 8.4; both p and E measured in units of mass). However, this was known before Einstein's relativity theory, both from Maxwell's theory of electromagnetic radiation and from direct observa-

tion of the pressure exerted by light on a mirror suspended in a vacuum. This measurement had first successfully been carried out by E. F. Nichols and G. F. Hull between 1901 and 1903. (By now the experiment has been so simplified and increased in sensitivity that it can be carried out in an elementary laboratory. See the reference for Exercise 8-3.)

Thus the radiation carries momentum and energy to the right while the box carries momentum and mass to the left. But the center of mass of the system, box plus radiation, cannot move. So the radiation must carry to the right not merely energy but mass. How much mass? To discover the answer is the object of these questions.

a What is the velocity of the box during the time of transit of the radiation?

b After the radiation is absorbed in the other end of the box, the system is once again at rest. How far has the box moved during the transit of the radiation?

c Now demand that the center of mass of the system be at the same location both before and after the flight of the radiation. From this argument, what is the mass equivalent of the energy that has been transported from one end of the box to the other?

Solution

a During the transit of the radiation the momentum of the box must be equal in magnitude and opposite in direction to the momentum p of the radiation. The box moves with a very low velocity v. Therefore the Newtonian formula Mv suffices to calculate its momentum:

$$Mv = -p = -E$$

From this relation we deduce the velocity of the box,

$$v = -E/M$$

b The transit time of the photon is very nearly

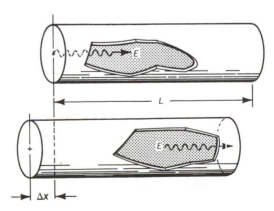

EXERCISE 8-5, first figure. *Transfer of mass by radiation.*

$t = L$ meters of light-travel time. In this time the box moves a distance

$$\Delta x = vt = -EL/M$$

c If the radiation transported no mass from one end of the box to the other, and if the box were the sole object endowed with mass, then this displacement Δx would result in a net motion of the center of mass of the system to the left. But, Einstein reasoned, an isolated system with its center of mass originally at rest can never set itself into motion nor experience any shift in its center of mass. Therefore, he argued, there must be some countervailing displacement of a part of the mass of the system. This transport of mass to the right can be understood only as a new feature of the radiation itself. Consequently, during the time the box is moving to the left, the radiation must transport to the right some mass m, as yet of unknown magnitude, but such as to ensure that the center of mass of the system has not moved. The distance of transport is the full length L of the box diminished by the distance Δx through which the box has moved to the left in the meantime. But Δx is smaller than L in the ratio E/M. This ratio can be made as small as one pleases for any given transport of radiant energy E by making the mass M of the box sufficiently great. Therefore it is legitimate to take the distance moved by the radiation as equal to L itself. Thus, with arbitrarily high precision, the condition that the center of mass shall not move becomes

$$M\Delta x + mL = 0$$

Calculate the mass m and find, using Δx from part **b**,

$$m = -\Delta x M/L = -(-EL/M)(M/L)$$

or, finally,

$$m = E$$

In conventional units, we have the famous equation

$$E_{conv} = mc^2$$

We conclude that the process of emission, transport, and reabsorption of radiation of energy E is equivalent to the transport of a mass $m = E$ from one end of the box to the other end. The simplicity of this derivation and the importance of the result makes this analysis one of the most interesting in all of physics.

Discussion: The mass equivalence of radiant energy implies the mass equivalence of thermal energy and — by extension — of other forms of energy, ac-cording to the following reasoning. The energy that emerges from the left wall of the box may reside there originally as heat energy. This thermal energy excites a typical atom of the surface from its lowest energy state to a higher energy state. The atom returns from this higher state to a lower state and in the course of this change sends out the surplus energy in the form of radiation. This radiant energy traverses the box, is absorbed, and is ultimately converted back into ther-mal energy. Whatever the details of the mechanisms by which light is emitted and absorbed, the net effect is the transfer of heat energy from one end of the box to the other. To say that mass has to pass down the length of the box when radiation goes from one wall to the other therefore implies that mass moves when thermal energy changes location. The thermal energy in turn is derived from chemical energy or the energy of a nuclear transformation or from electrical energy. Moreover, thermal energy deposited at the far end of the tube can be converted back into one or another of these forms of energy. Therefore these forms of energy — and likewise all other forms of energy — are equivalent in their transport to the transport of mass in the amount $m = E$.

How can one possibly uphold the idea that a pulse of radiation transports mass? One already knows that a photon has zero mass, by virtue of the relation (Section 8.4)

$$(\text{mass})^2 = (\text{energy})^2 - (\text{momentum})^2 = 0$$

Moreover, what is true of the individual photon is true of the pulse of radiation made up of many such photons: The energy and momentum are equal in magnitude, so that the mass of the radiation necessar-ily vanishes. Is there not a fundamental inconsistency in saying in the same breath that the mass of the pulse is zero and that radiation of energy E transports the mass $m = E$ from one place to another?

The source of our difficulty is some confusion be-tween two quite different concepts: (1) energy, the time component of the momentum–energy 4-vector, and (2) mass, the magnitude of this 4-vector. When the system divides itself into two parts (radiation going to the right and box recoiling to the left) the components of the 4-vectors of the radiation and of the recoiling box add up to identity with the compo-nents of the original 4-vector of the system before emission, as shown in the second figure. However, the magnitudes of the 4-vectors (magnitude = mass) are not additive. No one dealing with Euclidean geome-try would expect the length of one side of a triangle to be equal to the sum of the lengths of the other two sides. Similarly in Lorentz geometry. The mass of the system (M) is not to be considered as equal to the sum of the mass of the radiation (zero) and the mass of the recoiling box (less than M). But components of 4-vectors *are* additive; for example,

$$\begin{pmatrix} \text{energy of} \\ \text{system} \end{pmatrix} = \begin{pmatrix} \text{energy of} \\ \text{radiation} \end{pmatrix} + \begin{pmatrix} \text{energy of} \\ \text{recoiling box} \end{pmatrix}$$

Thus we see that the energy of the recoiling box is $M - E$. Not only is the energy of the box reduced by the emission of radiation from the wall; also its mass is reduced (see shortened length of 4-vector in diagram). Thus the radiation takes away mass from the wall of the box even though this radiation has zero mass. The inequality

$$\begin{pmatrix} \text{mass of} \\ \text{system} \end{pmatrix} \neq \begin{pmatrix} \text{mass of} \\ \text{radiation[zero]} \end{pmatrix} + \begin{pmatrix} \text{mass of} \\ \text{recoiling box} \end{pmatrix}$$

is as natural in spacetime geometry as is the inequality $5 \neq 3 + 4$ in Euclidean geometry.

What about the gravitational attraction exerted by the system on a test object? Of course the redistribution of mass as the radiation moves from left to right makes some difference in the attraction. But let the test object be at a distance r so great that any such redistribution has a negligible effect on the attraction. In other words, all that counts for the pull on a unit test object is the total mass M as it appears in Newton's formula for gravitational force:

$$\begin{pmatrix} \text{force per} \\ \text{unit mass} \end{pmatrix} = \frac{GM}{r^2}$$

Even so, will not the distant detector momentarily experience a less-than-normal pull while the radiation is in transit down the box? Is not the mass of the radiation zero, and is not the mass of the recoiling box reduced below the original mass M of the system? So is not the total attracting mass less than normal during the process of transport? No! The mass of the system —one has to say again— is not equal to the sum of the masses of its several parts. It is instead equal to the magnitude of the total momentum–energy 4-vector of the system. And at no time does either the total momentum (in our case zero!) or the total energy of the system change—it is an isolated system. Therefore neither is there any change in the magnitude M of the total momentum–energy 4-vectors shown in the second figure. So, finally, there is never any change in the gravitational attraction.

There is one minor swindle in the way this problem has been presented: The box cannot in fact move as a rigid body. If it could, then information about the emission of the radiation from one end could be obtained from the motion of the other end before the arrival of the radiation itself—this information would be transmitted at a speed greater than that of light! Instead, the recoil from the emission of the radiation travels along the sides of the box as a vibrational wave, that is, with the speed of sound, so that this wave arrives at the other end long after the radiation does. In the meantime the absorption of the radiation at the second end causes a second vibrational wave which travels back along the sides of the box. The addition of the vibration of the box to the prob-

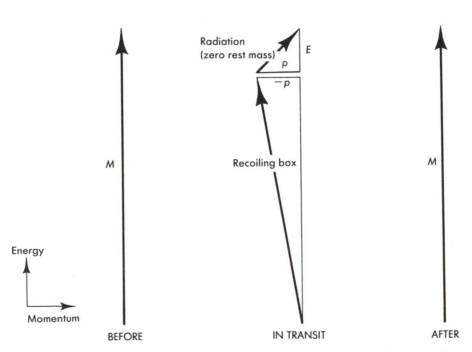

EXERCISE 8-5, second figure. *Radiation transfers mass from place to place even though the mass of the radiation is zero!*

lem requires a more complicated analysis but does not change in any essential way the results of the exercise.

References: A. Einstein, *Annalen der Physik,* Volume 20, pages 627–633 (1906). For a more careful treatment of the box, see A. P. French, *Special Relativity* (W. W. Norton, New York, 1968), pages 16–18 and 27–28.

8-6 gravitational red shift

Note: Exercises 8-6 and 8-7 assume an acquaintance with the following elementary facts of gravitation.

(1) A very small object—or a spherically symmetric object of any radius—with mass M attracts an object of mass m—also small or spherically symmetric—with a force

$$F = \frac{GMm}{r^2}$$

Here r is the distance between the centers of the two objects and G is the Newtonian constant of gravitation, $G = 6.67 \times 10^{-11}$ (meter)³/(kilogram-second²).

(2) The work required to move a test particle of unit mass from r to $r + dr$ against the gravitational pull of a fixed mass M is $GM(dr/r^2)$. Translated from conventional units of energy to units of mass this work is

$$dW_{conv} = \frac{GM}{c^2} \frac{dr}{r^2} = M^* \frac{dr}{r^2}$$

per unit of mass contained in the test particle.

(3) The symbol $M^* = GM/c^2$ in this formula has a simple meaning. It is the mass of the center of attraction translated from units of kilograms to units of meters. For example, the mass of Earth ($M_{Earth} = 5.974 \times 10^{24}$ kilograms) expressed in length units is $M^*_{Earth} = 4.44 \times 10^{-3}$ meters, and the mass of Sun ($M_{Sun} = 1.989 \times 10^{30}$ kg) is $M^*_{Sun} = 1.48 \times 10^3$ meters.

(4) Start the test particle at a distance r from the center of attraction of mass M and carry it to an infinite distance. The work required is $W = M^*/r$ in units of mass per unit of mass contained in the test particle.

So much for the minitutorial. Now to business.

a What fraction of your rest energy is converted to potential energy when you climb the Eiffel Tower (300 meters high) in Paris? Let g^* be the acceleration of gravity in meters/meter² at the surface of Earth:

$$g^* = \frac{GM_{Earth}}{c^2} \frac{1}{r^2_{Earth}} = \frac{M^*_{Earth}}{r^2_{Earth}} = \frac{g}{c^2}$$

b What fraction of one's rest energy is converted to potential energy when one climbs a very high ladder that reaches higher than the gravitational influence of Earth? Assume that Earth does not rotate and is alone in space. Does the fraction of the energy that is lost in either part **a** or part **b** depend on your original mass?

c Apply the result of part **a** to deduce the fractional energy change of a photon that rises vertically to a height z in a uniform gravitational field g^*. Photons have zero mass; one can say formally that they have only kinetic energy $E = K$. Thus photons have only one purse—the kinetic energy purse—from which to pay the potential energy tax as they rise in the gravitational field. Light of frequency f is composed of photons of energy $E = hf/c^2$ (see Exercise 8-31). Show that the fractional energy loss for photons rising in a gravitational field corresponds to the following fractional change in frequency:

$$\frac{\Delta f}{f} = -g^* z \quad \text{[uniform gravitational field]}$$

Note: We use f for frequency instead of the usual Greek nu, ν, to avoid confusion with v for speed.

d Apply the result of part **b** to deduce the fractional energy loss of a photon escaping to infinity. (To apply **b** for this purpose is an approximation good to one percent when this fractional energy loss itself is less than two percent.) Specifically, let the photon start from a point on the surface of an astronomical object of mass M (kilograms) or M^* (meters) = GM/c^2 and radius r. From the fractional energy loss, show that the fractional change of frequency is given by the expression

$$\frac{\Delta f}{f} = -\frac{M^*}{r} \quad \text{[escape field of spherical object]}$$

This decrease in frequency is called the **gravitational red shift** because, for visible light, the shift is toward the lower-frequency (red) end of the visible spectrum.

e Calculate the fractional gravitational red shifts for light escaping from the surface of Earth and for light escaping from the surface of Sun.

Discussion: The results obtained in this exercise are approximately correct for light moving near Earth, Sun, and white dwarf (Exercise 8-7). Only general relativity correctly describes the motion of light very close to neutron star or black hole (Box 9-2).

8-7 density of the companion of Sirius

Note: This exercise uses a result of Exercise 8-6.

Sirius (the Dog Star) is the brightest star in the heavens. Sirius and a small companion revolve about one another. By analyzing this revolution using Newtonian mechanics, astronomers have determined that the mass of the companion of Sirius is roughly equal to the mass of our Sun (M is about 2×10^{30} kilograms; M^* is about 1.5×10^3 meters). Light from the companion of Sirius is analyzed in a spectrometer. A spectral line from a certain element, identified from the pattern of lines, is shifted in frequency by a fraction 7×10^{-4} compared to the frequency of the same spectral line from the same element in the laboratory. (These figures are experimentally accurate to only one significant figure.) Assuming that this is a gravitational red shift (Exercise 8-6), estimate the average density of the companion of Sirius in grams/centimeter^3. This type of star is called a **white dwarf** (Box 9-2).

CREATIONS, TRANSFORMATIONS, ANNIHILATIONS

8-8   nuclear excitation

A nucleus of mass m initially at rest absorbs a gamma ray (photon) and is excited to a higher energy state such that its mass is now 1.01 m.

   a   Find the energy of the incoming photon needed to carry out this excitation.

   b   Explain why the required energy of the incoming photon is greater than the change of mass of the nucleus.

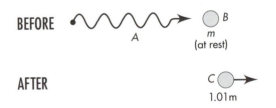

EXERCISE 8-8. *Excitation of a nucleus by a gamma ray.*

8-9   photon braking

A moving radioactive nucleus of known mass M emits a gamma ray (photon) in the forward direction and drops to its stable nonradioactive state of known mass m. Find the energy of the incoming nucleus (BEFORE diagram in the figure) such that the resulting mass m nucleus is at rest (AFTER diagram). The unknown energy E_c of the outgoing gamma ray should not appear in your answer.

BEFORE M ⟶ E_A (to be found)
A

AFTER m •~~~~~~~~⟶ E_C
D C (not known)
(at rest)

EXERCISE 8-9. *Stopping a nucleus by emission of a gamma ray.*

8-10   photon integrity

Show that an isolated photon cannot split into two photons going in directions other than the original direction. (Hint: Apply the laws of conservation of momentum and energy and the fact that the third side of a triangle is shorter than the sum of the other two sides. What triangle?)

8-11   pair production by a lonely photon?

A gamma ray (high-energy photon, zero mass) can carry an energy greater than the rest energy of an electron–positron pair. (Remember that a positron has the same mass as the electron but opposite charge.) Nevertheless the process

(energetic gamma ray) ⟶ (electron) + (positron)

cannot occur in the absence of other matter or radiation.

   a   Prove that this process is incompatible with the laws of conservation of momentum and energy as employed in the laboratory frame of reference. Analyze the alleged creation in the frame in which electron and positron go off at equal but opposite angles $\pm\phi$ with the extended path of the incoming gamma ray.

   b   Repeat the demonstration—which then becomes much more impressive—in the center-of-momentum frame of the alleged pair, the frame of reference in which the total momentum of the two resulting particles is zero.

8-12   photoproduction of a pair by two photons

Two gamma rays of different energies collide in a vacuum and disappear, bringing into being an electron–positron pair. For what ranges of energies of the two gamma rays, and for what range of angles between their initial directions of propagation, can this reaction occur? (Hint: Start with an analysis of the reaction at threshold; at threshold the electron and positron are relatively at rest.)

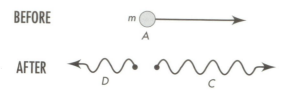

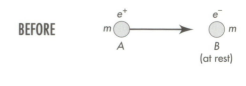

EXERCISE 8-13. *Decay of positronium in flight.*

8-13 decay of positronium

A moving "atom" called positronium (an electron and positron orbiting one another) of mass m and initial energy E decays into two gamma rays (high-energy photons) that move in opposite directions along the line of motion of the initial atom. Find the energy of each gamma ray, E_C and E_D, in terms of the mass m and energy E_A of the initial particle. Check that $E_C = E_D$ in the case that the initial particle is at rest.

8-14 positron—electron annihilation I

A positron e^+ of mass m and kinetic energy K is annihilated on a target containing electrons e^- (same mass m) practically at rest in the laboratory frame:

$$e^+(\text{fast}) + e^-(\text{at rest}) \longrightarrow \text{radiation}$$

a By considering the collision in the center-of-momentum frame (the frame of reference in which the total momentum of the initial particles is equal to zero), show that it is necessary for at least two gamma rays (rather than one) to result from the annihilation.

b Return to the laboratory frame, shown in the figure. The outgoing photons move on the line along which the positronium approaches. Find an expression for the energy of each outgoing photon. Let your derivation be free of any reference to velocity.

c Using simple approximations, evaluate the answer to part **b** in the limiting cases (1) very small K and (2) very large K. (Very small and very large compared with what?)

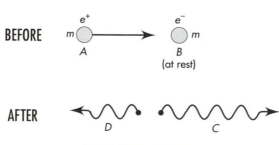

EXERCISE 8-14. *Positron—electron annihilation.*

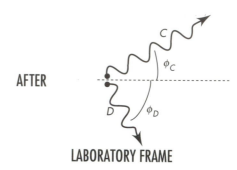

LABORATORY FRAME

EXERCISE 8-15, **first figure.** *Positron—electron annihilation.*

8-15 positron—electron annihilation II

A positron e^+ of mass m and kinetic energy K is annihilated on a target containing electrons e^- (same mass m) practically at rest in the laboratory frame:

$$e^+(\text{fast}) + e^-(\text{at rest}) \longrightarrow \text{radiation}$$

The resulting gamma rays go off at different angles with respect to the direction of the incoming positron, as shown in the first figure.

a Derive an expression for the energy of one of the gamma rays in the laboratory frame as a function of the angle between the direction of emergence of that gamma ray and the direction of travel of the positron before its annihilation. The gamma ray energy should be a function of only the energy and mass of the incoming positron and the angle of the outgoing gamma ray. (Hint: Use the law of cosines, as applied to the second figure.)

$$p_D{}^2 = p_A{}^2 + p_C{}^2 - 2p_A p_C \cos \phi_C$$

b Show that for outgoing gamma rays moving along the positive and negative x-direction, the results of this exercise reduce to the results of Exercise 8.14.

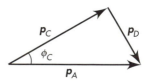

EXERCISE 8-15, **second figure.** *Conservation of vector momentum means that the momentum triangle is closed.*

8-16   creation of proton—antiproton pair by an electron

What is the threshold kinetic energy K_{th} of the incident electron for the following process?

electron (fast) + proton (at rest) ⟶
electron + antiproton + two protons

8-17   colliders

How much more violent is a collision of two protons that are moving toward one another from opposite directions than a collision of a moving proton with one at rest?

Discussion: When a moving particle strikes a stationary one the energy available for the creation of new particles, for heating, and for other interactions — or, in brief, the available interaction energy — is less than the initial energy (the sum of the rest and kinetic energies of the initial two particles). Reason: The particles that are left over after the reaction have a net forward motion (law of conservation of momentum), the kinetic energy of which is available neither for giving these particles velocity relative to each other nor for producing more particles. For this reason much of the particle energy produced in accelerators is not available for studying interactions because it is carried away in the kinetic energy of the products of the collision.

However, in the center-of-momentum frame, the frame in which the total momentum of the system is equal to zero, no momentum need be carried away from the interaction. Therefore the energy available for interaction is equal to the total energy of the incoming particles.

Is there some way that the laboratory frame can be made also the center-of-momentum frame? One way is to build two particle accelerators and have the two beams collide head on. If the energy and masses of the particles in each beam are respectively the same, then the laboratory frame is the center-of-momentum frame and all the energy in each collision is available interaction energy. It is easier and cheaper to achieve the same efficiency by arranging to have particles moving in opposite directions in the same accelerator. A magnetic field keeps the particles in a circular path, "storing" them at their maximum energy for repeated tries at interaction. Such a facility is called a **collider.** The figure on page 262 gives some details of a particular collider.

a   What is the total available interaction energy for each encounter in the laboratory frame of the Tevatron?

b   Now transform to a frame in which one of the incoming particles is at rest (transformation given in Exercise 7-5). This would be the situation if we tried to build an accelerator in which moving antiprotons hit a stationary target of, say, liquid hydrogen (made of protons and electrons). [Simplify: At 0.9 TeV what is the effective speed v of the proton? What is its momentum compared with its energy? What is the value of the time stretch factor $\gamma = E/m$?] If the target protons were at rest, what energy, in TeV, would the incoming antiproton need to have in order to yield the same interaction energy as that achieved in the Tevatron?

Wait a minute! You keep telling us that energy and momentum have different values when measured with respect to different reference frames. Yet here you assume the "interaction energy" is the same in the Tevatron laboratory frame as it is in the rest frame of a proton that moves with nearly the speed of light in the Tevatron frame. Is the energy of a system different in different frames, or is it the same?

There is an important distinction between the total energy of a system and the "available interaction energy," just as there is an important distinction between your money in the bank and money in the bank you can spend. If some of your money in the bank has been put in escrow for payment on a house you are buying, then you cannot spend that part of your bank money to buy a new car. Similarly, the total energy of the proton—antiproton system is much smaller in the Tevatron laboratory frame than in the frame in which the proton is initially at rest, but all of the Tevatron laboratory-frame energy can be spent — used to create new particles, for example. In contrast, only a minute fraction of the energy in the frame in which the proton is initially at rest can be spent to create new particles, since total momentum must be conserved; most of the total energy is kept "in escrow" for this purpose. The number and kinds of new particles created must be the same for all observers! Therefore the "available interaction energy" must be the same for all observers. The central point here is that the Tevatron collider design makes all of the energy in the proton—antiproton system "available" for use in the laboratory.

EXERCISE 8-17. *Top: Aerial view of the Tevatron ring at Fermi National Accelerator Laboratory in Batavia, Illinois. The ring is 6.3 kilometers in circumference.* **Bottom:** *View along the tunnel of the Tevatron. Protons (positive charge) and antiprotons (antiparticle of the proton: same mass, negative charge) circulate in separate beams in opposite directions in the same vacuum chamber in the lower ring of superconducting magnets shown in the photo. The upper ring of regular magnets accelerates protons from 8 GeV to 150 GeV. Some of these protons are injected into the lower set of magnets directly, rotating clockwise. Other protons strike a copper target and create antiprotons at a lower energy that are accumulated over approximately 15 hours in a separate ring (not shown) and then reaccelerated to 150 GeV and inserted into the lower ring, circulating counterclockwise. (Opposite charge, opposite motion yields same magnetic force toward the center, hence counterrotation around the same circle.) Then particles in both beams in the lower ring of magnets are accelerated at the same time from 150 GeV to a final energy of 0.9 TeV per particle. (1 teraelectron-volt $= 10^{12}$ electronvolts, or approximately 1000 times the rest energy of the proton or antiproton.) After acceleration, the beams are switched magnetically so that they cross each other at multiple intersection points around the ring, allowing protons and antiprotons to collide in the laboratory center-of-momentum frame. Detectors at the points of intersection monitor products of the collisions. Protons and antiprotons that do not interact at one intersection are not wasted; they may interact at another intersection point or on subsequent trips around the ring. The particles are allowed to coast around and around at full energy for as long as 24 hours as they interact. Question: Approximately how many revolutions around the ring does a given proton or antiproton make in 24 hours?* Photographs courtesy of Fermi Laboratory.

DOPPLER SHIFT

8-18 Doppler shift along the x-direction

Note: Recall Exercise L-5 in the Special Topic on Lorentz Transformation, following Chapter 3.

Apply the momenergy transformation equations (Exercise 7-5) to light moving in the positive x-direction for which $p_x = p = E$.

a Show that the relation between photon energy E' in the rocket frame and photon energy E in the laboratory frame is given by the equation

$$E = \gamma(1 + v)E' = \frac{(1 + v)E'}{(1 - v^2)^{1/2}}$$

$$= \frac{(1 + v)E'}{[(1 - v)(1 + v)]^{1/2}} = \left[\frac{1 + v}{1 - v}\right]^{1/2} E'$$

[photon moves along positive x-direction]

b Use the Einstein relation between photon energy E and classical wave frequency f, namely $E_{conv} = hf$ or $E = hf/c^2$ and $E' = hf'/c^2$, to derive the transformation for frequency

$$f = \left[\frac{1 + v}{1 - v}\right]^{1/2} f'$$

[wave motion along positive x-direction]

This is the Doppler shift equation for light waves moving along the positive x-direction.

Note: We use f for frequency instead of the usual Greek nu, ν, to avoid confusion with v for speed.

c Show that for a wave moving along the negative x-direction, the equation becomes

$$f = \left[\frac{1 - v}{1 + v}\right]^{1/2} f'$$

[wave motion along negative x-direction]

d Derive the corresponding equations that convert laboratory-measured frequency f to rocket-measured frequency f' for waves moving along both positive and negative x-directions.

8-19 Doppler equations

A photon moves in the xy laboratory plane in a direction that makes an angle ϕ with the x-axis, so that its components of momentum are $p_x = p \cos \phi$ and $p_y = p \sin \phi$ and $p_z = 0$.

a Use the Lorentz transformation equations for the momentum–energy 4-vector (Exercise 7-5) and the relation $E^2 - p^2 = 0$ for a photon to show that in the rocket frame, moving with speed v_{rel} along the

laboratory x-direction, the photon has an energy E' given by the equation

$$E' = E\gamma(1 - v_{rel} \cos \phi)$$

and moves in a direction that makes an angle ϕ' with the x'-axis given by the equation

$$\cos \phi' = \frac{\cos \phi - v_{rel}}{1 - v_{rel} \cos \phi}$$

b Derive the inverse equations for E and $\cos \phi$ as functions of E', $\cos \phi'$, and v_{rel}. Show that the results are

$$E = E'\gamma(1 + v_{rel} \cos \phi')$$

$$\cos \phi = \frac{\cos \phi' + v_{rel}}{1 + v_{rel} \cos \phi'}$$

c If the frequency of the light in the laboratory frame is f, what is the frequency f' of the light in the rocket frame? Use the Einstein relation between photon energy E and classical wave frequency f, namely $E_{conv} = hf$ or $E = hf/c^2$ and $E' = hf'/c^2$, to derive the transformations for frequency

$$f' = f\gamma(1 - v_{rel} \cos \phi)$$
$$f = f'\gamma(1 + v_{rel} \cos \phi')$$

This difference in frequency due to relative motion is called the **Doppler shift.**

Note: We use f for frequency instead of the usual Greek nu, ν, to avoid confusion with v for speed.

d For wave motion along the positive and negative x-direction, show that the results of this exercise reduce to the results of Exercise 8-18.

e Discussion question: Do the Doppler equations enable one to determine the rest frame of the source that emits the photons?

8-20 the physicist and the traffic light

A physicist is arrested for going through a red light. In court he pleads that he approached the intersection at such a speed that the red light looked green to him. The judge, a graduate of a physics class, changes the charge to speeding and fines the defendant one dollar for every kilometer/hour he exceeded the local speed limit of 30 kilometers/hour. What is the fine? Take the wavelength of green light to be 530 nanometers = 530 × 10⁻⁹ meter) and the wavelength of red light to be 650 nanometers. The relation between wavelength λ and frequency f for light is $f\lambda = c$. Notice that the light propagates in the negative x-direction ($\phi = \phi' = \pi$).

8-21   speeding light bulb

A bulb that emits spectrally pure red light uniformly in all directions in its rest frame approaches the observer from a very great distance moving with nearly the speed of light along a straight-line path whose perpendicular distance from the observer is b. Both the color and the number of photons that reach the observer per second from the light bulb vary with time. Describe these changes qualitatively at several stages as the light bulb passes the observer. Consider both the Doppler shift and the headlight effect (Exercises 8-19 and L-9).

8-22   Doppler shift at the limb of Sun

Sun rotates once in about 25.4 days. The radius of Sun is about 7.0×10^8 meters. Calculate the Doppler shift that we should observe for light of wavelength 500 nanometers $= 500 \times 10^{-9}$ meter) from the edge of Sun's disk (the **limb**) near the equator. Is this shift toward the red end or toward the blue end of the visible spectrum? Compare the magnitude of this Doppler shift with that of the gravitational red shift of light from Sun (Exercise 8-6).

8-23   the expanding universe

Note: Recall Exercise 3-10.

   a   Light from a distant galaxy is analyzed by a spectrometer. A spectral line of wavelength 730 nanometers $= 730 \times 10^{-9}$ meters is identified (from the pattern of other lines) to be one of the lines of hydrogen that, for hydrogen in the laboratory, has the wavelength 487 nanometers. If the shift in wavelength is a Doppler shift, how fast is the observed galaxy moving relative to Earth? Notice that the light propagates in a direction opposite to the direction of motion of the galaxy ($\phi = \phi' = \pi$).

   b   There is independent evidence that the observed galaxy is 5×10^9 light years away. Estimate the time when that galaxy parted company from our own galaxy — the Milky Way — using the simplifying assumption that the speed of recession was the same throughout the past (that is, not slowed down by the gravitational attractions between one galaxy and another). The astronomer Edwin Hubble discovered in 1929 that this time — whose reciprocal is called the Hubble constant, and which may itself therefore appropriately be called the Hubble time — has about the same value for all galaxies whose distances and speeds can be measured. Hence the concept of the expanding universe.

   c   Will allowance for the past effect of gravitation in slowing the expansion increase or decrease the estimated time back to the start of this expansion?

Reference: E. Hubble, *Proceedings of the U. S. National Academy of Sciences,* Volume 15, pages 168–173 (1929).

8-24   twin paradox using the Doppler shift

The Twin Paradox (Chapter 4 and Exercises 4-1 and 5-8) can be resolved elegantly using the Doppler shift as follows. Paul remains on Earth. His twin sister Penny travels at a high speed, v, to a distant star and returns to Earth at the same speed. Both Penny and Paul observe a distant variable star whose light gets alternately dimmer and then brighter with a frequency f in the Earth frame (f' in the rocket frame). This variable star is very much farther away than the length of Penny's path and is in a direction perpendicular to this path in the Earth frame. Both observers will count the same total number of pulsations of the variable star during Penny's round trip. Use this fact and the expression for the Doppler shift at the 90-degree laboratory angle of observation (Exercise 8-19) to verify that at the end of the trip described in Chapter 4, Penny will be only 20 years older while Paul will have aged 202 years.

Reference: E. Feenberg, *American Journal of Physics,* Volume 27, page 190 (1959).

8-25   Doppler line broadening

The average kinetic energy of a molecule in a gas at temperature T degrees Kelvin is $(3/2)kT$. (The constant k is called the Boltzmann constant and has the value 1.38×10^{-23} joules/degree Kelvin). Molecules of gas move in random directions. Calculate the average speed from the low-velocity approximation of Newtonian mechanics. Estimate the fractional change in frequency due to the Doppler shift that will be observed in light emitted from a molecule in a gas at temperature T. Will this shift increase or decrease the observed frequency of the emitted light? This effect, called Doppler broadening of spectral lines, is one reason why a given spectral line from a gas excited in an electric discharge contains a range of frequencies around a central frequency.

8-26 $E_{\text{rest conv}} = mc^2$ from the Doppler shift

Einstein's famous equation in conventional units, $E_{\text{rest conv}} = mc^2$, and the relativistic expression for energy can be derived from (1) the relativistic expression for momentum (derived separately, for example in Exercise 7-12), (2) the conservation laws, and (3) the

Doppler shift (Exercise 8-18). In conventional units, a photon has energy $E_{conv} = hf$, where h is Planck's constant and f is the frequency of the corresponding classical wave. (We use f for frequency instead of the usual Greek nu, ν, to avoid confusion with v for speed.) Divide by c^2 to convert to units of mass: $E = hf/c^2$. Expressed in units of mass, a photon has equal energy and momentum. Therefore the momentum of a photon is also given by the equation $p = hf/c^2$. Momentum does differ from energy, however, in that it is a 3-vector. In one dimensional motion, the *sign* of the momentum (positive for motion to the right, negative for motion to the left) is important, as in the analysis below.

A particle of mass m_{before} emits two photons in opposite directions while remaining at rest in the laboratory frame. Conservation of momentum requires these two photons to have equal and opposite momenta and therefore to correspond to the same classical frequency f. In consequence, they also have the same energy.

a   First result: Energy released $= \Delta m$. Now view this process from a rocket frame moving at speed $v = v_{conv}/c$ along the direction of flight of the two photons. The particle moves in this frame, but does not change velocity on emitting the photons. The photon emitted in the same direction as the rocket motion will be upshifted in energy (and in corresponding classical frequency) as compared with the energy observed in the laboratory; the other backward-moving photon will be downshifted. We can calculate this frequency shift using the Doppler formulas (Exercise 8-18). Use the expression $m\gamma v$ for momentum of a particle, equation (7-8), to state the conservation of momentum (notice the minus sign before the second photon term, representing the photon moving to the left):

$$m_{before}v\gamma = m_{after}v\gamma + \frac{hf}{c^2}\left[\frac{1+v}{1-v}\right]^{1/2}$$
$$-\frac{hf}{c^2}\left[\frac{1-v}{1+v}\right]^{1/2}$$

Simplify this expression to

$$m_{before} = m_{after} + 2hf/c^2 \qquad (1)$$

or

$$m_{before} - m_{after} = \Delta m = 2hf/c^2 = \text{energy released}$$

Conservation of momentum in both frames implies a change in particle mass equal to the total energy of the emitted photons. Multiply the mass-

units result by c^2 to convert to conventional units and the equation in the well-known form

$$\text{energy released (conventional units)} = (\Delta m)c^2$$

b   Second result: $E_{rest} = m$. Now add the condition that energy is conserved in the laboratory frame:

$$E_{before} = E_{after} + 2hf/c^2 \qquad (2)$$

Compare equations (1) and (2). These two equations both describe a particle at rest. Show that they are consistent if $E_{before} = m_{before}$ and $E_{after} = m_{after}$ and that therefore in general

$$E_{rest} = m$$

or, in conventional units,

$$E_{rest\ conv} = mc^2$$

c   Third result: At any speed, $E = m\gamma$. Next add the condition that energy be conserved in the rocket frame. Place primes on expressions for rocket-measured energy of the particle and use the Doppler equations to transform the classical frequency back to the laboratory value f. Show that the result is

$$E'_{before} = E'_{after} + (2hf/c^2)\gamma \qquad (3)$$

The salient difference between equations (2) and (3) is that in the rocket frame the particle is in motion. Deduce that the general expression for energy of a particle includes the stretch factor gamma:

$$E = m\gamma$$

or, in conventional units,

$$E_{conv} = m\gamma c^2$$

Reference: Fritz Rohrlich, *American Journal of Physics*, Volume 58, pages 348–349 (April 1990).

8-27   everything goes forward

"Everything goes forward" is a good rule of thumb for interactions between highly relativistic particles and stationary targets. In the laboratory frame, many particles and gamma rays resulting from collisions continue in essentially the same direction as the incoming particles.

The first figure (top) shows schematically the collision of two protons in the center-of-momentum frame, the frame in which the system has zero total momentum. A great many different particles are created in the collision, including a gamma ray (the fastest possible particle) that by chance moves perpendicular to the line of motion of the incoming particles: $\phi' = \pi/2$ radians.

The first figure (bottom) shows the same interaction in the laboratory frame, in which one proton is initially at rest. At what angle ϕ does the product gamma ray move in this frame?

a   From the Doppler equations (Exercise 8-19), show that the outgoing angle ϕ for the gamma ray in the laboratory frame is given by the expression

$$\cos\phi = v_{rel} \tag{1}$$

b   What is the speed v_{proton} of the rightward-moving proton in the laboratory frame? We define the laboratory frame by riding at speed v_{rel} on the leftward-moving proton in the center-of-momentum frame. Therefore the rightward-moving proton also moves with speed v_{rel} in the center-of-momentum frame. Use the law of addition of velocities to find the speed of the rightward-moving proton in the laboratory frame (Section L.7 and Exercise 3.11).

$$v_{proton} = \frac{2v_{rel}}{1 + v_{rel}^2} \tag{2}$$

c   In order to solve equation (1) for ϕ, we need to know the value of v_{rel}. Equation (2) is a quadratic in v_{rel}. Show that the solution is

$$v_{rel} = \frac{1}{v_{proton}}\left[1 - \frac{1}{\gamma_{proton}}\right] \tag{3}$$

Here γ_{proton} is the stretch factor γ using the proton velocity v_{proton}.

d   We are interested in finding the angle ϕ when the incoming proton is highly relativistic. In this case $v_{proton} \approx 1$. From the approximation for small angles (ϕ expressed in radians)

$$\cos\phi \approx 1 - \phi^2/2 \qquad |\phi| << 1$$

show that the angle ϕ is given approximately by the expression

$$\phi \approx \left[\frac{2}{\gamma_{proton}}\right]^{1/2} \tag{4}$$

e   What is the value of ϕ in radians and in degrees for incident protons of energy $E = 200$ GeV? For incident protons of energy 2×10^4 GeV? (1 GeV $= 10^9$ electron-volts. Mass of the proton is approximately 1 GeV.)

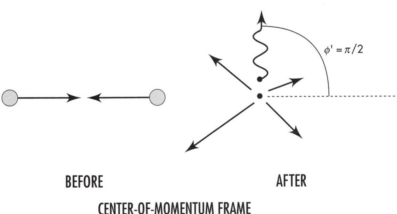

BEFORE   **AFTER**

CENTER-OF-MOMENTUM FRAME

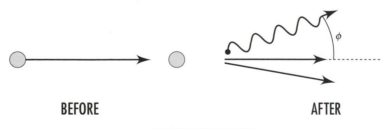

BEFORE   **AFTER**

LABORATORY FRAME

EXERCISE 8-27, first figure. *In the* **center-of-momentum frame** *two incoming protons collide, creating many particles, among them a gamma ray that moves perpendicular to the original line of motion. In the* **laboratory frame,** *in which one proton is initially at rest, in what direction does the gamma ray move?*

8-28 decay of π°-meson

A π° meson (neutral pi-meson) moving in the x-direction with a kinetic energy in the laboratory frame equal to its mass m decays into two photons. In the rocket frame in which the meson is at rest these photons are emitted in the positive and negative y'-directions, as shown in the figure. Find the energies of the two photons in the rocket frame (in units of the mass of the meson) and the energies and directions of propagation of the two photons in the laboratory frame.

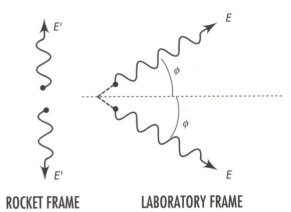

ROCKET FRAME   LABORATORY FRAME

EXERCISE 8-28. *Two photons resulting from the decay of a π°-meson, as observed in rocket and laboratory frames.*

COMPTON SCATTERING

8-29 Compton scattering

Analyze Compton scattering of an incident photon that collides with and recoils from an electron that is initially at rest. Compton scattering in one dimension was discussed in Section 8.4. Here we analyze Compton scattering in two dimensions. The goal is to determine the reduced energy of the photon that has been scattered with a change of direction measured by the

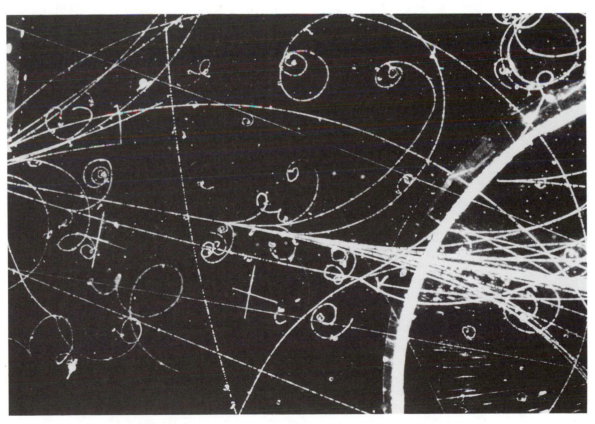

EXERCISE 8-27, second figure. *Forward spray of particles created in collisions near the middle of the picture. An incident particle, probably a charged π-meson, enters from the left with energy approximately 100 to 200 times its rest energy and strikes a nucleus of neon or hydrogen. Curving paths in the imposed magnetic field are probably knock-on electrons. These and the cascade of other particles move initially in the same direction as the incoming π-meson: "Everything goes forward!" Photograph courtesy of Fermi Laboratory.*

BEFORE

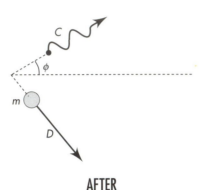

AFTER

EXERCISE 8-29, first figure. *Compton scattering of a photon from an electron initially at rest. The angle ϕ is called the scattering angle.*

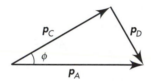

EXERCISE 8-29, second figure. *Conservation of vector momentum means that the momentum triangle is closed.*

angle ϕ. The angle ϕ is called the **scattering angle.** Use the notation in the first figure. Do not use frequency or wavelength or Planck's constant or speed in your analysis — only the laws of conservation of momentum and energy plus equations:

$$E^2 - p^2 = m^2 \qquad \text{[for an electron]}$$
$$E^2 - p^2 = 0 \qquad \text{[for a photon]}$$

Discussion: The conservation of momentum is a vector conservation law. This means that the vector sum of the momenta after the collision equals the momentum of the photon before the collision. In other words, the vectors form a triangle, as shown in the second figure. Apply the law of cosines to this figure:

$$p_D{}^2 = p_A{}^2 + p_C{}^2 - 2\,p_A\,p_C\cos\phi$$

a Now replace all momenta with energies (easy for photons, more awkward for the electron), com-

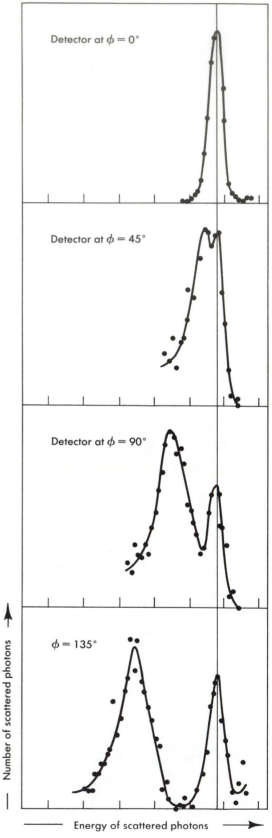

EXERCISE 8-29, third figure. *Results of the Compton experiment in which photons were scattered from the electrons in a graphite target. At each angle of the detector except $\phi = 0$ there are some photons scattered with loss of energy (electron recoils by itself) and other photons scattered with little or no loss of energy (electron and atom recoil as a unit).*

bine with the conservation of energy, and derive the Compton scattering formula:

$$E_{\text{scattered}} = \frac{E_{\text{incident}}}{1 + \dfrac{E_{\text{incident}}}{m}(1 - \cos \phi)}$$

Exercise 8-30 gives some examples of this result.

b Compton's original experiments showed that some photons were scattered without a measurable change of energy. These photons were scattered by electrons that did not leave the atom in which they were bound, so that the entire atom recoiled as a unit. Assume that the energy of the incoming photon is at most a few times the rest energy of the electron. In this case, show that the energy change is negligible for photons scattered by electrons tightly bound to an atom of average mass (say 10 × 2000 × mass of an electron). See the third figure.

Reference: A. H. Compton, *Physical Review*, Volume 22, pages 409–413 (1923).

8-30 compton scattering examples

a A gamma ray photon of energy equal to twice the mass of the electron scatters from an electron initially at rest. Provide the following answers in units of MeV. (Mass of the electron is 0.511 MeV.) From the Compton scattering formula find the energy of the scattered photon for scattering angles 0, 90, and 180 degrees. If you have access to a computer, calculate this energy at 10-degree increments between zero and 180 degrees and plot the resulting curve of energy vs. angle.

b In a new set of experiments, the incident gamma ray has energy equal to five times the rest energy of the electron. Repeat the calculations of part **a** for this case.

8-31 energy of a photon and frequency of light

Planck found himself forced in 1900 to recognize that light of frequency f (vibrations/second) is composed of quanta (Planck's word) or photons (Einstein's later word), each endowed with an energy $E = hf/c^2$ (energy in units of mass) where h is a universal constant of proportionality called Planck's constant. How can Planck's formula possibly make sense when—as we now know—not only E but also f depend upon the frame of reference in which the light is observed? (We use f for frequency instead of the usual Greek nu, ν, to avoid confusion with v for speed.)

a A photon moves along the positive x-axis. Results of Exercise 8-18 show the relation between the energy of this photon measured in the rocket frame and its energy measured in the laboratory frame. A classical electromagnetic wave moves along the positive x-axis. Results of Exercise L-5 (at the end of the Special Topic following Chapter 3) show the relation between the frequency of this wave measured in the rocket frame and its energy measured in the laboratory frame. Compare these two results to show that if we associate photons with a light wave in one coordinate system, this association will hold in all coordinate systems.

b The theory of relativity does not tell us the value of Planck's constant h in the formula $E = (h/c^2)f$ that relates photon energy (in units of mass) to classical wave frequency. Experiment shows the constant h to have the value 6.63×10^{-34} joule-second. Show that if energy is measured in conventional units, the relation between energy and frequency has the form

$$E_{\text{conv}} = hf \qquad \text{[energy in conventional units]}$$

c Show that the formula for Compton scattering (Exercise 8-29) becomes

$$f_{\text{scattered}} = \frac{f_{\text{incident}}}{1 + \dfrac{hf_{\text{incident}}}{mc^2}(1 - \cos \phi)}$$

In the 1920s there was great resistance to the idea that when the electron is "shaken" by the electric field of wave at one frequency it should scatter (reemit) this radiation at a lower frequency.

8-32 inverse Compton scattering

In Compton's original experiment an X-ray photon scattered with reduced energy from an electron initially at rest. In contrast, a photon scattered from a moving electron can increase the energy of the photon. Such an interaction is called **inverse Compton scattering.** The figure (page 270) shows an example.

When a high-energy electron collides head on with a low-energy photon, what is the energy of the outgoing photon? Answer this question using parts **a–e** or by some other method.

a Write down equations of conservation of energy and momentum, using subscripts A through D from the figure.

b Recall that the energy of a photon is equal to the magnitude of its momentum. Use this to simplify

BEFORE

AFTER

EXERCISE 8-32. *Inverse Compton scattering. A low-energy photon is scattered by a high-energy electron.*

the conservation equations, taking leftward momentum to be negative.

c We are not interested in the energy or the momentum of outgoing electron C. Therefore solve the energy equation for E_C and the momentum equation for p_C, square and subtract the two sides, and use $E_C{}^2 - p_C{}^2 = m$. What happens to $E_A{}^2$ and $p_A{}^2$ on the other side of the resulting equation? For now keep terms in the first power of p_A without substituting the awkward equivalent $p_A = (E_A{}^2 + m^2)^{1/2}$.

d Solve the resulting equation for the energy of the outgoing photon.

e Now consider an important special case in which the incoming electron is extremely energetic, with an energy of, say, thousands of times its rest energy as measured in the laboratory. Show that this case the incoming electron behaves in essential respects as a photon: $p_A \approx + E_A$. Simplify your equation of part **d** to show that under these circumstances the outgoing photon has the energy of the incoming electron *no matter what the energy of the incoming photon.*

TESTS OF RELATIVITY

Note: Exercises 8-33 through 8-39 form a connected tutorial on tests of relativity. Some of these exercises depend on each other and on earlier exercises, especially Exercise 8-6.

8-33   photon energy shift due to recoil of emitter

Note: This exercise uses the results of Exercise 8-25.

A free particle of initial mass m_0 and initially at rest emits a photon of energy E. The particle (now of mass m) recoils with velocity v, as shown in the figure.

EXERCISE 8-33. *Recoil of a particle that emits a photon.*

a Write down the conservation laws in a form that makes no reference to velocity. Consider the case in which the fractional change in mass in the emission process is very small compared to unity. Show that for this special case the photon has an energy $E_0 = m_0 - m$. For the general case show that

$$E = E_0 \left(1 - \frac{E_0}{2m_0} \right)$$

or

$$\frac{E - E_0}{E_0} = \frac{\Delta E}{E_0} = -\frac{E}{2m_0}$$

b Show that this shift in energy for visible light ($E_{0 \text{ conv}} \sim 3$ eV) emitted from atoms ($mc^2 \sim 10 \times 10^9$ eV) in a gas is very much less than the Doppler shift due to thermal motion (Exercise 8-25) even for temperatures as low as room temperature ($kT \sim 1/40$ eV).

8-34   recoilless processes

a A free atom of iron ^{57}Fe—formed in a so-called "excited state" by the radioactive decay of cobalt ^{57}Co—emits from its nucleus a gamma ray (high-energy photon) of energy 14.4 keV and transforms to a "normal" ^{57}Fe atom. By what fraction is the energy of the emitted ray shifted because of the recoil of the atom? The mass of the ^{57}Fe atom is about equal to that of 57 protons.

b That not all emitted gamma rays experience this kind of frequency shift was the important discovery made in 1958 by R. L. Mössbauer at the age of 29. He showed that when radioactive nuclei embedded in a solid emit gamma rays, some significant fraction of these atoms fail to recoil as free atoms. Instead they behave as if locked rigidly to the rest of the solid. The recoil in these cases is communicated to the solid as a whole. The solid being heavier than one atom by many powers of 10, these events are called **recoilless processes.** For gamma rays emitted in recoilless processes, the m_0 in Exercise 8-33 is the mass of the entire chunk in which the iron atoms are embedded. When this chunk has a mass of one gram, by what fraction is the frequency of the emitted ray shifted in this "recoilless" process?

c The gamma rays emitted from excited ^{57}Fe atoms do not have a precisely defined energy but are spread over a narrow energy range—or frequency range—or natural line width, shown as a bell-shaped curve in the figure. (The physical basis for this curve is explained by quantum physics.) The full width of this curve at half maximum is denoted by $\Delta \nu$. R. V. Pound and G. A. Rebka selected ^{57}Fe for experiments

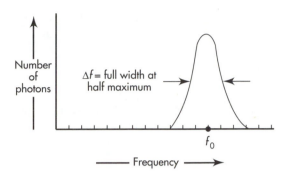

EXERCISE 8-34. *Natural line width of photons emitted from* ^{57}Fe.

with recoilless processes because the fractional ratio $\Delta f/f_o$ has the very small value 6×10^{-13} for the 14.4 keV gamma ray from ^{57}Fe. How much is the natural line width, Δf, of ^{57}Fe expressed in cycles/second? Compare the fractional natural line width with the fractional shift due to recoil of a free iron atom. And compare it with the fractional shift of a gamma ray from a recoilless process.

Reference: For a more detailed account of Mössbauer's discovery—for which the German scientist was awarded the Nobel prize in 1961—see S. DeBenedetti, "The Mössbauer Effect," *Scientific American*, Volume 202, pages 72–80 (April 1960). For the selection of ^{57}Fe, see R. V. Pound and G. A. Rebka, Jr., *Physical Review Letters*, Volume 3, pages 439–441 (1959).

Pound and Rebka's application of recoilless processes thus put into one's hands a resonance phenomenon sharp in frequency to the fantastic precision of 6 parts in 10^{13}. Exercise 8-35 deals with detection of this radiation. Exercise 8-36 uses motion (Doppler shift) as a means for producing controlled changes of a few parts in 10^{13} — or much larger changes — in the effective frequency of source or detector or both. To what uses can radiation of precisely defined frequency be put? There are many uses. For instance, the effect is the basis of important techniques in solid-state physics, molecular physics, and biophysics. One can detect the change in the natural frequency of radiation from ^{57}Fe atoms caused by other atoms in the neighborhood — and by external magnetic fields — and in this way analyze the interaction between the iron atom its surroundings. Here we aim at detection of various effects predicted by relativity.

8-35   resonant scattering

The nucleus of normal ^{57}Fe absorbs gamma rays at the resonant energy of 14.4 keV much more strongly than it absorbs gamma rays of any nearby energy. The energy absorbed in this way is converted to internal energy of the nucleus and transmutes the ^{57}Fe to the "excited state." After a time this excited nucleus drops back to the "normal state," emitting the excess energy in various forms in all directions. Therefore the number of gamma rays transmitted through a thin

sheet containing ^{57}Fe will be less at the 14.4 keV resonance energy than at any nearby energy. This process is called **resonant scattering.**

a   Show that when a gamma ray of the resonant energy E_o is incident on a free iron atom initially at rest then the free nucleus cannot absorb the gamma ray at its resonant energy, because the process cannot satisfy both the law of conservation of momentum and the law of conservation of energy.

b   Show that both conservation laws are satisfied when an iron atom embedded in a one-gram crystal absorbs such a gamma ray by a recoilless process, in which the entire crystal absorbs the momentum of the incident gamma ray. ("Satisfied"? For momentum, yes; for energy, no. However, the fractional discrepancy in energy — equivalent to the fractional discrepancy in frequency — is less than 6 parts in 10^{13} and therefore small enough so that the iron nucleus is "unable to notice" the discrepancy and therefore absorbs the gamma ray.)

8-36   measurement of Doppler shift by resonant scattering

In the experimental arrangement shown in the figure, a source containing excited ^{57}Fe nuclei emits (among other radiations) gamma rays of energy E_o by a recoilless process. An absorber containing ^{57}Fe nuclei in the normal state absorbs some of these gamma rays by another recoilless process and reemits this energy in various forms in all directions. Thus the counting rate on a gamma ray counter placed as shown is less for an absorber containing normal ^{57}Fe than for an equivalent absorber without normal ^{57}Fe. Now the source is moved toward the absorber with speed v.

a   What must be the velocity of the source if the gamma rays are to arrive at the absorber shifted in frequency by 6 parts in 10^{13}? Express your answer in centimeters/second.

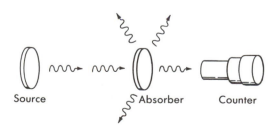

EXERCISE 8-36. *Resonant scattering of photons.*

b Will the counting rate of the counter increase or decrease under these circumstances?

c What will happen to this counting rate if the source is moved away from the absorber with the same speed?

d Make a rough plot of counting rate of the counter as a function of the source velocity toward the absorber (positive velocity) and away from the absorber (negative velocity).

e **Discussion question:** Does this method allow one to measure the "absolute velocity" of the source, in violation of the Principle of Relativity (Chapter 3)?

8-37 test of the gravitational red shift I

A 14.4-keV gamma ray emitted from ^{57}Fe without recoil travels vertically upward in a uniform gravitational field. By what fraction will the energy of this photon be reduced in rising to a height z (Exercise 8-6)? An absorber located at this height must move with what speed and in what direction in order to absorb such gamma rays by recoilless processes? Calculate this velocity when the height is 22.5 meters. Plot the counting rate as a function of absorber velocity expected if (a) the gravitational red shift exists, and (b) there is no gravitational red shift. A frequency shift of $\Delta f/f_o = (2.56 \pm 0.03) \times 10^{-15}$ was determined in an experiment conducted by R. V. Pound and J. L. Snider. You will notice that this shift is very much smaller than the natural line width $\Delta f/f_o = 6 \times 10^{-13}$ (see the figure for Exercise 8-34). Therefore the the result depended on a careful exploration of the shape of this line and was derived statistically from a large number of photon counts.

References: Original experiment: R. V. Pound and G. A. Rebka, Jr., *Physical Review Letters,* Volume 4, pages 337–341 (1960). Improved experiment: R. V. Pound and J. L. Snider, *Physical Review,* Volume 140, pages B788–B803 (1965).

8-38 test of the gravitational red shift II

On June 18, 1976, a Scout D rocket was launched from Wallops Island, Virginia, carrying an atomic hydrogen-maser clock as the payload. It achieved a maximum altitude of 10^7 meters. By means of microwave signals, its clock was compared with an identical clock at the surface of Earth. The experiment used continuous comparison of these two clocks as the payload rose and fell. Simplifying (and somewhat misrepresenting) the experiment, we report their result as a fractional frequency red shift at the top of the trajectory due to gravitational effects of $\Delta f/f = 0.945 \times 10^{-10} \pm 6.6 \times 10^{-15}$.

Modify the analysis of Exercise 8-6 to make a prediction about this experiment and compare your prediction with the results of the Scout D rocket experiment.

References: Description of experiment and preliminary results: R. F. C. Vessot and M. W. Levine, *General Relativity and Gravitation,* Volume 10, Number 3, pages 181–204 (1979). Final results: R. F. C. Vessot, M. W. Levine, and others, *Physical Review Letters,* Volume 45, pages 2081–2084 (1980). Popular explanation: Clifford M. Will, *Was Einstein Right?* (Basic Books, New York, 1986), pages 42–64.

8-39 test of the twin paradox

For Penny to leave her twin brother Paul behind in the laboratory, go away at high speed, return, and find herself younger than stay-at-home Paul is so contrary to everyday experience that it is astonishing to find that the experiment has already been done and the prediction upheld! Chalmers Sherwin pointed out that the twins can be identical iron atoms just as well as living beings. Let one iron atom remain at rest. Let the other make one forth-and-back trip. Or many round trips. The percentage difference in aging of the twin atoms is the same after a million round trips as after one round trip—and it is easier to measure. How does one get the second atom to make many round trips? By embedding it in a hot piece of iron, so that it vibrates back and forth about a position of equilibrium (thermal agitation!). How does one measure the difference in aging? In the case of Penny and Paul the number of birthday firecrackers that each sets off during their separation are counted. In the experiment with iron atoms one compares not the number of flashes of firecrackers up to the time of meeting but the frequency of the photons emitted by recoilless processes, and thus—in effect—the number of ticks from two identical nuclear clocks in the course of one laboratory second. In other words, one compares the effective frequency of INTERNAL nuclear vibrations (not to be confused with the back-and-forth vibration of the iron atom as a whole!) as observed in the laboratory for (a) an iron nucleus at rest and (b) an iron nucleus in a hot specimen.

It is difficult to obtain an iron nucleus at rest. Therefore the actual experiment compared the effective internal nuclear frequency two crystals of iron with a difference of temperature ΔT. R. V. Pound and G. A. Rebka, Jr., measured that a sample warmed up by the amount $\Delta T = 1$ degree Kelvin underwent a fractional change in effective frequency of $\Delta f/f_o = (-2.09 \pm 0.24) \times 10^{-15}$ (fewer vibrations; fewer clock ticks; fewer birthdays; more youthful!). (We use f for frequency instead of the usual Greek nu, ν, to avoid confusion with v for speed.)

To simplify thinking about the experiment, go back to the idea that one iron atom is at rest and the other is in thermal agitation at temperature T; predict the fractional lowering in number of internal vibrations in the hot sample per laboratory second; and compare with experiment.

Discussion: The figure compares the effective "ticks" of the two "internal nuclear clocks" in the laboratory time dt. Note that the speed of thermal agitation is about 10^{-5} the speed of light. What algebraic approximation suggests itself for the discrepancy factor $1 - (1 - v^2)^{1/2}$? How much is the deficit in number of "ticks" (for hot atom versus atom at rest) in the lapse of laboratory time dt? Show that the cumulative deficit in number of "ticks" from the hot atom in one second is $f_o(v^2/2)_{avg}(1\text{ second})$ where $(v^2)_{avg}$ means "the time average value of the square of the atomic speed" (relative to the speed of light). Note that the mean kinetic energy of thermal agitation of a hot iron atom (mass $m_{Fe} = 57\,m_{proton}$) is given by the classical kinetic theory of gases:

$$1/2\ m_{Fe}(v^2)_{avg}c^2 = 3/2\ kT$$

Here k is Boltzmann's factor of conversion between two units of energy, degrees and joules (or degrees and ergs); $k = 1.38 \times 10^{-23}$ joule/degree Kelvin ($k = 1.38 \times 10^{-16}$ erg/degree Kelvin). How does the experimental result of Pound and Rebka compare with the result of your calculation?

References: Chalmers W. Sherwin, *Physical Review*, Volume 120, pages 17–21 (1960). R. V. Pound and G. A. Rebka, Jr., *Physical Review Letters*, Volume 4, pages 274–275 (1960).

FREE-FOR-ALL!

8-40 momentum without mass?

A small motor mounted on a board is powered by a battery mounted on top of it, as shown in the figure on page 274. By means of a belt the motor drives a paddlewheel that stirs a puddle of water. The paddlewheel mechanism is mounted on the same board as the motor but a distance x away. The motor performs work at a rate dE/dt.

a How much mass is being transferred per second from the motor end of the board to the paddlewheel end of the board?

b Mass is being transferred over a distance x at a rate given by your answer to part **a**. What is the momentum associated with this transfer of mass? Since this momentum is small, Newtonian momentum concepts are adequate.

c Let the mounting board be initially at rest and supported by frictionless rollers on a horizontal table. The board will move! In which direction? What happens to this motion when the battery runs down? How far will the board have moved in this time?

d Show that an observer on the board sees the energy being transferred by the belt; an observer on the table sees the energy being transferred partly by the belt and partly by the board; an observer riding one way on the belt sees the energy being transferred partly by the belt moving in the other direction and partly by the board. Evidently it is not always possible

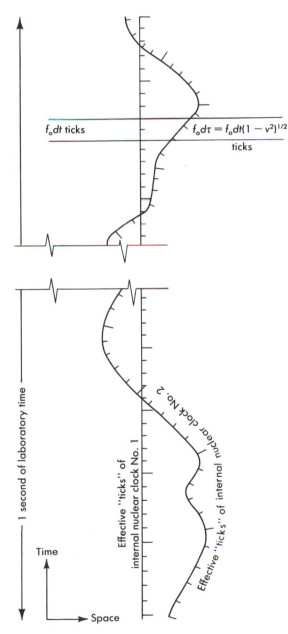

EXERCISE 8-39. *Comparison of nuclear clock at rest with nuclear clock in thermal motion.*

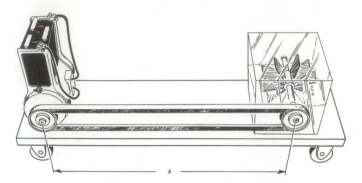

EXERCISE 8-40. *Transfer of mass without net transfer of particles or radiation.*

to make a statement satisfactory to all observers about the path by which energy travels from one place to another or about the speed at which this energy moves from one place to another!

8-41   the photon rocket and interstellar travel

The "perfect" rocket engine combines matter and antimatter in a controlled way to yield photons (high-energy gamma rays), all of which are directed out the rear of the rocket. Suppose we start with a spaceship of initial mass M_o, initially at rest. At burnout the remaining spaceship moves with speed v and has a mass equal to the fraction f of the original mass. For a given fraction f, we want to know the final rocket speed v or, better yet, the time stretch factor $\gamma = 1/(1 - v^2)^{1/2}$. (**Note:** Here, f is *not* frequency.)

a   What is the total energy of the system initially? Let E_{rad} stand for the total energy of radiation after burnout. Find an expression for the total energy of the system after burnout and set up the conservation of energy equation.

b   Similarly, set up the conservation of momentum equation. What is the total momentum of the system initially? The momentum of the radiation at burnout? The momentum of the spaceship at burnout?

c   Eliminate E_{rad} between the two conservation equations. Show that the result can be written

$$\gamma f + \gamma v f = 1$$

d   From the definition of γ, show that $\gamma v =$

$(\gamma^2 - 1)^{1/2}$ and hence that the equation of part **c** can be written in the form

$$f^2 - 2\gamma f + 1 = 0$$

e   What is the value of the fraction $f =$ (final spaceship mass)/(initial spaceship mass) for a time stretch factor $\gamma = 10$? In your opinion, is it possible to construct a spaceship whose shell and payload is this small a fraction of takeoff mass?

f   Substitute the result of part **e** into the conservation of energy equation in part **a**. Show that the total energy of emitted radiation is less than the mass of fuel consumed. Why?

g   Does your analysis apply to takeoff from Earth's surface? From Earth orbit? From somewhere else? What safety precautions apply to the backward blast of gamma rays?

h   You are the astronaut assigned to this spaceship. Do you want to stop at your distant destination star or fly past at high speed? Do you want to return to Earth? Do you want to stop at Earth on your return or merely wave in passing? Must all fuel for the entire trip be on board at takeoff or can you refuel at your destination star? From your answers to these questions, plan your trip and find the resulting fractions of spaceship mass to initial mass for different stages of your trip.

i   **Discussion question:** From your results for this exercise, what are your conclusions about the technical possibilities of human flight to the stars?

References: Adapted from A. P. French, *Special Relativity* (W.W. Norton, New York, 1968), pages 183–184. See also J. R. Pierce, *Proceedings of the IRE,* Volume 47, pages 1053–1061 (1959).

CHAPTER 9

GRAVITY: CURVED SPACETIME IN ACTION

9.1 GRAVITY IN BRIEF

the mutual grip of mass and spacetime

Gravity, as we see it today, does not count as a foreign force transmitted *through* space and time. Gravity manifests the curvature *of* spacetime.

Ten years after his special relativity, Einstein gave us his 1915 battle-tested and still standard theory of gravitation. Its message comes in a single simple sentence: *Spacetime grips mass, telling it how to move; and mass grips spacetime, telling it how to curve.*

The grip of spacetime on mass enforces a central principle of special relativity: conservation of energy and momentum in a smash (Figure 9-1). The coupling of mass and spacetime geometry, far from being the weakest force in nature, is the strongest.

Now for the back-reaction, the grip mass exerts on spacetime! What curvature does that grip impose on spacetime? And how does that curvature give an account of gravity unrivaled for scope and accuracy? 🖝

Spacetime tell mass how to move

Mass to spacetime: "Curve!"

9.2 GALILEO, NEWTON, AND EINSTEIN

Only historical judgment liberates the spirit from the pressure of the past; it maintains its neutrality and seeks only to furnish light.
—Benedetto Croce

Galileo and Newton viewed motion as properly described with respect to a rigid Euclidean reference frame that extends through all space and endures for all time. This

275

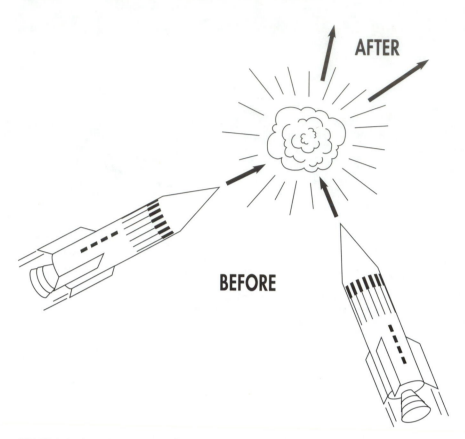

FIGURE 9-1. *Spacetime grips mass, keeping an object moving straight when free. By its power, it enforces conservation of energy and momentum in a smash.*

**Newton: One global frame.
Einstein: Many local frames.**

supposed reference frame stands high above the battles of matter and energy. Within this ideal space of Galileo and Newton there acts a mysterious force of gravity, an interloper from the world of physics, a foreign influence not described by geometry.

In contrast, Einstein says that there exists no mysterious "force of gravity," only the structure of spacetime itself. Climb into an unpowered spaceship, he says, and see for yourself that there is no gravity there. Physics is locally gravity-free (Chapter 2). Every free particle moves in a straight line at uniform speed. In a free-float (inertial) frame, physics looks simple. But such a frame rates as free-float in only a limited region of spacetime (Section 2.3) — a fact emphasized here by repeated use of the word "local" in describing a free-float frame.

Complications arise in describing the relation between (1) the direction of motion of a particle in one local frame and (2) the direction of motion of the same particle as observed from a nearby local frame. Any difference between the two directions is described in terms of the "curvature of spacetime," Einstein tells us. The existence of this curvature destroys the possibility of describing motion with respect to a single ideal Euclidean reference frame that pervades all space. What is simple is the geometry in a region small enough to look flat.

How did the views of Galileo, Newton, and Einstein develop? And what is the concrete substance of the strange phrase "curvature of spacetime"? ◄═

9.3 LOCAL MOVING ORDERS FOR MASS

moving orders from the local commander, spacetime!

Navigation satellites near Earth drift away from "perfect" orbits because thin air and solar radiation pressure affect their motion. Figure 9-2 shows an experimental satellite that carries a "conscience" designed to assure that the same motion will be maintained

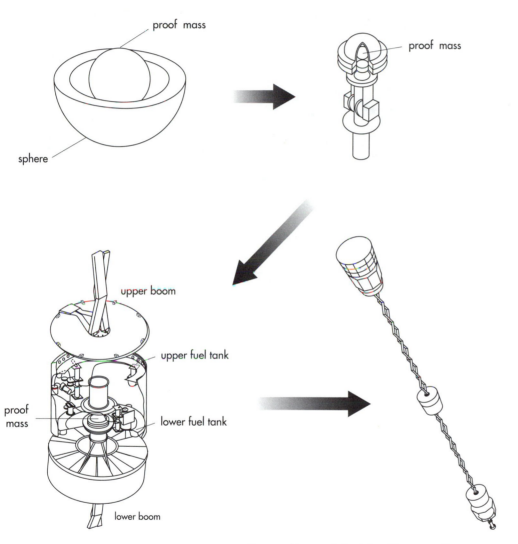

proof mass

sphere

proof mass

upper boom

upper fuel tank

proof mass

lower fuel tank

lower boom

FIGURE 9-2. *"Conscience-guided" satellite. A satellite in orbit around Earth is subject to small accelerations due to solar radiation pressure and residual atmospheric drag. Uncorrected, these accelerations are between $10^{-6}g$ and $10^{-9}g$, where g is the acceleration of gravity at Earth's surface. The acceleration was reduced to $5 \times 10^{-12}g$ for more than a year in orbit by use of a conscience or proof mass and the Disturbance Compensation System (DISCOS) mounted on a TRIAD U.S. Navy satellite. The conscience, a gold–platinum sphere 2.2 centimeters in diameter, floats freely inside a spherical housing. Any nongravitational force results in an incremental velocity change. The floating proof mass continues in its original state of motion in an ideal friction-free environment. Observing the proof mass through capacitor sensing devices, the satellite becomes aware that it is not keeping up with the motion demanded by the proof mass. An opposite vernier rocket fires long enough to bring the spaceship back into concord with its proof mass—its conscience. To reduce gravitational effects of the satellite itself on the proof mass, fuel for the vernier rockets is stored in donut-shaped tanks placed symmetrically above and below the proof mass; power supply and radio transmitter are each held at the end of a boom 2.7 meters long on either side of the control unit. For an Earth-based microgravity environment, recall Figure 2-3. (Used with permission of AIAA. Journal of Spacecraft.)*

ISAAC NEWTON

Woolsthorpe, December 25, 1642—Kensington (London), March 20, 1727

"The marble index of a mind forever
Voyaging through strange seas of thought, alone."—*Wordsworth*

★   ★   ★

"I do not know what I may appear to the world; but to myself I seem to have been only like a boy, playing on the sea-shore, and diverting myself, in now and then finding a smoother pebble or a prettier shell than ordinary, whilst the great ocean of truth lay all undiscovered before me."—*Newton*

★   ★   ★

"Why do I call him a magician? Because he looked on the whole universe and all that is in it as a *riddle*, as a secret which could be read by applying thought to certain evidence, certain mystic clues which God had laid about the world to allow a sort of philosopher's treasure hunt to the esoteric brotherhood. He believed that these clues were to be found partly in the evidence of the heavens and in the constitution of elements (and that is what gives the false suggestion of his being an experimental natural philosopher), but also partly in certain papers and traditions handed down by the brethren in an unbroken chain back to the original cryptic revelation in Babylonia. He regarded the universe as a cryptogram set by the Almighty—just as he himself wrapt the discovery of the calculus in a cryptogram when he communicated with Leibnitz. By pure thought, by concentration of mind, the riddle, he believed, would be revealed to the initiate."—*Keynes*†

———————

†Reprinted by permission of the publisher, Horizon Press, from *Essays in Biography* by John Maynard Keynes. Copyright 1951.

when it encounters these disturbances as when it moves through perfect emptiness. The "conscience" — called a **proof mass** — is a separate sphere that floats inside the larger ship. The proof mass undergoes no acceleration relative to the ship as long as the ship moves freely. When relative motion does occur, the error in the tracking must be due to the satellite. By small rockets the satellite gives itself a brief spurt of acceleration and comes back into step with the inner proof mass — the satellite's conscience. Though resistance is present, the rocket thrust overcomes it. The satellite takes the same course it would have taken had both resistance and thrust been absent.

As satellite and proof mass come to empty space, they fly through it in perfect step, without use of rockets or sensing devices. What a remarkable harmony they present! The inner proof mass does not see outer space. It does not touch, feel, or see the ship that surrounds it on every side. Yet it faithfully tracks the ship's route through spacetime. Moreover, this tracking is as perfect when the proof mass is made of aluminum as when it is made of gold. How do proof masses — of whatever atomic constitution and whatever construction — know enough to follow a standard world-line? Where does mass get its moving orders?

Locally, answers Einstein. From a distance, answers Newton.

Einstein says that the proof mass gets its information in the simplest way possible. It responds to the structure of spacetime in its immediate vicinity. It moves on a straight line in the local free-float frame. No simpler motion and no straighter motion can be imagined.

Newton says that the inner proof mass gets its information about how to move from a distance, via a "force of gravity." Motion relative to what? Motion relative to an ideal, God-given, never-changing Euclidean reference frame that spans all of space and endures for all time. He tells us that the proof mass would have moved along an ideal straight line in this global frame had not Earth deflected it. How can this ideal line be seen? How sad! There is nothing, absolutely nothing, that ever moves along this ideal line. It is an entirely imaginary line. But it nevertheless has a simple status, Newton tells us, in this respect: Every satellite and every proof mass, going at whatever speed, is deflected away from this ideal line at the same acceleration (Figure 9-3).

Einstein says: Face it; there is no ideal background Euclidean reference frame that extends over all space. And why say there is, when even according to Newton no particle, not even a light ray, ever moves along a straight line in that ideal reference frame. Why say spacetime is Euclidean on a large scale when no evidence directly supports that hypothesis? To try to set up an all-encompassing Euclidean reference frame and attempt to refer motion to it is the wrong way to do physics. Don't try to describe motion relative to faraway objects. *Physics is simple only when analyzed locally.* And locally the worldline that a satellite follows is already as straight as any worldline can be. Forget all this talk about "deflection" and "force of gravitation." I'm inside a spaceship. Or I'm floating outside and near it. Do I feel any "force of gravitation?" Not at all. Does the spaceship "feel" such a force? No. Then why talk about it? Recognize that the spaceship and I are traversing a region of spacetime free of all force. Acknowledge that the motion through that region is already ideally straight.

How can one display the straightness of the motion? Set up a local lattice of meter sticks and clocks, a local free-float (inertial) reference frame — also called a Lorentz reference frame (Chapter 2). How does one know the frame is free-float? Watch every particle, check every light ray, test that they all move in straight lines at uniform speed relative to this frame. And having thus verified that the frame is free-float, note that the proof mass too moves at a constant speed in a straight line — or remains at rest — relative to this local free-float frame. What could be simpler than the moving orders for mass: "Follow a straight line in the local free-float reference frame." Does a proof mass have to know the location of Earth and Moon and Sun before it knows how to move? Not at all! Surrounded on all sides by the black walls of a satellite, it has only to sense the local structure of spacetime — right where it is — in order to follow the correct track. ✒

"Conscience-guided" satellite. What guides the conscience?

Physics is simple only when analyzed locally

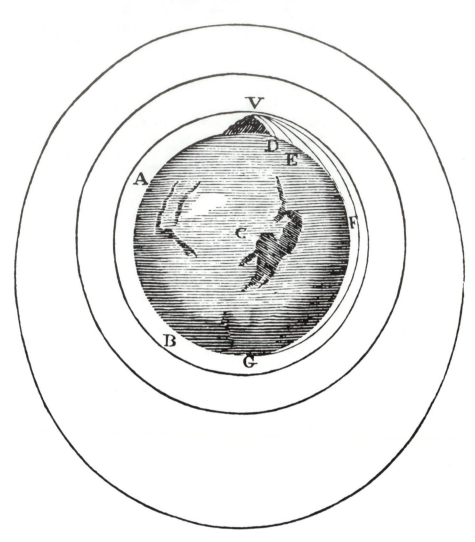

FIGURE 9-3. *In Newtonian mechanics different particles going at different speeds are all deflected away from the ideal straight line with equal acceleration. In this respect there is no difference in principle between the fall of a projectile and the motion of a satellite. In this picture of Newton's published in 1686, cannon of successively greater power mounted on a mountaintop fire out their balls horizontally. The more powerful cannon launches a satellite. The outer two curves show other possible satellite orbits. In brief, Newton has one global reference frame, but within this reference frame no satellite is ever gravity-free, and no particle ever moves in a straight line at constant speed. Einstein, in contrast, makes use of many local regions in each of which the geometry is Lorentzian (as in special relativity); the laws of gravitation arise from the lack of ideality in the relation between one local region and the next (gravitation; spacetime curvature; general relativity).*

9.4 SPACETIME CURVATURE

not one but two particles witness to gravitation

Splendid! And also simple! But isn't Einstein's view of motion *too* simple? We started out interested in the motion of a spaceship around Earth and in "gravitation." We seem to have ended up talking only about the motion of the satellite — or the proof mass — relative to a strictly local inertial reference frame, a trivially simple straight-line motion. Where is there any evidence of "gravitation" to be seen in that? Nowhere.

This is the great lesson of Einstein: Spacetime is always and everywhere locally Lorentzian. No evidence of gravitation whatsoever is to be seen by following the motion of a single particle in a free-float frame.

One has to observe the relative acceleration of two particles slightly separated from each other to have any proper measure of a gravitational effect. Separated by how much? That depends on the region of spacetime and the sensitivity of the measuring equipment. Two ball bearings with a *horizontal* separation of 20 meters, dropped from a height of 315 meters above Earth's surface with 0 initial relative velocity, hit the ground 8 seconds later (24×10^8 meters of light-travel time later) with a separation that has been *reduced* by 10^{-3} meter (Section 2.3). Two ball bearings with a *vertical* separation of 20 meters, dropped from a height of 315 meters with 0 initial relative velocity, in the same 8 seconds *increase* their separation by 2×10^{-3} meter. To measuring equipment unable to detect such small relative displacements the ball bearings count as moving in one and the same free-float reference frame. No evidence for gravitation is to be seen. More sensitive apparatus detects the **tide-producing action** of gravity—the accelerated shortening of horizontal separations parallel to Earth's surface, the accelerated lengthening of vertical separations. Each tiny ball bearing still moves in a straight line in its own local free-float reference frame. But now—with the new precision—the region of validity of the one free-float reference frame does not reach out far enough to give a proper account of the motion of the other steel ball. The millimeter or two discrepancy is the way "gravity" manifests itself.

Tidal acceleration displays gravity as a local phenomenon. No mention here of the distance of the steel balls from the center of Earth! No mention here of acceleration relative to that center! The only accelerations that come into consideration are those of nearby particles relative to each other, the tidal accelerations described in the preceding paragraph.

These relative accelerations double when the separations are doubled. The true measure of the tide-producing effect has therefore the character of an acceleration per unit of separation. Let the acceleration be measured in meters of distance per meter of light-travel time per meter of light-travel time; that is, in units meters/meter^2 or 1/meter [$x = 1/2 at^2$, so $a = 2x/t^2$]. Then the measure of the tide-producing effect (different for different directions) has the units (acceleration/distance) or (1/meter^2). In the example, in the two horizontal directions this quantity has the value [$2(-0.001$ meter)/$(24 \times 10^8$ meter)^2]/20 meter $= -17.4 \times 10^{-24}$ meter^{-2} and in the vertical direction twice the value and the opposite sign: $+34.6 \times 10^{-24}$ meter^{-2}. The tide-producing effect is small but it is real and it is observable. Further, it is a locally defined quantity. And Einstein tells us that we must focus our attention on locally defined quantities if we want a simple description of nature.

Einstein says more: This tide-producing effect does not require for its explanation some mysterious force of gravitation, propagated through spacetime and additional to the structure of spacetime. Instead, it can and should be described in terms of the geometry of spacetime itself as the **curvature of spacetime.**

Though Einstein speaks of four-dimensional spacetime, his concepts of curvature can be illustrated in terms of two-dimensional geometry on the surface of a sphere.

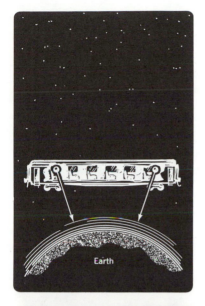

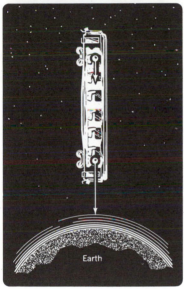

Einstein's railway coach in free fall.

9.5 PARABLE OF THE TWO TRAVELERS

space curvature on a sphere accounts for relative acceleration of travelers

One traveler, A, stands at the equator, ready to travel straight north. A's companion B, standing against him shoulder to shoulder, wheels 90 degrees and marches straight

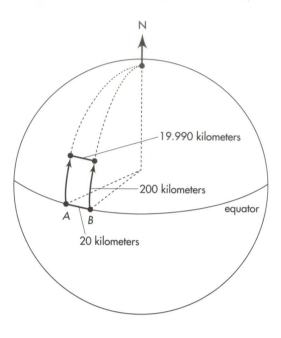

FIGURE 9-4. *Travelers A and B, starting out parallel and deviating neither to the left nor to the right, nevertheless find themselves approaching each other after they have traveled some distance. Interpretation 1: Some mysterious force of "gravitation" is at work. Interpretation 2: They are traveling on a curved surface. Figure not drawn to scale.*

east. She paces off 20 kilometers along the equator. There she again turns a sharp 90 degrees and faces straight north. Both travelers now start north and travel 200 kilometers (Figure 9-4). In the beginning their tracks are strictly parallel. Moreover, no travelers could be more conscientious than they are in continuing precisely in their original directions. Each of them deviates neither to the right nor to the left. Yet an umpire sent out to measure their separation after their 200-kilometer treks finds it to be less than the original 20 kilometers. Why? We know perfectly well: The surface of the globe is curved. If they continue north, their paths will meet at the north pole.

Already at this early stage of their trip the travelers are approaching each other, although they had started out not approaching at all. Initially their velocity relative to one another was zero; now they move toward one another with a small relative velocity. In this sense they are slowly accelerating toward each other.

The travelers accelerate toward each other as surely as two tiny ball bearings in a free-fall horizontal railway coach accelerate toward each other (Figure 9-5). We ascribe the relative acceleration of ball bearings in the railway coach to the "tidal" effects of nonuniform gravitation near Earth. To be sure, the relevant picture for the travelers is the two-dimensional curved space of the surface of Earth, whereas what counts for the ball bearings is curvature of spacetime. This parallelism between the geometrical concept of curvature and the gravitational concept of tide-producing effect foreshadows Einstein's geometrical interpretation of gravity.

The two travelers, who started out so conscientiously on parallel tracks and deviated neither to the left nor to the right, have been told by the umpire of distances that despite all precautions they are now slowly accelerating toward one another. They blame this development on the existence of some mysterious "gravitational force" that deflects their paths. They explore the nature of this "gravitational force." Repeating the travel with bicycles, motorcycles, light cars, and heavy trucks all moving northward with the same speed, they find always the same relative acceleration toward one another. They conclude that the "gravitational force" leads to the same acceleration of all objects, no matter what they are made of or how massive they are.

Learned would-be pundits analyze the motion of travelers. They say, in words utterly mysterious to us, "See here. You find the same acceleration for every vehicle

Curvature of Earth demonstrated by change in separation of two originally parallel paths

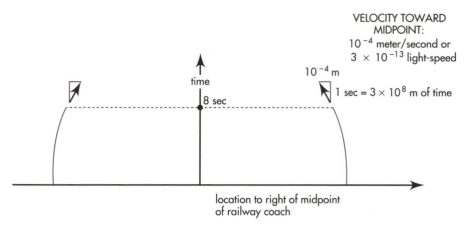

VELOCITY TOWARD
MIDPOINT:
10^{-4} meter/second or
3×10^{-13} light-speed

10^{-4} m

1 sec = 3×10^8 m of time

time

8 sec

location to right of midpoint
of railway coach

BALL BEARINGS FALLING "DOWN" IN RAILWAY COACH

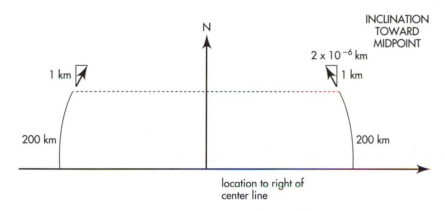

N

INCLINATION
TOWARD
MIDPOINT

2×10^{-6} km

1 km

1 km

200 km

200 km

location to right of
center line

TRAVELERS HEADED "NORTH" ON EARTH

FIGURE 9-5. *Comparison of the paths of northward travelers on Earth's surface with the worldlines of ball bearings released side by side from rest near Earth's surface. In both cases the "path" of each "traveler" starts parallel with that of the second traveler (zero initial relative velocity). In both cases this "path" gradually inclines toward the centerline ("relative acceleration"). In both cases the paths can be accounted for in terms of the local curvature of geometry (curvature of Earth's surface for the travelers; curvature of spacetime geometry — gravitation! — for the ball bearings). In each diagram, vertical distances are drawn — for vividness — to a different scale than horizontal distances. Both diagrams suffer from this additional imperfection: they attempt to show, on the flat Euclidean surface of this page, trajectories that can be correctly represented only in terms of a curved geometry.*

you try. This means that the ratio of gravitational mass to inertial mass is the same for all sorts of objects. You have made a great discovery about mass."

All this time we and our space-traveler friends are looking down from on high. We see the many treks. We watch the many measurements of distance. Through our intercommunication system we hear and approve as our friends on the ground interpret distance shortening as relative acceleration — and relative acceleration as "gravitation." But then they get into weighty discussions. They start speaking of "gravitation" as action at a distance. We smile. What is at issue — we know — is not action at a distance at all, but the geometry of curved space. All this talk about the identity of "gravitational mass" and "inertial mass" completely obscures the truth. Curvature and nothing more is all that is required to describe the increasing rate at which *A* and *B* approach each other. ✎

Curvature alone accounts for
relative acceleration

9.6 GRAVITATION AS CURVATURE OF SPACETIME

spacetime curvature accounts for tidal accelerations of objects

Spacetime curvature demonstrated by change in separation of two originally parallel worldlines

Einstein smiles, too, as he hears gravitation described as action at a distance. Curvature of spacetime and nothing more, he tells us, is all that is required to describe the millimeter or two change in separation in 8 seconds of two ball bearings, originally 20 meters apart in space above Earth, and endowed at the start with zero relative velocity. Moreover, this curvature completely accounts for gravitation.

"What a preposterous claim!" is one's first reaction. "How can such minor — and slow — changes in the distance between one tiny ball and another offer any kind of understanding of the enormous velocity with which a falling mass hits Earth?" The answer is simple: Many local reference frames, fitted together, make up the global structure of spacetime. Each local Lorentz frame can be regarded as having one of the ball bearings at its center. The ball bearings all simultaneously approach their neighbors (curvature). Then the large-scale structure of spacetime bends and pulls nearer to Earth (Figure 9-6). In this way many local manifestations of curvature add up to give the appearance of long-range gravitation originating from Earth as a whole.

Acceleration toward Earth: Totalized effect of relative accelerations, each particle toward its neighbor, in a chain of test particles that girdles globe

In brief, the geometry used to describe motion in any local free-float frame is the flat-spacetime geometry of Lorentz (special relativity). Relative to such a local free-float frame, every nearby electrically neutral test particle moves in a straight line with constant velocity. Slightly more remote particles are detected as slowly changing their velocities, or the directions of their worldlines in spacetime. These changes are described as tidal effects of gravitation. They are understood as originating in the local curvature of spacetime.

From the point of view of the student of local physics, gravitation shows itself not at all in the motion of one test particle but only in the change of separation of two or more nearby test particles. "Rather than have one global frame with gravitational forces we have many local frames without gravitational forces." However, these local dimension changes add up to an effect on the global spacetime structure that one interprets as "gravitation" in its everyday manifestations.

In contrast, Newton supposed the existence of one ideal overall reference frame. For him, "Absolute space, in its own nature, without relation to anything external, remains always similar and immovable." The ball bearing or spaceship is regarded by Newton as actually accelerated with respect to this ideal frame. The "gravitational force" that accelerates it acts mysteriously across space and is produced by distant objects. That the man in the spaceship finds no evidence either of the acceleration or the force is an accident of nature, according to the Newtonian view. Pundits used to interpret this accident of nature as the fortuitous equality of "gravitational mass" and "inertial mass" or in other "learned" ways.

In conversations with one of the authors of this book at various times over the years, Einstein emphasized his great respect for Newton and, in particular, his admiration for Newton's courage and judgment. He stressed that Newton was even better aware than his seventeenth-century critics of the difficulties with the ideas of absolute space and time. To postulate those ideas was nevertheless the only practical way to get on with the task of describing motion in Newton's century. In effect, Newton chopped the problem of motion into two parts: (1) space and time and their meaning: ideas that were puzzling but usable and that were destined to be clarified only 230 years later and (2) the laws of acceleration with respect to that idealized spacetime: laws that Newton gave the world.

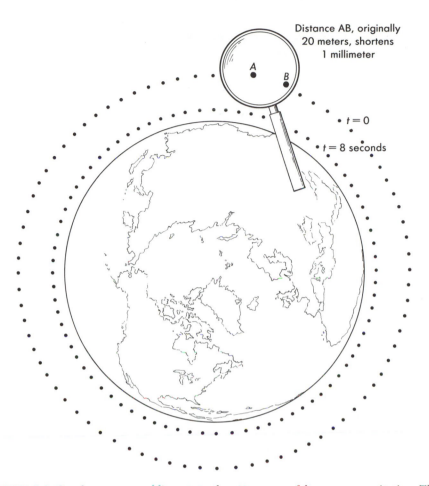

Distance AB, originally
20 meters, shortens
1 millimeter

A
B

t = 0

t = 8 seconds

FIGURE 9-6. *Local curvature adding up to the appearance of long-range gravitation.* The shortening of distance between any one pair, A and B, of ball bearings is small when the distance itself is small. However, small separation between each ball bearing and its partner demands many pairs to encompass Earth. The totalized shortening of the circumference in any given time—the shortening of one separation times the number of separations—is independent of the fineness of the subdivision. That totalized pulling in of the circumference carries the whole necklace of masses inward. This is free fall, this is gravity, this is a large scale motion interpreted as a consequence of local curvature. Example:

Original separation between A *and* B—*and every other pair: 20 meters*
Time of observation: 8 seconds
Shortening of separation in that time: 1 millimeter
Fractional shortening: 1 millimeter/20 meters = 1/20,000
Circumference of Earth (length of airy necklace of ball bearings): 4.0030×10^7 meters
Shrinkage of this circumference in 8 seconds: $1/20,000 \times 4.0030 \times 10^7$ meters = 2001.5 meters
Decrease in the distance from the center of Earth (drops by the same factor 1/20,000):
$$1/20,000 \times 6.371 \times 10^6 \text{ meters} = 315 \text{ meters}.$$

This apparently large-scale effect is caused—in Einstein's picture—by the addition of a multitude of small-scale effects: the changes in the local dimensions associated with the curvature of geometry (failure of B *to remain at rest as observed in the free-float frame associated with* A).

What is the source of the curvature of spacetime? Momenergy is the source. In Chapter 8 we saw the primacy of momenergy in governing interactions between particles. Crash of mass on mass, no matter how elastic or how destructive, leaves the total momenergy of the system quite unaltered. By what miracle does this come about? Education of momenergy from birth onward to good behavior? Goodness of heart?

BOX 9-1

HOW SPACETIME CURVATURE CARRIES INFLUENCE FROM ONE MASS TO ANOTHER

The necklace of ball bearings (Figure 9-6) as they approach Earth, examined more closely, reveals a remarkable feature of spacetime curvature outside a great, essentially uniform, essentially isolated sphere of mass. The curvature in its character is totally "tide-producing," totally "noncontractile."

An array of test masses covering the surface of a hollow sphere freely floating above the Earth's surface will shrink in two dimensions and lengthen in one. The volume remains constant; only the shape changes. This change is evidence of the noncontractile, tide-driving spacetime curvature outside Earth.

What do these descriptive terms mean, and how do we verify that they apply? We look at a cluster of ball bearings dotted here and there over the surface of an imaginary small sphere, all momentarily at rest relative to each other and relative to Earth. That shape, however, as the seconds tick by, changes from sphere to ellipsoid. How come? First let's look at the two dimensions of the sphere that lie perpendicular to each other but parallel to Earth's surface. Both these dimensions of the sphere shrink as the ball bearings converge toward Earth's center. The up-down dimension of the pattern, however, lengthens, and twice as much. Why? Newton says because of the greater gravitational acceleration of the one nearer Earth. Einstein says because two-percent stretch in that dimension compensates one-percent shrinkage in the other two dimensions and keeps the volume of the pattern unchanged. Spacetime curvature,

yes; but a totally noncontractile curvature. Einstein's famous equation, stated in simple terms, tells us how spacetime curvature responds to mass:

$$\begin{pmatrix} \text{appropriate measure of} \\ \text{spacetime contractile curvature} \\ \text{at any place, any time,} \\ \text{in any Lorentz frame} \end{pmatrix}$$

$$= \begin{pmatrix} \text{a universal} \\ \text{constant} \end{pmatrix} \times \begin{pmatrix} \text{density of energy} \\ \text{at that locale} \\ \text{perceived in that} \\ \text{Lorentz frame} \end{pmatrix}$$

Outside, no mass, no energy, a spacetime curvature that is totally noncontractile. Inside Earth, however, there is mass, therefore there is energy — or in a mov-

Spacetime controls momenergy

Obedience to the eyes of a corps of bookkeepers? No, Einstein taught us. The enforcing agency does not lie far away. It's close at hand. It's the geometry of spacetime right where the crash takes place. Not only does spacetime grip isolated mass, telling it how to move. In addition, in a crash it sees to it that the participants neither gain nor lose momenergy. But there is more! Spacetime, in so acting, cannot maintain the

ing frame, energy plus energy flow—and therefore spacetime curvature there has a contractile character. The ball bearings—when shafts are drilled for them so that not one of them encounters any obstacle to free-float motion—start to converge vertically as well as horizontally. The volume shrinks. That, overlooking details, is what we mean when we say that "mass grips spacetime, telling it how to curve."

There is no Earth mass out at Moon's orbit. How then does Einstein's spacetime geometry account for Moon's motion? Answer: Earth's mass imposes on spacetime a contractile curvature throughout Earth's interior, as a jumper's feet impose a contractile curvature on a trampoline. That contractile curvature, where the feet push, forces on the surrounding nontear fabric a corresponding lateral stretch. That effect transmits itself in ever more dilute measure to the ever more remote regions of the trampoline.

Likewise spacetime does not tear. Its fabric just above Earth's surface experiences the same lateral contractility as it does just below the surface. Not so with the curvature in the two-dimensional domain defined by time and by direction perpendicular to Earth's surface. In that one plane, curvature within Earth is contractile but suddenly jumps just above Earth's surface to the opposite character. Hence the tide-producing character of spacetime curvature outside Earth. A point twice as far from Earth's center lies on an imaginary Earth-centered sphere that encompasses eight times the volume. There the tide-producing curvature experiences eight times the dilution and has one eighth the strength. Despite this rapid dilution of tide-producing power with distance, it has strength enough at Moon, 60 Earth radii away from Earth's center, to deform Moon from sphere to ellipsoid, 1738.35 kilometers in radius along the Earth–Moon direction, 1738.15 kilometers in radius for each of the other two perpendicular directions.

Easy as it is to regard Earth as running the whole show, Moon too has its part. Like an infant standing on the trampoline some distance from its mother, it imposes its own small curvature on top of the curvature evoked by Earth. That additional curvature, contractile in Moon's interior, has tide-driving character outside. Were the Earth an ideal sphere covered by an ideal ocean of uniform depth, then Moon would draw that ocean's surface 35.6 centimeters higher than the average in two domains, one directly facing Moon, one directly opposite to it—simultaneously lowering those waters 17.8 centimeters below the average on the circle of points midway between the two. (These low figures show how important are funneling and resonant sloshing in determining heights of actual ocean tides on Earth.)

The local contractile curvature of spacetime at Moon's location added up along Moon's path yields the appearance of long-range gravitation, similar to that illustrated in Figure 9-6. Box 2-1 tells a little of the many influences that have to be taken into account in any fuller treatment of the tides.

The deformation of the nontear trampoline fabric under the jumper's feet and elsewhere is analogous to the nontear curvature of spacetime geometry inside Earth and elsewhere.

perfection assumed in textbooks of old. To every action there is a corresponding reaction. *Spacetime acts on momenergy, telling it how to move; momenergy reacts back on spacetime, telling it how to curve.* This "handshake" between momenergy and spacetime is the origin of momenergy conservation—and the source of spacetime curvature that leads to gravitation (Box 9-1). ✒️

Momenergy tells spacetime how to curve

9.7 GRAVITY WAVES

gravitational energy moving at light speed

Gravity waves from
collapsing matter

In the depths of an ill-fated, collapsing star, billions upon billions of tons of mass cave in and crash together. The crashing mass generates a wave in the geometry of space—a wave that rolls across a hundred thousand light-years of space to "jiggle" the distance between two mirrors in our Earthbound gravity-wave laboratory.

A cork floating all alone on the Pacific Ocean may not reveal the passage of a wave. But when a second cork is floating near it, then the passing of the wave is revealed by the fluctuating separation between the two corks. So too for the separation of the two mirrors. There is, however, this great difference. The cork-to-cork distance reveals a momentary change in the two-dimensional geometry of the surface of the ocean. The

BOX 9-2

COMPACT STELLAR OBJECTS

Three kinds of astronomical objects exist comparable in mass to Sun but very much smaller. Two of these have been observed; the third seems an inevitable result of Einstein's theory.

A **white dwarf star** is a star of about one solar mass, with radius about 5000 kilometers. (The radius of Earth is 6371 kilometers.) This gives the white dwarf a density of approximately 10^9 kilograms/meter^3 (or one metric ton per cubic centimeter). As of 1990, approximately 1500 white dwarfs had been identified.

White dwarfs were observed and studied astronomically long before they were understood theoretically. Today we have come to recognize that a white dwarf is a star that quietly used up its fuel and settled gently into this compact state. The electrons and nuclei that make up the body of a white dwarf are not separated into atoms. Instead, the electrons form a gas in which the nuclei swim. The pressure of this "cold" electron gas keeps the white dwarf from collapsing further.

S. Chandrasekhar calculated in 1930 that no white dwarf can be more massive than approximately 1.4 solar masses ("Chandrasekhar limit") without collapsing under its own gravitational attraction. His analysis assumed the mix of electrons and nuclei to be unaltered under compression by a load so heavy, an assumption that had to be modified in later years. Today we recognize that enormous compressions squeeze electrons into combining with protons to make neutrons. At compressions near the Chandrasekhar limit, the electron gas transforms into a neutron gas, the interior of the star becomes a giant nucleus, and the whole nature of the compact object changes to that of a neutron star.

A **neutron star** has roughly the same density as an atomic nucleus, of the order of 10^{17} kilograms/meter^3, or one Earth mass per cube of edge length 400 meters. The radius of a neutron star is approximately 10 kilometers.

mirror-to-mirror distance reveals a momentary change in the three-dimensional geometry of space itself.

The idea of extracting energy from ocean waves is old. After all, the ability of a water wave to change a distance lets itself be translated into the ability to do work. The same reasoning applies to a gravity wave. Because it can change distance, it can do work. It carries energy. Energy once resident as mass in the interior of a star has radiated out to us and to all the universe.

Of all the workings of the grip of gravity, none is more fascinating or opens up for exploration a wider realm of ideas than a gravity wave. None pushes to a higher pitch the art of detecting a small effect, and none gives more promise of providing an unsurpassable window on cataclysmic events deep inside troubled stars. Nevertheless, no other great prediction of Einstein's geometric theory of gravity stands today so far from triumphant exploitation. As of this writing, not one of the nine ingenious

How often is a neutron star formed? Towards answering this still open question we have one important lead: In our own galaxy we see one supernova explosion on average about every 300 years [most recent supernova in the Large Magellanic Cloud, a satellite of our galaxy, on February 23, 1987; one seen by Kepler, October 13, 1604; one seen by Tycho Brahe, November 6, 1572; earlier ones: 1181 A.D.; July 4, 1054 A.D.; 1006 A.D. (the brightest); 185 A.D.; and two possibles in 386 A.D. and 393 A.D.]. In such an event a star teetering on the edge of instability finally collapses. The Niagara Falls of infalling mass in some cases go too far and overcompress the inner region of the star. That region thereupon acts like a spring, or explosive charge, and drives off the outer portions of the star. This explains the spectacular luminosity that is such a prominent feature of a supernova. The core that remains becomes a neutron star in some events, it is believed, in others a black hole.

Neutron stars were predicted in 1934 but not observed until 1968. Many neutron stars spin rapidly — with a period as short as a few milliseconds. A neutron star typically has an immense magnetic field. When that field is aligned at an angle relative to the axis of spin of the star (as in Earth, for example), it sweeps around like a giant whisk brush through the plasma in the space around the star. The periodic shock to the electrons of the plasma from the periodic arrival of this field excites those electrons to radiate periodic pulses of radio waves and visible light — both observed on Earth. Because of this behavior, such neutron stars are called **pulsars.** As of 1990, nearly 500 pulsars were identified.

A **black hole** is an object created when a star collapses to a size so small that strong spacetime curvature prevents it from communicating outward with the external universe. Even light cannot escape from a black hole, whence its name. No one who accepts general relativity has found any way to escape the prediction that black holes must exist in our galaxy. Strong evidence for the existence of black holes has been found, but it is not yet convincing to all astrophysicists. A black hole can have a mass as small as a few times the mass of our Sun. A black hole of three solar masses would have a ''radius'' of about 9 kilometers. There is no theoretical upper limit to its mass.

detectors built to this day has proved sensitive enough to secure any generally agreed detection of an arriving gravity wave.

Does any truly simple line of reasoning assure us that gravity will inescapably carry energy away from two masses that undergo rapid change in relative position? Yes is the conclusion of a little story that savors of mythology. The Atlas of our day, zooming through space in free float, insists as much as ever on maintaining physical fitness. He pumps iron, not by raising iron against the pull of Earth's gravity, but by throwing apart two identical great iron spheres, Alpha and Beta. He floats between those minor moons and plays catch with them. Each time they fall together under the influence of their mutual gravity, he catches them, absorbs their energy of infall in his springlike muscles, and flings them apart so that they always travel the same distance before returning. It's an enchanting game, but Atlas finds that it's a losing game. When the masses fly back together, they never yield up to him as much energy as he must supply to throw them apart again. Why not?

Gravity waves result from time delay

Say the central point in two words: time delay. Like any force that makes itself felt through the emptiness of space, the force of gravity cannot propagate faster than the speed of light. This limitation imposes a delay on the attraction between the two iron spheres. Alpha, on each little stretch of its outbound path, feels a pull that originated from Beta when the two were a tiny bit *closer* than they are now. The actual force that's slowing Alpha is therefore a tiny bit bigger than we would judge from thinking of them as stationary at their momentary separation. On its return trip inbound along the same little stretch of path, Alpha experiences a helping pull that originated from Beta when the two had a separation slightly *greater* than its present value. The actual force that's speeding Alpha inward is therefore a tiny bit less than we would judge from thinking of them as stationary at their momentary separation. In each stretch of their outbound trip, the two masses have to do more work against the pull of gravity than they get back — in the form of work done on them by gravity — on the same stretch of path inbound. A calculable amount of energy disappears from the local scene on each out-in cycle of Atlas's exercise. Yet the total energy must somehow be conserved. Therefore the very gravity that steals energy from Atlas and his iron, or from any two

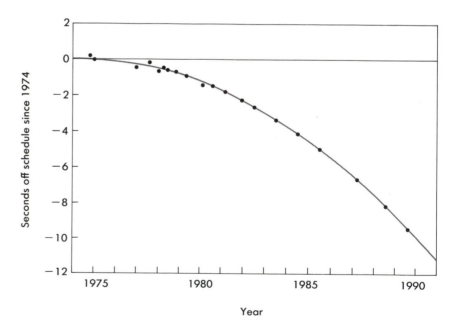

FIGURE 9-7. *Two whirling neutron stars furnish a giant clock, whose time-keeping hand is the line, ever-turning, that separates the centers of those two stars. That hand does not today keep the "slow" schedule (straight horizontal line) one might have expected from its timing as measured in 1974. The downward sloping curve shows gravity-wave theory's prediction of the shortening in the time required to accumulate any specified number of revolutions. The dots show the actual observed shortening in that time.*

masses that rapidly change their relative position, must somehow all the time be transporting the stolen energy to the far-away. That inescapable theft of energy is in its quality, its directional distribution, and its magnitude none other than what Einstein had treated long before under the head of *gravity radiation* and what we now call gravity waves.

Atlas couldn't "see" those gravity waves. Neither have we today yet succeeded in detecting directly the gravity waves we feel sure must be radiating from sources dotted here and there in the galaxy and in the universe. However, we have an exciting indirect confirmation that gravity waves exist — not through their action on any receptor, but through the energy they carry away from a whirling pair of neutron stars. That particular "binary pulsar" first revealed itself to Joseph H. Taylor, Jr., and Russell A. Hulse by periodic pulses of radio waves picked up on the huge disklike antenna at Arecibo in Puerto Rico. As one of these neutron stars spins on its axis, its magnetic field spins with it, giving timing comparable in accuracy to the best atomic clock ever built (Box 9-2). Thanks to this happy circumstance, Taylor and his colleagues have been able to follow the ever-shortening separation of the two stars and the ever-higher speed they attain as they slowly spiral in toward an ultimate catastrophe some 400 million years from now. The timing of the orbits gives us a measure of energy lost as the stars spiral in. No reasonable way has ever been found to account for the thus observed loss of energy except gravitational radiation. As of September 1989, 14 years after first observation, this loss of energy agrees with the rate predicted by theory to better than one percent (Figure 9-7).

Gravity waves and pulses of gravity radiation are sweeping over us all the time from sources of many kinds out in space. Detecting them, however, we are no better than the primitive jungle dweller unable to detect and even totally unaware of the radio waves that carry past her every minute of the day music, words, and messages. However, experimentalists are working out ingenious technology and building detector instrumentation of ever-growing sensitivity (Figures 9-8 and 9-9). Few among them have any doubt of their ability to detect pulses of gravity radiation from one or another star catastrophe by sometime in the first decade of the twenty-first century.

Gravity waves steal energy from orbiting neutron stars

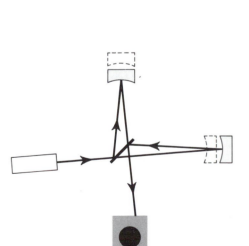

FIGURE 9-8. *The proposed MIT – Caltech gravity-wave detector will (1) use the beam from a laser (left), (2) split it by a device (center) analogous to a half-silvered mirror, (3) send one half-strength beam to one faraway mirror (top) and the other to the other faraway mirror (right), (5) allow these beams to undergo many many reflections (not shown), and (6) recombine them at the detector (bottom). A gravity-wave incident on Earth will slightly shorten the 4-kilometer distance to the one mirror and slightly lengthen the 4-kilometer distance to the other mirror. This relative alteration in the path length of the laser beams, if big enough, amplified enough, and picked up by detectors sensitive enough, will reveal the passage of the gravity wave.*

FIGURE 9-9. *Prototype gravity-wave detector, California Institute of Technology, Pasadena. The laser beam is tailored (lower right) for entry into the beam splitter (located where the two long light pipes meet, just to the left of center in the photograph). The mirrors at the ends of these two evacuated light pipes lie outside the boundary of the photograph.*

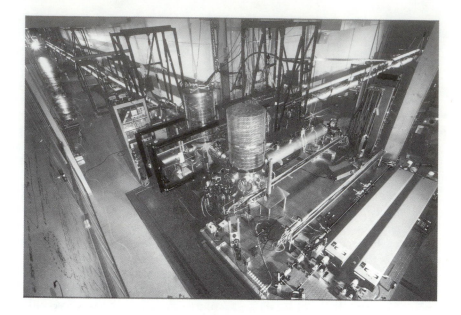

Astronomy uses signals of many kinds — light, radio waves, and X-rays among them — to reveal the secrets of the stars. Of all signals from a star, none comes out from deeper in the interior than a gravity wave. Among all violent events to be probed deeply by a gravity wave, none is more fascinating than the dance of death of two compact stars as they whirl around each other and undergo total collapse into . . . a black hole! 🖋

9.8 BLACK HOLE

over the edge with a scream of radiation

A black hole is a domain whose mass is so tightly compacted that nothing can escape from it, not even light. Everything that falls in is caught without hope of escape (Figure 9-10).

"Escape velocity c" implies black hole

To fire a missile from Moon's surface so that it escapes that satellite's attraction demands a speed of 2.38 kilometers per second or greater. The critical speed for escape from Earth — in the absence of drag from the atmosphere — is 11.2 kilometers per second. When the object does not rotate and is so compact that even light cannot escape, the "effective radius" or so-called "horizon radius" is

$$(\text{effective radius}) = \frac{\left(\begin{array}{c}\text{circumference of region} \\ \text{out of which} \\ \text{light cannot escape}\end{array}\right)}{2\pi}$$

$$= 2 \times (1.47 \text{ kilometers}) \times \left(\begin{array}{c}\text{mass of black hole} \\ \text{expressed in} \\ \text{number of Sun masses}\end{array}\right)$$

Black hole still exerts "pull" of gravity

When a star or cloud of matter collapses to a black hole it disappears from view as totally as the Cheshire cat did in Alice in Wonderland. The cat, however, left its grin behind; and the black hole — via the effect of spacetime curvature that we call gravity — exerts as much "pull" as ever on normal stars in orbit around it. They are like participants in a formal dance with lights turned low. Only the white dress of the girl is visible as she whirls around in the arms of her black-suited companion. From the

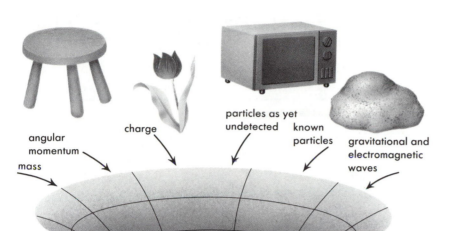

FIGURE 9-10. *Whatever objects fall into a black hole, they possess at the end — as seen from outside — only mass, angular momentum, and electric charge. Not one other characteristic of any in-falling object remains to betray its past — not a hair. This leads to the saying, "A black hole has no hair."*

speed of the girl and the size of the circle in which she swirls, we know something of the mass of the invisible companion. By such reasoning it was possible to conclude by 1972 that that the optically invisible companion of one long-known star has a mass of the order of 9.5 solar masses.

This remarkable object came first to attention because in December 1971 the Uhuru orbiting X-ray observatory detected X-ray pulsations with time scales from one tenth to tens of seconds from an object located in the Cygnus region close to the known star. Why does it give off X-rays? And why does the intensity of the X-rays vary rapidly from instant to instant? The gas wind from the visible companion varies from instant to instant like the smoke from a factory chimney. This gas, falling on a compact object, gets squeezed. To picture the how and why of this squeeze, look from a low-flying plane at the streams of automobiles converging from many directions on a football stadium for a Saturday afternoon game. The particles and the gas are pushed together as surely as the cars in the traffic. The compression of the traffic raises the temper of the driver, and the compression of the gas raises its temperature as air is heated when pumped in a bicycle pump. However, because the gas falls from an object of millions of kilometers in size to one a few kilometers across, the compression is so stupendous that the temperature rises far above any normal star temperature, and X-rays come off.

The time scale of the fluctuations in X-ray intensity depends on the size of the object that is picking up the star smoke, a size less by a fantastic factor than that of any normal star. Could the object be a white dwarf (Box 9-2)? No, because such a star would be

Cygnus X-1: A black hole?

TABLE 9-1

BLACK HOLES FOR WHICH THERE WAS SUBSTANTIAL EVIDENCE AS OF SEPTEMBER 1989

(Uncertainties in masses are of the order of 20 to 50 percent.)

Astronomical designation of black hole	Mass (in solar masses)
Cygnus X-1	9.5
LMC X-1	2.6
AO 620-00	3.2
LMC X-3	7.0
SS 433	4.3
Black hole at center of our galaxy	3.5×10^6

visible. A neutron star? No, because even matter compressed so tightly that it is transformed to neutrons cannot support itself against gravity if it has a mass much over two solar masses. No escape has been found from concluding that Cygnus X-1 is a black hole. This great discovery transformed black holes from pencil-and-paper objects into a lively and ever-growing part of modern astrophysics (Table 9-1).

Black hole at center of our galaxy?

Much attention went in the 1980s to a presumptive black hole with a mass of about three and a half million times the solar mass and a horizon radius of about ten million kilometers. It floats at the center of our galaxy, the Milky Way. Around it buzz visible stars of the everyday kind, most of them fated to fall eventually into that black hole and increase its mass and size. That stars close to the center of our galaxy go around as fast as they do is one of the best indicators we have for the presence, and one of the best measures we have for the mass, of the central black hole, which is itself invisible.

Quasar energy output from matter swirling into black hole?

In contrast to dead solitary black holes, the most powerful source of energy we know or conceive or see in all the universe is a black hole of many millions of solar masses, gulping down enormous amounts of matter swirling around it. Maarten Schmidt, working at the Mount Palomar Observatory in 1956, was the first to uncover evidence for these quasistellar objects, or quasars, starlike sources of light located not billions of kilometers but billions of *light-years* away. Despite being far smaller than any galaxy, the typical quasar manages to put out more than a hundred times as much energy as our own Milky Way, with its hundred billion stars. Quasars, unsurpassed in brilliance and remoteness, we call lighthouses of the heavens.

High-efficiency conversion of gravitational energy to radiation

Observation and theory have come together to explain in broad outline how a quasar operates. A black hole of some hundreds of millions of solar masses, itself built by accretion, accretes more mass from its surroundings. The incoming gas, and stars-converted-to-gas, does not fall in directly, any more than the water rushes directly down the bathtub drain when the plug is pulled. Which way the gas swirls is a matter of chance or past history or both, but it does swirl. This gas, as it goes round and round, slowly makes its way inward to regions of ever-stronger gravity. Thus compressed, and by this compression heated, the gas breaks up into electrons — that is negative ions — and positive ions, linked by magnetic fields of force into a gigantic accretion disk. Matter little by little makes its way to the inner boundary of this accretion disk and then, in a great swoop, falls into the black hole, on its way crossing the horizon, the surface of no return. During that last swoop, hold on the particle is relinquished. Therefore, the chance is lost to extract as energy the full 100 percent of the mass of each infalling bit of matter. However, magnetic fields do hold onto the ions effectively enough for long enough to extract, as energy, several percent of the

Hal McIntosh. Courtesy of *The Saturday Review.*

ALBERT EINSTEIN
Ulm, Germany, March 14, 1879 — Princeton, New Jersey, April 18, 1955

"Newton himself was better aware of the weaknesses inherent in his intellectual edifice than the generations which followed him. This fact has always roused my admiration."

★   ★   ★

"Only the genius of Riemann, solitary and uncomprehended, had already won its way by the middle of the last century to a new conception of space, in which space was deprived of its rigidity, and in which its power to take part in physical events was recognized as possible."

★   ★   ★

"All of these endeavors are based on the belief that existence should have a completely harmonious structure. Today we have less ground than ever before for allowing ourselves to be forced away from this wonderful belief."

---

mass. In contrast, neither nuclear fission nor nuclear fusion is able to obtain a conversion efficiency of more than a fraction of a percent. Of all methods to convert bulk matter into energy, no one has ever seen evidence for a more effective process than accretion into a black hole, and no one has even been able to come up with a more feasible scheme for one.

Of all the features of black hole physics in action, none is more spectacular than a quasar. And no lighthouse of the skies gives more dramatic evidence of the scale of the universe. ✦

9.9 THE COSMOS

a final crunch?

Expanding universe: Evidence for big bang beginning

The more distant quasars and galaxies are, the greater the speed with which they are observed to be receding from us. This expansion argues that somewhere between ten and twenty billion years ago the universe began with a big bang, a time before which there was no time.

We see around us relics of the big bang, not only today's rapidly receding galaxies but also today's abundance of the chemical elements—some among them still radioactive, the "still warm ashes of creation" (V. F. Weisskopf)—and today's greatly cooled but still all-pervasive "primordial cosmic fireball radiation." We now believe that in the first instants of its life, the entire universe filled an infinitesimally small space of enormous density and temperature where matter and energy fused in a homogeneous soup. Immediately the universe began expanding. After about 10^{-6} seconds it had cooled enough that subatomic particles condensed from the matter–energy soup. In the first three minutes after the big bang, neutrons and protons combined to make heavier elements. Eons later stars and galaxies formed. Never since has the universe paused in its continual spread outward.

"Open" universe expanding forever? Or "closed" universe that recontracts to crunch? An open question!

Will the universe continue expanding forever? Or will its expansion slow, halt, and turn to contraction and crunch (Table 9-2), a crunch similar in character but on a far larger scale than what happens in the formation of a black hole? Great question! No one who cares deeply about this question can fail to celebrate each week that week's astrophysical advances: instruments, observations, conclusions.

We have come to the end of our journey. We have seen gravity turned to float, space and time meld into spacetime, and spacetime transformed from stage to actor. We have examined how spacetime grips mass, telling it how to move, and how mass grips spacetime, telling it how to curve. Of all the indications that existence at bottom has a simplicity beyond anything we imagine today, there is none more inspiring than the unsurpassed simplicity of gravity as we now see it. ✦

REFERENCES

Extended portions of this chapter were copied (and sometimes modified) from John Archibald Wheeler, *Journey Into Gravity and Spacetime* (Scientific American Library, a division of HPHLP, New York, 1990).

For details of Galileo's views on motion, see Galileo Galilei, *Dialogues Concerning Two New Sciences,* originally published March 1638; one modern translation is by Henry Crew and Alfonso de Salvio (Northwestern University Press, Evanston, Ill., 1950).

How Newton came only in stages to the solution of the problem of fall is told nowhere with such care for the fascinating documentation as in Alexander Koyre, "A Documentary History of the Problem of Fall from Kepler to Newton," *Transactions of the American Philosophical Society,* Volume 45, Part 4 (1955).

Keynes quotation under Newton portrait: Reprinted by permission of the publisher, Horizon Press, from *Essays in Biography* by John Maynard Keynes, copyright 1951.

⟨ **TABLE 9-2** ⟩

A CLOSED-MODEL UNIVERSE
COMPATIBLE WITH OBSERVATION

Radius at phase of maximum expansion	18.9×10^9 light-years or 1.79×10^{26} meters
Time from start to maximum size	29.8×10^9 years or 2.82×10^{26} meters
Radius today	13.2×10^9 light-years
Time from start to today's size	10.0×10^9 years
Time it would have taken from start to today's size if the entire expansion had occurred at today's slowed rate of expansion	20.0×10^9 years
Present expansion rate	An extra increment of recession velocity of 15.0 kilometers/second for every extra million light-years of remoteness of the galactic cluster
Fraction of the way around the 3-sphere universe from which we can in principle receive light today	$\dfrac{113.2 \text{ degrees}}{180 \text{ degrees}} = 62.9\%$
Fraction of the matter in the 3-sphere universe that has been able to communicate with us so far	74.4%
Number of new galaxies that come into view on average every three days	One!
Average mass density today	14.8×10^{-27} kilogram/meter^3
Average mass density at phase of maximum expansion	5.0×10^{-27} kilogram/meter^3
Rate of increase of volume today	1.82×10^{62} meters^3/second
Amount of mass	$M_{conv} = 5.68 \times 10^{53}$ kilograms In geometric units: $M = GM_{conv}/c^2 = 4.21 \times 10^{26}$ meters
Equivalent number of suns like ours	2.86×10^{23}
Equivalent number of galaxies like ours	1.6×10^{12}
Equivalent number of baryons (neutrons and protons)	3.39×10^{80}
Total time, big bang to big crunch	59.52×10^9 years

Figure 9-2: Figure and data from *Journal of Spacecraft,* Volume 11 (September 1974), pages 637–644, published by the American Institute of Aeronautics and Astronautics. Data also from D. B. De Bra, *APL Technical Digest,* Volume 12: pages 14–26.

Figure 9-3 from *Philosophiae Naturalis Principia Mathematica* (Joseph Streater, London, July 5, 1686); Motte translation into English revised and edited by Florian Cajori and published in two paperback volumes (University of California Press, Berkeley, 1962). This is also the source of the quote in Section 9.6: "Absolute space, in its own nature, without relation to anything external, remains always similar and immobile."

Three quotations under the Einstein picture come from Albert Einstein, *Essays in Science* (Philosophical Library, New York, 1934).

Quotation in Section 9.6: "Rather than have one global frame with gravitational forces we have many local frames without gravitational forces." Steven Schutz, in January 1966 final examination in course in relativity, Princeton University.

For an exciting and readable overview of the experimental proofs of Einstein's general relativity theory, see Clifford M. Will, *Was Einstein Right? Putting General Relativity to the Test* (Basic Books, New York, 1986). In particular (Chapter 10, pages 181–206), he describes at some length the emission of gravity waves by the binary pulsar system studied by Joseph H. Taylor, Jr., and Russell A. Hulse.

ANSWERS TO ODD-NUMBERED EXERCISES AND PROBLEMS

chapter 1

1-1a 10.2 meters   **b** 270 meters   **c** 10^3 meters   **d** 10^4 kilometers ≈ 2 times Boston–San Francisco distance   **1-3a** 2.6×10^{13} meters   **b** 5.3×10^{-6} second   **c** 1.85×10^{-10} hours   **d** 52 weeks   **e** 5.4×10^9 furlongs   **1-5a** 4 years   **b** 4/5 the speed of light $= 2.4 \times 10^8$ meters/second   **1-7a** 4 meters   **b** $\sqrt{7}$ meters $= 2.65$ meters   **c** $\sqrt{15}$ meters $= 3.87$ meters   **d** 2 meters   **e** 4 meters (same as part **a**)   **1-9a** 2×10^5 years   **b** $v = 0.995$ **c** 6.33×10^4 years, $v = 0.9995$ **d** $v = 1 - 5 \times 10^{-11} = 0.99999999995$ **1-11a** 2×10^{-4} second   **b** 133 half-lives; $(1/2)^{133} \approx 10^{-40}$ **c** 3 half-lives   **d** zero space separation (creation and decay occur at the same place in rocket frame)   **e** 3 half-lives $= 4.5 \times 10^{-6}$ second

chapter 2

2-1a hit the ceiling   **b** same answer   **c** Rider cannot tell when elevator reaches top.   **2-3** Set clock to 10 meters of time, start when reference flash arrives.   **2-5a** Experiment in progress for $1/0.96 = 1.04$ meters of time. In this time, test particle falls 6×10^{-17} meters, about 10^{-2} diameter of a nucleus.   **b** 3×10^{-4} second, 10^5 meters   **2-7** 3.6 millimeters; 19.7 seconds. Spacetime region: 20 meters $\times$ 20 meters $\times$ 20 meters of space $\times$ 59 $\times 10^8$ meters of time   **2-9a** decrease (think of each ball bearing in an elliptical orbit around the center of Earth)   **b** apart   **c** No, you cannot distinguish rising from falling. At the top you notice nothing inside the coach.   **2-11** $v_{\max} = 0.735$ the speed of light.   **2-13a** Effective time of fall: 4.67 seconds. Net velocity of fall: 1284 meters/second.   **b** Angle of deflection: 4.3×10^{-6} radian $= 2.5 \times 10^{-4}$ degree $= 0.88$ second of arc

chapter 3

3-1a 60 seconds   **b** 45 seconds against the current, 22.5 seconds with the current, 67.5 seconds round trip   **c** No   **3-3** If different kinds of clocks ran at different rates in a free-float rocket frame, then this difference could be used to detect the relative velocity of the laboratory from inside the rocket, which violates the Principle of Relativity. This does not mean that the common rate of rocket clocks will be the same as measured in rocket and laboratory frames.   **3-5a** 11.5 light-years   **b** 9.43 years   **c** $v = 0.6$ **d** 8 years $=$ the interval between the two events.   **3-7** The bullet misses. Coincidence of A and A' (event 1) and firing of the bullet at the other end of spaceship 0 (event 2) cannot be simultaneous in both rocket reference frames. The right panel of the figure is wrong. Consistent with the Train Paradox (Section 3.4), spaceship $0'$ (standing in for the train frame) will observe the bullet to be fired before coincidence of A and A', thus accounting for the fact that bullet misses.   **3-9a** $\sin \psi = v_{\text{Earth}}$ (in meters/meter)   **b** $\sin \psi \approx \psi \approx 10^{-4}$ radian $= 21$ seconds of arc   **c** $\sin \psi$ and $\tan \psi$ are both approximately equal to ψ for small ψ. Therefore the difference between the two predictions cannot be used to distinguish between relativistic and nonrelativistic predictions.   **d** in a direction 0.524 radians $= 30$ degrees ahead of transverse   **3-11g(1)** $v_{\text{rel}} = 10^{-7}$, $v'_{\text{bullet}} = 2 \times 10^{-6}$. Their product is 2×10^{-13}, very small compared with 1; therefore we expect v_{bullet} to be the sum of v'_{bullet} and v_{rel}, the form verified in everyday experience at low speeds. **(2)** $v_{\text{bullet}} = 24/25 = 0.96$ **(3)** $v_{\text{bullet}} = v_{\text{light}} = +1$ **(4)** $v_{\text{bullet}} = v_{\text{light}} = -1$. Yes, expected from the Principle of Relativity.   **3-13a** 0.32 meters $= 1.1$ nanosecond   **b** 6.0×10^5 periods   **c** No shift would imply the speed of light is the same for the frame of Earth going one way around Sun as compared with frame of Earth going the opposite direction around Sun.   **d** $dc = -(2/n^2)(\Delta l/T)dn$ and $dc/c = -2$

dn/n For $dn = 3 \times 10^{-3}$ and $n = 6.0 \times 10^5$, we have the maximum value of $dc/c = 1 \times 10^{-8}$ (sign not important). Hence $dc \approx 3$ meters/second is the maximum change in the speed of light that could have escaped detection in this very sensitive experiment. **3-15a** visual distance apart $= v\Delta t$; time lapse between images $= (1 - v)\Delta t$; visual speed of approach $= v_{approach} = v/(1 - v)$; $v_{approach} = 4$ when $v = 4/5$; $v_{approach} = 99$ when $v = 0.99$ **b** visual distance apart $= v\Delta t$; time lapse between images $= (1 + v)\Delta t$; visual speed of recession $= v_{recede} = v/(1 + v)$; for $v_{approach} = 4$ when $v = 4/5$, then $v_{recede} = 4/9 = 0.44$; for $v_{approach} = 99$ when $v = 0.99$, then $v_{recede} = 0.497$ **3-17a** Light leaves E one meter of time earlier than light from G in order to enter the eye at the same time. In this time the cube moves v meter of distance, equal to x in the top right figure. **b** The angle ϕ is given by the expression $\sin \phi = v$. For $v \rightarrow 0$, this visual angle of rotation goes to zero, as we experience in everyday life. For $v \rightarrow 1$, this visual angle of rotation goes to 90 degrees, and the cube shows us its back side as it passes overhead. **c** The word "really" is not appropriate; each mode of observation is valid; some will be more useful than others for different applications. (Requested speech to each observer not included here.) **d** The "cube" will look sheared, with top EF pulled backward a distance x with respect to bottom GH in the left panel.

special topic

L-1a $v_{rel} = 3/17 = 0.176$ for speed of Super 6 times speed of light **b** $v_{rel} = 1/3 = 0.333$ for infinite speed of Super **L-3b** 128 days **e(1)** 0.1 meter of time; too small for either wristwatch or electronic clock **(2)** about 10^4 meters of time; too small for wristwatch but easily detected by electronic clock **(3)** distance is 10^{12} meters, or about 6.7 times the Earth-Sun distance. **L-5d** $v_{rel} = 0.944$ **L-11** The manhole is tilted, so it passes over the meter stick without collision. **L-13a** At the beginning and the end of their trip (and all during their trip), Dick and Jane are separated by 12 light-years as measured in the Earth frame. Final velocity: $v = 3/4$. Aging of each astronaut $=$ proper time along either worldline $=$ sum of the spacetime intervals along each segment of either worldline $= \sqrt{15} + \sqrt{12} + \sqrt{7}$ years $= 10$ years. **b** Yes. Yes. **c(1)** Jane stops accelerating 13.6 years earlier than Dick. **(2)** 30 years **(3)** 30 years **(4)** 43.6 years **(5)** Dick: 50 years old. Jane: 63.6 years old. **(6)** 18.1 lightyears, which is just $\gamma = 1.51$ times their 12-light-year separation measured in the Earth frame by Mom and Dad. **(d)(1)** Yes **(2)** Yes Yes **(3)** Jane's **(4)** Yes. No. **(5)** It's the old story: relativity of simultaneity, in this case the fact that Dick and Jane stop accelerating simultaneously only in the Earth frame. **e** Then, by symmetry, Dick will be older than Jane in their final rest frame. All the numbers will otherwise be the same. **f** Then they will start and stop simultaneously in Earth frame and also in the final rocket frame; they will be the same age at these stopping events in both frames. **L-15c** For the extreme relativistic case when $v_{rel} \rightarrow 1$, then $v_{t=t'} \rightarrow 1$ also.

chapter 4

4-1a 11.6 years **b** 18.6 years **c** 30.2 years **d** 7.67 years **e** 14.67 years **f** 22.34 years **g** 5.75 light-years **h** 7.67 years, 5.07 years **i** 14.67 years, $30.2 - 5.1 = 25.1$ years **4-3a** The engineer is wrong. **b** Frequency of oscillation increases by $\sqrt{2}$ when voltage doubles. **c** frequency in cycles/second $= f = (qV_o/8mL^2)^{1/2}$, where m and q are mass and charge of the electron, V_o is the voltage applied, and L is the width of the box $= 1$ meter. **d** Minimum round-trip time across box at the speed of light is $2L/c$ so $f_{max} = c/2L$. **e** For the Newtonian region, $f/f_{max} = [qV_o/(2mc^2)]^{1/2}$. For the extreme relativistic region, $f/f_{max} = 1$. The quantity qV_o is a measure of electron potential energy at the wall or electron kinetic energy at the screen.

We expect the Newtonian analysis to be correct when this energy of motion is very much less than the rest energy mc^2. The extreme relativistic analysis will be correct when qV_o is very much greater than mc^2. The crossover occurs (the two dashed curves intersect) where $qV_o \approx 2mc^2$ or $V_o \approx 10^6$ volts. **f** For low speeds, the ratio f_{proper}/f_{max} will follow the Newtonian curve. At extreme relativistic speeds, the proper time for one period $\rightarrow 0$ and the proper frequency $\rightarrow$ infinity.

chapter 5

5-1a(1) 1 year **(2)** 1.94 years **(3)** 0.87 year **(4)** 3.81 years **b** 5.20 years **c** solid-line traveler will be younger   **5-3a** event A is at $(x, t) = (0, 0)$; event B is at $(0, 1)$; event C is at $(1.5, 3.5)$ or $(-1.5, 3.5)$; event D is at $(3, 6)$ or $(-3, 6)$ **b** event D is at $(x, t) = (0, 0)$; event C is at $(0, -2)$; event B is at $(0, -4)$; event A is at $(-0.75, -5.25)$ or $(+0.75, -5.25)$ **c** $v_{rel} = \pm 0.6$ **d** Yes   **5-5d** 3136 cycles/second **e** 31.4 cycles/second   **5-7** Hint: As with most paradoxes in relativity, the solution has to do with the relativity of simultaneity.

chapter 6

6-1a Events 1 and 2: **(1)** Proper time: 4 meters **(2)** Yes **(3)** Yes **(4)** No Events 1 and 3: **(1)** Proper distance: 4 meters **(2)** No **(3)** No **(4)** Yes **Events 2 and 3:** **(1)** zero **(2)** Yes **(3)** No **(4)** No **b** $v_{rel} = 3/5$ in $+x$-direction for both   **6-3a** Set $t' = 0$ in the first inverse Lorentz transformation equation (L-11) and solve for v_{rel}. **b** Set $x' = 0$ in the second equation (L-11) and solve for v_{rel}. (Why does the result look so funny?)   **6-5a** Yes, explosion. (Sorry!) **b** No change in prediction. (The impact at A and activation of the detonation switch are spacelike events; the laser pulse cannot connect them.)

chapter 7

7-1a $[5m, \sqrt{24}m, 0, 0]$ **b** $[m, 0, 0, 0]$ **c** $[\sqrt{10}m, 0, 0, 3m]$ **d** $[5m, 0, -\sqrt{24}m, 0]$ **e** $[10 \ m, 2.66 \ m, 5.32 \ m, 7.98 \ m]$.   **7-3a** 0.05 milligram **b** 0.1 milligram **7-7a** wristwatch time: 32 seconds; Earth time: 1000 centuries **b** $E/m \approx 10^{36}$. 1.7 million metric tons.   **7-9a** $E_B = 9$ units **b** $p_B = \sqrt{32}$ units $= 5.66$ units **c** $m_B = 7$ units **d** greater: $m_C = 15$ units $> m_A + m_B = 9$ units   **7-11a** proton: 938 MeV; electron: 0.511 MeV **b** $v_{limit} \approx 0.12$. Proton kinetic energy at limit ≈ 6 MeV. Electron kinetic energy at limit $\approx 3.4 \times 10^{-3}$ MeV $= 3.4$ keV. Yes, designer of color TV tubes (electron kinetic energy ≈ 25 keV) must use special relativity.

chapter 8

8-1a approximately 35×10^{-9} kilograms $= 35$ micrograms **b** approximately 600 kilograms. More. **c** approximately 6×10^{13} seconds or about 2 million years! Chemical burning in Eric's body produces large quantities of waste products. Elimination of these products carries away mass enormously faster than mass is carried away as energy.   **8-3a** Force is approximately 3×10^{-9} newtons, or the weight of 3×10^{-10} kilograms. You should not be able to feel it. **b** pressure on a perfectly absorbing satellite $\approx 5 \times 10^{-6}$ newton/meter^2; on a perfectly reflecting satellite $\approx 9 \times 10^{-6}$ newton/meter^2; somewhere in between for a partially absorbing surface. Total energy absorbed/meter^2, not color of the incident light, determines pressure. **c** acceleration approximately 10^{-9} g **d** particle radius approximately 10^{-6} meter, independent of the distance from Sun   **8-7** density approximately 5×10^{10} kilograms/meter$^3 = 5 \times 10^7$ grams/centimeter^3, or 50 million times the density of water!   **8-9** $E_A = (M^2 + m^2)/(2m)$ **8-11a** From conservation equations, show that $\cos \phi > 1$, which is impossible.   **b** If the total momentum is zero after the collision, it must be zero before the collision. But the alleged single photon before the collision cannot have zero momentum. Therefore the reaction is impossible.   **8-13** $2E_C = E_A + (E_A^2 - m^2)^{1/2}$ and $2E_D = E_A - (E_A^2 - m^2)^{1/2}$. If the particle is at rest, then $E_A = m$ and $E_C = E_D = m/2$.   **8-15a** $E_C = m(E + m)/[E + m - (E^2 - m^2)^{1/2} \cos \phi_C]$ **8-17a** 1.8 TeV   **b** $E \approx 1.7 \times 10^6$ TeV   **8-19e** No   **8-21** When the bulb is seen way ahead, its light is very intense and radically blue-shifted. While still seen ahead, there is an angle of observation (depending on the speed) at which the light is red, but dim. As the bulb is seen to pass the observer, its light is infrared and very dim. As the bulb is seen to retreat into the distance, its light is extremely dim and radically red-shifted.   **8-23a** $v = 0.38$ **b** 13×10^9 years   **c** Allowance for gravitational slowing will *decrease* the estimated time back to the start of the expansion.   **8-25** $\Delta f/f \approx [3kT/(mc^2)]^{1/2}$. The observed frequency will increase for molecules approaching the observer and decrease for molecules receding from the observer. The overall effect — at temperatures for which Newtonian expressions are valid — is to produce a spread of frequencies approximated by the expression above ("Doppler line broadening").   **8-27** $E' = m/2$, $E = m$, $\phi = 30$ degrees.   **8-35a** The incident gamma ray (with excitation energy E) imparts a small kinetic energy K to the iron atom, for which Newtonian expression is valid: $K = p^2/2m = E^2/2m$, since $p = E$ for the gamma ray. Then (energy of recoil)/(energy for excitation) $= K/E \approx E/(2m) \approx 1.4 \times 10^{-7}$. But fractional resonance width (6×10^{-13}) is smaller than this by a factor of almost a million, so the iron nucleus cannot accept the gamma ray and conserve energy.   **b** One gram is about 10^{22} atoms. If the m in the above equation increases by the factor 10^{22}, then the energy of recoil is a factor 10^{22} smaller, and the nucleus will not notice the residual mismatch in energy.   **8-37** $\Delta f/f = - gz/c^2$, $v = 0.7 \times 10^{-6}$ meter/second towards emitter   **8-39** $\Delta f/(f_o \Delta T) \approx (3/2)k/(mc^2) \approx 1.2 \times 10^{-15}$ per degree.

INDEX